Global Risk Agility and Decision Making

Daniel Wagner • Dante Disparte

Global Risk Agility and Decision Making

Organizational Resilience in the Era of Man-Made Risk

palgrave
macmillan

Daniel Wagner
Country Risk Solutions
USA

Dante Disparte
Risk Cooperative
USA

ISBN 978-1-349-94859-8   ISBN 978-1-349-94860-4   (eBook)
DOI 10.1057/978-1-349-94860-4

Library of Congress Control Number: 2016946025

Cover illustration: © NASA photo/Alamy stock photo

Printed on acid-free paper

This Palgrave Macmillan imprint is published by Springer Nature
The registered company is Macmillan Publishers Ltd. London

This book is dedicated to everyone who devotes their time and energy to managing the nexus between man-made and natural risk. It is you who ultimately control our collective long-term destiny.

Preface

This is a book about survival. Having the ability to discern between perception and reality can make the difference between making the right moves on the global business chessboard and moving forward, or making a critical mistake. Gone are the days when the job of successfully managing the turbulent waters of international business was the domain of just a risk manager. In a world that has become increasingly defined by unpredictable events with severe and lasting consequences—and where those consequences can impact an organization's ability to survive—it is *every* manager's job to be a risk '*navigator*'.

What is required to be an effective risk navigator is not a degree from a good business school, completing the latest round of continuing education, or relying on twenty years of experience in a chosen field. Rather, what is required today is the ability to know a *lot* about how the world works—how it *really* works—and to be able to *think*, not just about the impact of an event on this quarter's bottom line, but, more importantly, about an event's current and long-term impact on a trading, investing, lending, or operating landscape.

Doing so is, of course, no simple task. It requires a commitment to being informed, to understanding the implication of an event or action, and to constantly thinking "outside the box" about what it all means. The idea is not to be reactive, but rather, to be *proactive* and *anticipatory*. A typical international organization does not practice "*Anticipatory Risk Management*". Instead, it relies on information from a third party that must first be delivered before it even knows there is a problem. Then, it may not have either the personnel or tools in-house to interpret the information. Finally, it may not have a decision-making process in place that effectively incorporates the result of that interpretation, so that it is transformed into meaningful action.

If your firm is a small or medium-sized organization and fits this mold, you are not alone. Some of the largest and most sophisticated organizations on the planet also suffer from this malady. Operating in more than 100 countries and having hundreds of thousands of employees does not necessarily translate into being prepared to operate effectively internationally. It is often the case that these firms believe themselves to be invulnerable as a result of their global presence. As we will see, they are indeed vulnerable, on many levels. There is also an inherent vulnerability by remaining on the sidelines of engaging, investing, and operating around the world. The masters of *risk-ready* enterprises that stand to dominate the twenty-first century will take to the turbulent waters of global trade and integration, recognizing all the risk implied in this endeavor.

The Illusion of Effective Risk Management

To illustrate just how vulnerable, consider the state of security in commercial aviation today. Since the beginning of the War on Terror, many billions of dollars have been spent (and continue to be spent) on putting into place a security protocol that methodically screens passengers to identify threats before a plane pulls back from the gate. Of course, this has greatly reduced the *perceived* risk of a terrorist threat on a typical commercial flight. However, it has not eliminated the threat, nor has it prevented unanticipated threats from emerging, nor does the system function as it should.

Despite all the time, effort, and money thrown at the problem, in 2015 there remained some truly significant gaps in the system. Some 14 years *after* the War on Terror began, the United States (U.S.) Department of Homeland Security's (DHS) Office of the Inspector General was able to successfully transfer mock explosives or banned weapons through airport security screening systems in the U.S. 95 percent of the time, by deploying teams intending to do just that.[1] It was not until 2009 that the Transportation Security Administration (TSA) required that just 50 percent of the cargo placed inside commercial aircraft in the U.S. be screened for explosives[2] (today it is 100 percent).

As a result of logistics and the cost implied, in 2015 only two major U.S. airports made comprehensive employee screening part of their operational

---

[1] J. Date et al., "Undercover DHS Tests Find Security Failures at U.S. Airports," ABC News, accessed June 1, 2015, accessed November 24, 2015, http://abcnews.go.com/ABCNews/exclusive-undercover-dhs-tests-find-widespread-security-failures/story?id=31434881.

[2] C. Berrick, *Transportation Security Administration Has Strengthened Planning to Guide Investments in Key Aviation Security Programs, but More Work Remains*, U.S. Government Accountability Office, accessed July 24, 2008, accessed November 24, 2015, http://www.gao.gov/assets/130/120883.html.

protocol,[3] meaning that all other major airports in the country do not do so. Also in 2015, another DHS investigation found that the TSA had failed to identify 73 employees of airlines, airport vendors, and other employers with active clearance badges who had links to terrorism.[4] In other words, the TSA was sanctioning the employment of individuals with terrorist links to work in roles that could endanger the public.

Hundreds of billions of dollars, tens of thousands of people, and numerous congressional mandates have only created the *illusion* of effective airport security in the U.S. today. If someone wanted to bring an aircraft down—inside the U.S.—he or she could. That is exactly what happened in 2015 with the Russian Metrojet crash in the Sinai Peninsula, the result of a bomb having been placed in the aircraft at the Sharm El-Sheik (Egypt) airport[5]—the third successful bombing of Russian commercial aircraft post-9/11 (the two previous bombings occurring in Russia in 2004).[6] If the devotion of these kinds of resources has resulted in such gross inadequacies in the system, does a business stand a chance of creating a better mousetrap? Some risk is never 'managed' out of existence, but merely tamed, creating a placebo effect—as in the minds of the traveling public that the system is in fact safer than it really is.

The Unforeseen Versus the Unknown

One thing no one can predict, and therefore no one can control, is the unknown. It is worth distinguishing between the 'unforeseen' and the 'unknown'. The difference is that what is *unforeseen* is not necessarily something that was unknown, but rather was not predicted or anticipated. By contrast, what is *unknown* is ambiguous and a mystery, so it cannot be predicted or anticipated. We cannot do much about what is unknown, precisely because there is no way to quantify or understand it. The most we can hope to do is 'manage' the risks associated with the unforeseen, by anticipating those risks. Herein lies the distinction between risk and uncertainty. Risk can always be

---

[3] S. Kerry, "Closing an Airport Security Gap: Airport Screening," *Wall Street Journal*, accessed January 27, 2015, http://www.wsj.com/articles/closing-an-airport-security-gap-employee-screening-1422396346.

[4] T. McCay, "The TSA Has Been Letting a Lot of People with Links to Terrorism Work at Airports," news. mic, June 9, 2015, accessed November 24, 2015, http://mic.com/articles/120393/the-tsa-has-been-letting-a-lot-of-people-with-links-to-terrorism-work-at-airports.

[5] The bombing occurred even though the following security advisory had been issued in March 2015 by the U.S. Federal Aviation Administration, strongly advising against traveling into, out of, or over the Sinai below 26,000 feet. See Federal Aviation Administration, *Advisory for Egypt Sinai Peninsula*, accessed November 24, 2015, https://www.faa.gov/air_traffic/publications/us_restrictions/media/FDC_5-9155_Egypt-Sinai_Advisory_NOTAM.pdf.

[6] *Timeline of Airline Bombing Attacks*, Wikipedia, 2015, accessed November 24, 2015, https://en.wikipedia.org/wiki/Timeline_of_airliner_bombing_attacks.

measured and generally understood, while uncertainty cannot be measured, and is the domain along the risk spectrum that paralyzes markets and causes bank runs.

Sticking with the aviation theme, recall the terrible tragedy of Germanwings Flight 9525, which crashed in 2015. The passengers and cargo had been successfully screened for explosives and weapons, which, ordinarily, would imply a safe and successful flight. Yet, the flight's co-pilot, Andreas Lubitz, intentionally crashed the plane into the Alps, killing all on board. Germanwings' owner, Lufthansa, learned during Lubitz's Flight School training in 2009 that he suffered from severe depression, and, subsequently, that he had suicidal tendencies, yet it still allowed him to fly commercial aircraft.[7] It cannot be argued that there was *not* a potentially substantial risk associated with allowing Lubitz to fly. Such risk was neither unknown to the airline, nor could his action be considered unforeseen. That he was allowed to continue to fly was criminally negligent, and the risk could certainly have been managed by preventing him from continuing to fly.

The simple mitigation strategy to this known risk, which was compounded by the secured cockpit doors mandated post-9/11, would have been to conduct mental health screenings with greater frequency. Germanwings, Lufthansa, and the airline industry exposed the traveling public to the risk, however remote, that pilots would attempt to take their own lives along with everyone on board. The fatal error was assuming that a sound mental checkup resulting in no known signs of suicidal tendencies at the pilot onboarding process was a constant variable in the system.

Contrast this example with that of Greece's default on its debt in 2015. A prudent lender would say that it is unwise to continue to lend money to a country (or person or business) that is heavily indebted and clearly unable to repay its debt. And yet, the lending to Greece continued, and continued, and continued, prior to and throughout the Great Recession, until it reached the point of absurdity. Even when it reached the point of absurdity, the European Central Bank (ECB) and a consortium of commercial banks wanted to *continue* to lend to Greece, knowing the money could not possibly be repaid. Was that an unknown risk? No. Was Greece's default unforeseen? No.

So, what was the reason the ECB and banks wanted to continue to lend to Greece? To keep the 'House of Cards' from falling down. Yet, by anticipating (and accepting) the risks associated with continuing to lend, the risk of

---

[7] J. Ewing and N. Kulish, "Lufthansa Says Germanwings Pilot Reported Deep Depression," *New York Times* (March 31, 2015), accessed November 24, 2015, http://www.nytimes.com/2015/04/01/world/europe/lufthansa-germanwings-andreas-lubitz.html.

default *could* have been managed by changing course, stopping the imposition of excruciating austerity measures, and working with the Greek government to find a more acceptable, and workable, solution. In doing so, the ECB and banks would have stood a better chance of being at least partially repaid, Greece could have been rescued, and the future of the European Union (EU) need not have been put in jeopardy.

These two examples serve to emphasize several things. First, some risks that are thought to be unknown, are not unknown. Second, with some foresight and critical thought, some risks that at first glance may seem unforeseen, can in fact be foreseen. Third, with the right set of tools, procedures, knowledge and insight, light can be shed on variables that lead to risk, allowing us to manage them. That is a key message we will be exploring further in this book.

Acknowledgements

No book can be written without inspiration and guidance—personally and professionally. A great number of individuals have inspired and guided both of us over the course of our lives and careers, and we wish to acknowledge them here.

Dante Disparte

No worthwhile endeavor is possible without a great reason to persevere. My reasons to try and *tame* risk in the twenty-first century are my incredible children and the coming generation, who inherit a more turbulent world. My loving wife, who has endured all my follies and misadventures, deserves all the credit for this effort, as do all the helping hands that wove a safety net behind me enabling me to embrace risk. My family, the one I was born into and those I inherited, receive my gratitude for their belief. The Berrios family in particular defined safety and security for me. I would also like to acknowledge my co-author, who has given me a voice on current affairs and risk. He is a great thinker, mentor, and friend and nothing about him is conventional.

Over the course of my career I have had the great privilege to see the world for all its diversity, inequities, and promise. One thing stands in the way of this promise and it is the fascinating subject of risk. For all its permutations and how it continually stays ahead of mankind's ability to manage this remarkable process, I must also acknowledge risk for inspiring this body of work. There are countless colleagues, friends and mentors to whom I owe a great debt of gratitude. I am particularly grateful to the incredible academics at NYU Stern, who honed my understanding of risk and how it shapes our world.

Daniel Wagner

I have been fortunate to work for a number of esteemed institutions that enabled me to interact with some of the best minds in the insurance, risk management, finance and development arenas. I am grateful to have had the opportunity to work for AIG, GE, the Asian Development Bank and the World Bank Group, and with the many fine people there, who gave me the opportunity to learn so much about the world through their eyes. I want to thank the many people throughout my career who encouraged me to be inquisitive and turn the pyramid upside down for perspective.

I would also like to give a special thanks to my family, for always being so encouraging, supportive, and a constant source of love. In particular, I want to thank my mother, who was a Holocaust survivor and intellectual historian, for showing me what willpower, determination and perseverance—against all odds—really means. And my father, who came from modest roots and taught me never to forget who I am and where I came from, to treat all people with dignity and respect, and to always try to see the other person's perspective. I thank Fidel for standing right by my side through thick and thin, and for never losing his sense of humor. Finally, I wish to thank Dante Disparte for suggesting we write this book together and for being a soul mate. It is rare to find someone who instinctively turns the pyramid upside down. Dante is a naturally agile global risk manager and decision maker.

Contents

About the Authors

Dante Disparte Dante Disparte is the Founder and CEO of Risk Cooperative, a strategy, risk and capital management firm focusing on mid-market opportunities, market expansion and equity investments on a global scale.

Prior to forming Risk Cooperative, Dante served as the Managing Director of Clements Worldwide, a leading insurance brokerage with customers in more than 170 countries. He is a specialist in strategy and risk reduction through the design and delivery of comprehensive risk solutions of worldwide scope. Dante is credited with designing the world's first card-based life insurance program for the United Nations, a plan that has placed more than a half billion USD of risk with the markets in more than 150 countries. This innovation was heralded as one of the top product innovations of 2011 by the MENA Insurance Review.

He was formerly the Managing Director of Land Rover's operations in 32 Sub-Saharan African markets and held numerous general management roles in Denmark, where he developed applied skills in social entrepreneurship. Dante is credited with developing a humanitarian fleet management solution that is proven to reduce the economic, environmental, and social impact of humanitarian operations. This body of work is profiled in a business case published by INSEAD's Social Innovation Centre.

He served on the board of directors of Kjaer Group A/S, which is ranked as one of the top ten workplaces in Europe and the top workplace in Denmark for four years and he currently serves on the board of Communities in Schools Nation's Capital, where he chairs the fund development committee. He is conversant in six languages and has published numerous articles on the subject of risk, strategy and business effectiveness. A graduate of Harvard Business School's Program for Leadership Development, Dante holds a degree in International and Intercultural Studies from Goucher College and an MSc. in risk management from New York University's Stern School of Business.

Dante serves chairs the board of the Harvard Business School Club of Washington, D.C., and sits on Harvard Business School's global alumni board. He also serves as the chair of the Business Council for American Security and as a board member with the American Security Project.

Daniel Wagner Daniel Wagner is the Founder and CEO of Country Risk Solutions (CRS), a cross-border risk advisory firm based in Connecticut, U.S.

Prior to founding CRS, Daniel was Senior Vice President of Country Risk at GE Energy Financial Services where he was part of a team making annual investments of billions of dollars into global energy projects. He was responsible for advising senior management on a variety of country risk-related issues, strategic planning, and portfolio management. Daniel created a Center of Excellence for country risk analysis in GE and led a team that produced a comprehensive automated country risk rating methodology.

He began his career underwriting Political Risk Insurance (PRI) at AIG in New York and subsequently spent five years as Guarantee Officer for the Asia Region at the World Bank Group's Multilateral Investment Guarantee Agency in Washington, D.C. During that time he was responsible for underwriting PRI for projects in a dozen Asian countries. After then serving as Regional Manager for Political Risks for Southeast Asia and Greater China for AIG in Singapore, Daniel moved to Manila where he was Guarantee and Risk Management Advisor, Political Risk Guarantee Specialist, and Senior Guarantees and Syndications Specialist for the Asian Development Bank's Office of Co-financing Operations. Over the course of his quarter-century-long career he has also held senior positions in the PRI brokerage business in London, Dallas and Houston.

Daniel has published more than 500 articles on risk management and current affairs, is a non-resident scholar at the Institute for Near East and Gulf Military Analysis, and a regular contributor to the *Huffington Post*, *National Interest*, *South China Morning Post*, and a plethora of other publishing platforms. His editorials have been published in such notable newspapers as the *International Herald Tribune* and the *Wall Street Journal*. His two previous books were *Political Risk Insurance Guide* and *Managing Country Risk*.

Daniel holds master's degrees in International Relations from the University of Chicago and in International Management from the American Graduate School of International Management (Thunderbird) in Phoenix. He received his bachelor's degree in Political Science from Richmond College in London.

Praise for *Global Risk Agility and Decision Making*

"You have to manage risks before they manage you. Global Risk Agility provides a practical platform for greater risk governance at all layers of an organization and society. With this compelling work, we can no longer plead ignorance to the era of man-made risk, nor can we afford a slow search for causality. It is time to mount our defenses to the effects of cyber risk, terrorism, climate change and other interconnected threats, which Wagner and Disparte masterfully deconstruct in this book."

Tom Ridge, first U.S. Secretary of Homeland Security, 43rd Governor of Pennsylvania and Chairman of Ridge Global (U.S.)

"Globalization's 'interconnectedness' is moderating our tribal savage instincts; Conflict has become a growth industry. This book reveals what diplomats know—if you accept a risk environment for its hidden opportunities, your interests will be greater, and better served. The rest is up to us—to be able to identify risks and realize those opportunities. The authors' lens adds clarity to this risk/opportunity landscape to illustrate that agility, not fear, will capture those opportunities. Wagner and Disparte split the atom on risk, and their book shows how to harness it."

Hugh T. Dugan, U.S. Department of State career diplomat, senior advisor to ten U.S. ambassadors to the United Nations, and Distinguished Visiting Scholar of UN Studies, School of Diplomacy and International Relations, Seton Hall University (U.S.)

"*Global Risk Agility and Decision Making* is a must-read for government officials (either elected or career), for business men and women, and for all those who value their economic future. Sun Tzu famously said in his epic book *The Art of War,* "if you know the enemy and know yourself, you need not fear the result of a hundred battles." That book is a must read at every military school. Wagner and Disparte's book gives you the knowledge and ability to know about the enemy—risk—and what the threats are. They address innovation, agility, and preparation, along with a treasure trove of other recommendations. Put this book at the top of your reading list!"

Steve Cheney, Brigadier General, United States Marine Corps (Retired) and CEO of the American Security Project (U.S.)

"A true treasure chest of insights that provides the reader with a clear understanding of how risk can be managed. *Global Risk Agility* and Decision Making is a wonderful tour de force across the shifting landscapes of risk, providing deep insights about how to navigate our globalized and interconnected world. A must-read for business executives and anyone who has been entrusted with risk preparedness."

Georg Kell, Chief Architect and former CEO of the United Nations Global Compact (U.S.)

"Given the frequency with which international crises are occurring, and their consequences, it is important not only to read the right book about how to manage them, but to do so at the right time. It certainly is the right time for political and business leaders alike to read *Global Risk Agility and Decision Making*, a book that provides great insight into how to manage man-made risks, and survive them in our increasingly unpredictable and dangerous world."

Guihong Zhong, Chief Risk Officer, Orient Minerva Asset Management Co. Ltd. (China).

"What I particularly valued about this book is the emphasis the authors place on the importance of achieving risk agility. Agile entrepreneurs always anticipate, innovate, and where possible, avoid headwinds, by changing tack and direction. Professional risk managers must do the same, and this book drives that point home very effectively."

Tony O. Elumelu, Chairman of Heirs Holdings & Founder of The Tony Elumelu Foundation (Nigeria).

"Globalization offers enormous opportunities, but also exposes companies to a host of risks that span the continuum from currency volatility and geopolitics to cyber security and climate change. Risk management is complex, layered and multifaceted, and requires interdisciplinary coordination and insight. Threats abound, and their capacity to undermine customer trust and market confidence can level incalculable damage. *Global Risk Agility and Decision Making* thoughtfully examines these threats and makes the case that containing risk is not only a strategic imperative, but essential to organizational survival in the 21st century."

Wanda Felton, Vice Chair & First Vice President, The Export-Import Bank of the United States (U.S.)

"If you believe that risk management is a boring topic for accountants, this book will change your mind. The authors offer a fresh perspective on the discipline, giving abundant evidence that accurate risk management is the cornerstone for the long-term growth and prosperity for any business or enterprise. A must-read."

Luigi de Pierris, Senior Consultant, Initiative for Risk Mitigation in Africa, African Development Bank (Ivory Coast).

"*Global Risk Agility and Decision Making* is a fascinating, accessible and important book, full of compelling insights and original perspectives. In casting their discerning eyes over the future of risk, Wagner and Disparte firmly put tomorrow's issues on to today's agenda and give the reader the skills needed to systemically analyze risk and react with agility—to look beyond headlines and think through the tertiary effects of any event to anticipate and identify new opportunities."

Nicholas Wyman, CEO of the Institute for Workplace Skills and Innovation and author of award-winning book *Job U: How to Find Wealth and Success by Developing the Skills Companies Actually Need* (U.S.)

"This book effectively tackles the complex 'man-made' risk factors that must inform our collective risk perspectives. The authors lay out a compelling argument that these are the filters we must apply to fully understand today's current risk environments and residual impacts. Specific case studies are included to demonstrate beneficial risk management approaches designed to help private and public entities attack today's emerging risks."

Yvette K. Connor, Managing Director, Alvarez & Marsal Risk Advisory Services (U.S.)

List of Abbreviations

AA	Arthur Andersen
ACA	Affordable Care Act
ADB	Asian Development Bank
AIG	American International Group
ARC	Agile Risk Control
ARM	Anticipatory Risk Management
ASEAN	Association of Southeast Asian Nations
BCM	Business Continuity Management
BP	British Petroleum
CDOs	Collateralized Debt Obligations
CEO	Chief Executive Officer
CFO	Chief Financial Officer
CIO	Chief Information Officer
CMO	Chief Mobility Officer
CNPC	China National Petroleum Corporation
CO_2	Carbon Dioxide
CRO	Chief Risk Officer
CSR	Corporate Social Responsibility
Daesh	IS
Dodd-Frank	The Dodd-Frank Wall Street Reform and Consumer Protection Act
DHS	U.S. Department of Homeland Security
ECB	European Central Bank
ERM	Enterprise Risk Management
Ex-Im	Export-Import Bank
EU	European Union
FDI	Foreign Direct Investment
FDIC	Federal Deposit Insurance Corporation

FEMA	U.S. Federal Emergency Management Agency
FIFA	Fédération Internationale de Football Association
GAAP	Generally Accepted Accounting Principles
GDP	Gross Domestic Product
GE	General Electric
GM	General Motors
GoP	Guardians of Peace
GRI	Global Reporting Initiative
GSP	Global Sullivan Principles
HHI	Herfindahl—Hirschman Index
HIV	Human immunodeficiency virus
HPAA	Health Policy and Administration
IAS	International Accounting Standards
ICJ	International Court of Justice
ILO	International Labor Organization
IMF	International Monetary Fund
IRD	International Relief and Development
IS	Islamic State
ISO	International Organization for Standardization
KSA	Kingdom of Saudi Arabia
KwH	Kilowatt hour
LIBOR	London Interbank Offered Rate
LNG	Liquefied Natural Gas
MDBs	Multilateral Development Banks
MENA	Middle East and North Africa
MIGA	Multilateral Investment Guarantee Agency
MNE	Multinational Enterprise
NATO	North Atlantic Treaty Organization
NFC	Near-field communication
NGOs	Non-governmental Organizations
NINJA	No Income, No Job and No Assets
NKW	NKW Holdings
NOAA	U.S. National Oceanic and Atmospheric Administration
NSA	U.S. National Security Agency
OECD	Organization of Economic Cooperation and Development
OPIC	Overseas Private Investment Corporation
PCA	Permanent Court of Arbitration
PNG	Papua New Guinea
PRI	Political Risk Insurance
PRI	Principles for Responsible Investment
PRISM	Clandestine surveillance program of the NSA
PSU	Pennsylvania State University
SDGs	Sustainable Development Goals

SDRs	Slowly Developing Risks
SEC	U.S. Securities and Exchange Commission
SIFI	Systemically important financial institution
SOE	State-owned enterprise
SPVs	Special Purpose Vehicles
SWIFT	Society for Worldwide Interbank Financial Telecommunication
TPP	Trans-Pacific Partnership
TRIA	Terrorism Risk Insurance Act
TRM	Transactional Risk Management
TSA	U.S. Transportation Security Administration
TTIP	Transatlantic Trade and Investment Partnership
UAE	United Arab Emirates
UN	United Nations
UN Norms	The UN Norms on the Responsibilities of Transnational Corporations and other Business Enterprises with regard to Human Rights
UNCLOS	United Nations Convention on the Law of the Sea
UNCTAD	United Nations Conference on Trade and Development
UNSC	UN Security Council
U.S.	United States of America
VaR	Value at risk
VW	Volkswagen
WEF	World Economic Forum
WHO	World Health Organization
WTO	World Trade Organization
YPF	Yacimientos Petroliferos Fiscales

List of Illustrations

List of Tables

Part I

The Risk Management Conundrum

1

Risk Management in a Global World

Declaring Battle

It should be abundantly clear to anyone who manages risk that the rules of engagement for conducting international business have changed since the advent of globalization, instant communication, the War on Terror, the new Cold War, the rise of the Islamic State (IS), the Great Recession, and multiple simultaneous sovereign defaults. The risks associated with cross-border transactions and risk aversion are in general high, while the margin for errors is usually low. It is only natural, given all that has happened in rapid succession in the new millennium that international businesses would think more carefully about assuming and managing cross-border risk. Many have done just that, but doing so successfully has become increasingly difficult.

The new risk management 'normal' includes a paradigm shift. The rulebook has changed and, in a very short period of time, the pyramid has been turned upside down. As a result of the combination of a decoupling in growth patterns between the developed and developing worlds, ever-growing income disparity across the globe, the rise of the emerging economies, and the fact that global economic crises have emanated from both the developed and developing worlds, it has become difficult to distinguish between North and South, and East and West. This new risk management paradigm is *global*, and it is here to stay.

The driving forces behind this new paradigm are not only permanent, but gaining momentum. Technology continues to develop and change at a lightning pace. Socioeconomic forces are shifting. There is a growing collec-

© The Author(s) 2016
D. Wagner, D. Disparte, *Global Risk Agility and Decision Making,*
DOI 10.1057/978-1-349-94860-4_1

tive revulsion at the levels of income inequality. Concern about human rights, climate change and the environment are prominent issues in global forums and legislative bodies. Natural resources are being depleted at an astonishing rate. Population growth is accelerating at an unprecedented pace in the developing world. All these issues—and many others—are contributing to *a fundamental shift in how we should be thinking about the world*, and, by extension, *risk management*.

The role of government in managing economies has been declining for decades. As a result of privatization, public/private partnerships, regulatory reform and deregulation, policy is increasingly being influenced and made by non-state actors—multilateral organizations, nongovernmental organizations (NGOs), think-tanks and corporations. Here again, a paradigm shift is under way. The 'debate' over these issues is not limited to individual nations or organizations, but rather, the *global* community. As a result of social media, any 'event' can become an instant sensational news story. Public opinion has become *global* public opinion.

The transparency and interconnectedness of information, the mobility of goods, services, and people, and the seemingly constant nature of change in the twenty-first century makes for a truly breathtaking landscape from which risk managers and decision makers *must* view the world. In this increasingly complex environment, we have no choice but to be informed, shake off our preconceived notions about what constitutes operational 'normality', embrace change, and do so being prepared for battle—because to succeed in this kind of environment, a battle must be waged on how risk managers and decision makers think and act.

The Global Risk Landscape

It may be trite and simplistic to say that the global risk landscape is perilous and is becoming more so every day, but in fact, it is. Given how interconnected the world is—on so many levels—a challenge in one country or region can quickly have implications on the other side of the world. Global risks transcend borders and are inextricably linked with each other. It does not matter whether it is political, economic, financial, socio-cultural, or environmental in nature—a risk in one country can easily impact its next door neighbor, the region, and even the rest of world.

Consider, for example, the Ebola pandemic of 2014. What started out as a West African phenomenon quickly became a global risk, by virtue of porous borders, global travel, inadequate resources, poor medical treatment, local burial customs, and poor government response. Any one of these contributing factors could, by themselves, have made the problem much worse; in combination, they were almost guaranteed to make Ebola a pandemic. Each global line of defense quickly succumbed to Ebola's march. It was not until the response was militarized that the world regained its footing in the battle to contain the disease.

When it first began, at the end of December 2013 in a rural village in Guinea, the virus had been restricted to a few neighboring villages. By March 2014, it had spread to Liberia and Sierra Leone, and in the coming months it would spread to the Democratic Republic of Congo, Italy, Senegal, Nigeria, Mali, Spain, the United Kingdom, and the U.S. By September of that year, global concern had risen to such a degree that travel restrictions from the three most severely impacted countries were imposed by dozens of countries. Early in 2015 such restrictions remained in effect in nearly 40 countries around the world.

Another example was the birth and growth of the IS, which served to re-emphasize how a political movement can turn into a global phenomenon. What started as a disparate group of Islamic extremists in Iraq and Syria morphed into a regional, and then global force to be reckoned with. After only one year in existence, the IS came to define the core of the modern Sunni/Shi'ia fissure in the Middle East, and quickly spread to North Africa.

The threat of 'home grown' terrorism soon became associated with IS devotees either having traveled to Iraq or Syria to fight on the front lines and then returning to their home countries, or that of impressionable individuals who never left their home countries but became potential IS 'warriors' while sitting at home behind their laptops. Almost instantaneously the IS became synonymous with global terrorism across the world, and triggered the notion that radical fringe groups such as Boko Haram in Nigeria can and will operate as quasi-nation states, filling in the governance void, collecting revenues and applying a virulent judicial system over their subjects.

In the economic and finance arena, the Great Recession was a reminder of the extremely fragile nature of the global economic system. The Recession brought the world to its knees. Its cause—a combination of old-fashioned greed, predatory lending practices, a nepotistic system of checks and balances, insufficient regulatory structures, and an inadequate enforcement apparatus—was evidence of systemic risk management failure.

Regrettably, even after legal 'safety mechanisms' were put in place in the U.S., lobbyists from the banking industry in Washington, D.C. were busy trying to dismantle them piece by piece. The asset 'bubbles' that existed prior to 2008 had re-inflated just six years later, setting the stage for the *next* bubble to pop, and the next recession. However, with the majority of governments' 'tools' having been deployed and spent to prevent the Great Recession from turning into a global depression, fiscal policy and law makers around the world were left with fewer choices to 'manage' the next downturn.

Consider the impact that population change will have on our planet over the next 35 years. The United Nations (UN) predicts that by 2050 there will be 9.6 billion people living on the Earth.[1] It projects that the population of the developed world will remain largely the same, at approximately 1.3 billion, while the population of the 49 least developed countries will double—to approximately 1.8 billion. This has profound implications on natural resource utilization, the potential for a significant rise in global poverty (which is contrary to the long-term trend), migration patterns, and the possibility of political unrest.

The UN also notes that, globally, the number of people over the age of 60 is expected to more than *double* to 2 billion by 2050, and that by that time, for the first time ever, those over the age of 60 will outnumber children in the world.[2] Just think of the implications on national budgets, entitlement programs, and health care. What pension and insurance actuaries call 'longevity risk'—the effect of people outliving financial models—is a very real global challenge, widening the gap in most developed nations' tattered safety nets.

Water is becoming increasingly scarce globally. More than half of the known freshwater sources available for human consumption have already been used, and if current trends continue, by 2030 the OECD predicts that 90 percent will have been used. If consumption patterns remain the same, approximately two-thirds of the world's population will be living in water shortage conditions by 2025. This has profound implications for a host of issues, ranging from food production capabilities to poverty to the spread of disease to the ability of businesses to function to a rise in the propensity for water wars.

There are numerous environmental issues worthy of being mentioned in this context—such as the debate about global warming, climate change, expansive droughts, and the growing intensity of hurricanes. All of these impact more than just a city, state, or country; they are *global* issues. Depending on which side of the debate you may find yourself—either that global warming and climate change are man-made and preventable, or that they are simply acts

---

[1] United Nations Department of Economic and Social Affairs, *World Population Projected to Reach 9.6 Billion by 2050*, June 13, 2013, accessed November 24, 2015, http://www.un.org/en/development/desa/news/population/un-report-world-population-projected-to-reach-9-6-billion-by-2050.html.

[2] United Nations Department of Economic and Social Affairs, *World Population Ageing 2013*, 2013, accessed November 24, 2015, http://www.un.org/en/development/desa/population/publications/pdf/ageing/WorldPopulationAgeing2013.pdf.

of God and are not preventable—they can be addressed proactively (through risk management) or reactively (through policy implementation and/or post-crisis clean-up and rebuilding). As long as 'the debate' lingers, and there is no majority agreement about cause and effect, there is little that can be done from a risk management perspective to address these issues.

Also worthy of mention are the growth of the global middle class, creating hundreds of millions of new consumers who collectively demand, consume, and impact national economies and environments. There is also mass migration as a result of conflict, war, famine, and natural disasters. When hundreds of thousands—or even millions—of people change location, they impact food and water supplies, national economies, and political and security landscapes. And there is the ever increasing threat of terrorism, cybercrime, and computer viruses that have come to dominate news headlines, with more significant global impact with each year that passes.

There can be no doubt, as a result of these forces all occurring at the same time, that the nature of *systemic* risk is itself changing. Our world is interconnected and one event or circumstance half-way across the world can and does impact populations on the other side of the globe. Politics, economics, and financial issues are inextricably connected *globally*. An earthquake can unleash a tsunami thousands of miles away. A computer virus can spread at the speed of light. Our world is not only globalized, it is *fragile*.

The need for twenty-first-century risk management designed to tackle these issues is absolutely essential, yet risk management protocols are playing a game of constant catchup. As a result of the dynamic and constantly changing global landscape, risk managers and decision makers are struggling to respond, and they have been forced into a state of *perpetual reactivity*. While it is absolutely true that some of the examples presented simply cannot be predicted or prevented, others can, but even those require a change of mindset, standards, and protocols in order to address them proactively.

Evolving Perceptions of Risk

Every year the World Economic Forum (WEF) publishes a noteworthy study that discusses the changing nature of the global risk landscape. In its 2015 report[3] a comparison is made of the top perceived risks since 2007, in terms of both their likelihood and their impact. Asset price collapse was the most

---

[3] World Economic Forum, *Global Risks 2015*, 2015, accessed November 24, 2015, http://www.weforum. org/reports/global-risks-report-2015.

likely risk perceived in 2008, 2009 and 2010, but as soon as the worst of the Great Recession was over, focus quickly shifted to storms and cyclones in 2011, severe income disparity in 2012, 2013, and 2014, and (given the rise of the IS) intestate conflict with regional consequences in 2015.

At first glance, one may conclude that the likelihood of perceived risk appears to change according to what is most prominent in the news at any particular time. If income disparity was not such a high-profile aspect in the news during those years (which it was), is it likely that 'perceptions' would have been so focused on it? Perhaps not. When we look at the WEF study to learn what were the top perceived risks in terms of *impact*, asset price collapse was the top concern again in 2008, 2009, and 2010, but fiscal crises and the risk of major systemic failure remained on people's minds from 2011 through to 2014. While the statistics told us that the crisis had eased (based on consistently lower unemployment figures and growing Gross Domestic Product (GDP)), the impacts of the Recession on people's *psyche* still lingered.

Psychology is, of course, an important component of how risk is perceived, assessed, and, ultimately, acted upon. If we believe the risk of an event happening is low, we act differently than if the opposite is perceived to be true. But what happens if we believe the risk of an event occurring is high (or if the risk environment is high) and we know there is little or nothing we can do about it? Many of the perceived risks of today did not exist a decade or even a year ago (the IS did not exist in 2013, for example, but by 2015 it was front-page news on a daily basis). Terrorism on a global scale is a relatively new phenomenon, as is the widespread nature and high impact of a broad spectrum of cyber risk—from petty cyber crime to state sponsored cyber terrorism and extortion.

An individual can easily manage the risk that he/she will be kidnapped by a terrorist by simply not traveling to a country where terrorism or kidnapping is prominent. The same cannot be said about becoming a victim of 'homegrown' terrorism, which is random in nature and can occur in any place at any time. By the same token, an organization can avoid the risk of political violence by deciding not to operate in a dangerous environment, but that may not prevent it from becoming a target of, to take two random examples, cyber crime or a tornado. The list of threats which can impact an organization is definitely growing, as is the list of perils outside of our control.

The top five global risks of highest concern over the next ten years, according to the WEF, are interstate conflict, unemployment/underemployment, extreme weather events, water crises and cyber-attacks (in other words, political, economic, environmental and technological risks). None of these is within the control of an individual or organization, and some are not even in the control of a government. Any 'Chinese Wall' that may have been perceived to

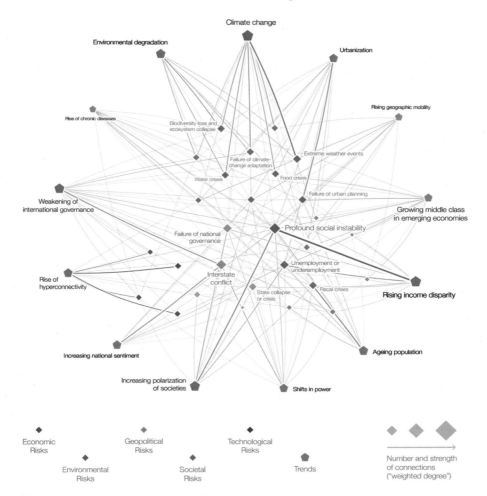

Illustration 1.1   Interconnected global risks

have existed between these categories no longer exists—either in terms of perceived likelihood or impact. So, put another way, there is fact and fiction, perception and reality, and what appears at the surface versus what lay beneath. The WEF produced the above Global Risk Map[4] (see Illustration 1.1), which is useful visualizing just how interconnected perceived risks are.

It is abundantly clear that a fundamental re-examination of how risk managers and decision makers think about risk—both in general terms and in *global* terms—is in order. We no longer have the luxury of assuming that an

---

[4] World Economic Forum, *Global Risks Report 2015*, 2015, accessed November 24, 2015, http://reports. weforum.org/global-risks-2015/part-1-global-risks-2015/fragile-societies-under-pressure/#frame/607b4.

event in another part of the world will not impact our ability to function at home, or elsewhere abroad. We now know that is simply not the case. The age of global shocks is upon us and risk agility is our best countermeasure.

Preparing for Global Shocks

To be considered a global shock, an event must affect more than one continent; it need not affect the entire world. Global shocks are, by their nature, also rapidly onset. So while climate change does not qualify as a global shock, a pandemic does. It is the rapid nature of such onset that makes global shocks difficult to manage. Most are unpredictable (when, how, and how severely a financial crisis may occur), but by the same token, most *types* of global shocks have happened before. It is unusual for a totally new form of shock to arise (i.e. we have seen pandemics, financial crises, and acute social unrest before).

What distinguishes a global shock from that on a local, regional, or national level are the types of interconnections. This implies the need for risk managers and decision makers to take a more systematic view of risk that links the causes of risk with their potential impacts and costs. For example, a local conflict in a coal-producing region of Indonesia affects the ability of a Chinese coal-fired power plant to operate, which results in a power shortage in a production facility based in China. Or a drought in Brazil decimates the soy bean crop in Brazil, preventing a firm from meeting its production targets for animal feed.

Although global shocks are becoming more frequent, they are *infrequent* when compared with a variety of other forms of risk (such as a one-hundred-year storm or a volcanic eruption that can darken the Earth for months or years). This makes it more difficult to produce criteria and models to predict and test them, relegating risk managers to incomplete or even experimental data. If there is any good news about the existence of global shocks, however, it is that because they appear to be increasing in frequency, we are gaining more data and insights into how and why they occur. The growing interdependence and concentration of triggers, and the mobility of capital, goods, people and information, all help us to better understand how and why some global shocks can occur. On this basis, then, what are some of the things risk managers can do to prepare for global shocks before they occur?

1. Create databases that contain the information that will allow for the identification of critical points of exposure, paying particular attention to People at Risk. Where is your organization vulnerable? What new forms of vulnerability exist today that did not exist five years ago?

2. Generate an assessment of conditions or events that would result in a breakdown in the risk management value chain, and identify the likelihood that such events would occur. What critical systems would be impacted if this were to occur, and what would be the likely impact?

3. Consider what safeguards can be built into the risk management matrix. There may be little that can actually be done here, but simply going through the process of considering your options can produce useful feedback.

4. Think about the monetary impact of a global shock on your organization. Would it result in a complete cessation of operation? Does your firm have sufficient financial resources to resume operation without relying on external funding? Is that funding available on short notice? If not, what options, if any, would you have?

5. What about direct and indirect consequences? Decision makers should consider how their chosen course of action may actually create new risks that are unforeseen. What additional liabilities might be produced? What are the unintended consequences?

These kinds of questions are easier to answer in a local environment, but with global operations there can be many manifestations of actions taken half a world away. Has the risk manager or decision maker ever been to all the locations where the business is operating? If so, was it more than a year ago? If so, you may not have the information, insight, and tools needed to prepare for a global shock. While many global organizations embed 'native' risk managers in their global or regional operations, many challenges remain in balancing risk culture and appetite. Thus, despite all the global decision-making frameworks available, few examples of truly agile enterprises exist.

The gaps in data, vulnerability assessments, and effective action plans can be daunting. Information from a remote or distant source may not even be reliable, which can skew the entire decision-making process. That is why assessing the risk of global shocks requires a more sophisticated blend of information, expertise, procedures, and cooperation. It is necessary not only to understand how the various components of the risk matrix blend together, but also the strengths or weaknesses embedded in those interdependencies.

Simply put, the larger and more sprawling the enterprise, the more likely that critical risk signals will get lost, go entirely unheard or get misinterpreted. This is known as the signal-to-noise ratio. When the noise is communicated in multiple languages, it is filtered through differing degrees of risk tolerance and travels across the world, exposing even the best-prepared firms. The other challenge is latency in risk related information—when company culture dictates that one should hesitate to sound the alarm, meaning vital information travels more slowly than it should.

By their very nature, global shocks are fraught with uncertainty and incomplete information. In many parts of the world, it is simply not possible to get accurate data. While some countries are transitioning from developing to newly industrialized countries, it will be many years before they create the types of information gathering and reporting regimes that allow for the collection and transmission of accurate data. Risk management systems must be adapted to account for such gaps and inconsistencies.

Global shocks are, by definition, large, high-profile, often-irreversible and with long-lasting consequences. The source of the problem may not be immediately known, your organization may not have previous experience of dealing with the problem, and a solution may require reliance on parties you may have never dealt with before. Cultural and language differences can inhibit effective communication, which can have a knock-on effect on getting the issues resolved. The resources needed to solve the problem may not be readily available, or not available at all.

Managing global shocks therefore implies the need for a range of inputs and resources that include the ability to integrate resources and information from the private and public sector together with that of the academic and NGO world, to produce the best possible planning tools. Clearly, government is an essential component of post-crisis management, yet coordinating with a foreign government sometimes means having to do things you may not otherwise ordinarily do (such as providing 'facilitation payments'). Doing so can have legal and policy implications. In short, the complexity of problem solving for global shocks is unprecedented and requires adopting a broader view of risks and potential solutions.

Taking a Broader View of Risk

Against this dynamic backdrop, where change is the one constant, and where interdependencies imply inherent uncertainty, past experience is not necessarily helpful in determining the best course of future action. What may have been true yesterday is, almost by definition, not going to be the case tomorrow. By default, organizations must adopt forward-thinking and forward-looking approaches to managing risk. In today's world, it is the forces that drive change—and the ability to identify and embrace them—that can matter the most.

While a broad range of quantitative and qualitative tools of course exist to help manage risk—and it is often best to integrate both types into an organizational toolkit—finding the right mix of tools and determining the most efficient ways to use them can make the difference between success and failure. Determining who in an organization should even assume an appropriate risk management role, what an appropriate role should be, and how it can be usefully integrated into an organization can also be daunting challenges. There are times when a Chief Executive Officer (CEO), Chief Financial Officer (CFO) or Treasurer, or mid-level manager must become a de facto risk manager, and when a risk manager must become the decision maker. Nevertheless, there is no substitute for an independent Chief Risk Officer (CRO), empowered as a *c-level decision maker and reporting to an independent board level risk oversight committee*. We will expand on this risk architecture later in this book.

Taking into account the wide range of factors that must be considered in choosing how to navigate in this kind of risk management landscape, risk prevention and mitigation are naturally important components of the risk management value chain, but so is the extent to which risk management is integrated into the corporate and strategic planning process. So often, organizations place risk management in the post-planning stage, when they would ultimately be much better off making it a central component of the planning process.

For example, adapting organizational processes to encompass information gathering, early warning, vulnerability identification and ongoing monitoring as a standard element of the planning process may require that additional resources be devoted to the planning process, but will ultimately save time, effort, and presumably money, in the long run. Given the prominence of terrorism, cyber-attacks, and natural catastrophes, organizations that fail to incorporate risk management into the planning process could suffer significant consequences that would otherwise be either preventable, or at least more easily managed in the process.

Adopting a broader view of risk that places an emphasis on having specialized knowledge inside an organization to specifically address issues that are, or will prove to be, critical to the organization is no longer just nice to have—it is essential. Ensuring policy consistency, prioritizing risks, exchanging information transparently, and making *best* practices the *standard* practice are also essential to effectively managing risk today. Moreover, the ability to understand what 'the process' tells a decision maker is equally important.

If risk management were based solely—or even primarily—on past data and experience, and if risk managers and decision makers were unable to adapt to the underlying conditions that define risk, the game would be over before it began. By the same token, adopting a 'reactive' approach clearly will not work. What is required is a proactive approach, which entails *closely monitoring the forces that drive change, analyzing their relationship to risk, and adapting the strategies required to manage risk flexibly.*

Global Risk Agility

This is what we call *risk agility*. Part of the challenge of adapting risk management to the evolving new normal is the tendency for the discipline to be backward looking. From the earliest pioneers of statistical methods and probabilistic science to the quants who mastered and came to define Wall Street, so much of risk management today is not based on anticipating tomorrow or next year, but on reading the tea leaves proffered by deep data sets of past observations. Once upon a time this careful, methodical backward-looking approach to risk was much more forgiving—akin to crossing the street while looking backwards. In the 1800s this foolhardy endeavor may have spared the pedestrian, as the sound of oncoming horse-drawn carriages may have signaled an emerging risk.

Today, navigating the risk landscape using backward-looking tools would be like trying to cross a busy superhighway while blindfolded. While we must integrate the lessons of history into the risk analysis process, risk agility is all about *harnessing those lessons and combining them with rigorous instinct.* An emerging class of risk-ready firms would not shun data and analysis from risk traditionalists. Instead, they would elevate the findings, depending on their materiality, to an interdisciplinary, multigenerational body that can challenge data-driven assumptions by benefitting from the lessons of history to form the general direction of a firm's strategy.

Risk agility is not all about downside risk—quite the contrary. Risk-ready organizations often have outsized financial results, with an innate ability to read the market and gain a first mover advantage. Agile enterprises can neutralize bad news and have a propensity to turn bad information and risk into an upside. They do this by confronting risk head-on. Their orientation is not the 'analysis paralysis' that can take hold of larger, more cautious organizations. Since doing nothing is often more dangerous than taking risks, agile enterprises are, by definition, entrepreneurial and instill a culture of bounded risk taking at all organizational levels. This is seen in innovation labs, high-performing innovative teams, or the Skunk Works concept popularized by Lockheed Martin.

In order to survive and thrive in the new emerging and highly interconnected risk landscape, even the largest organization must embrace the spirit and lithe response of the entrepreneur. While many may criticize entrepreneurs as having the propensity to be reckless and lacking bearing between their goals and their resources, we believe that entrepreneurs and their mastery of the creative–destructive cycle are the best agile risk managers. Each day a new entrepreneurial enterprise makes life and death decisions.

Miles away from where the rubber meets the road, corporate and political leaders of large and complex organizations operate as if they were impermeable to risk. There is often no perceived sense of finality in corporate board rooms, or in state houses and capitols around the world, as if the consequences of action or inaction are someone else's problem. Yet embracing a sense of urgency, finality, and consequences is healthy, and helps temper our approach and response to risk. It also helps temper the managerial and decision-making bravado most evident in high finance that continually threatens to take the global economy over the edge. While some would argue that not 'owning' the consequences of your action may help you sleep at night, we would argue just the opposite.

Global risk agility is as much about being bold and *taking responsibility* as it is about having both temerity and deference to the unprecedented upsurge in man-made risk. The agile risk manager is in many respects a new renaissance leader—who knows as much about quantitative as well as qualitative risk, is a student of history as well as a master of current events, and understands the long-term impact of actions being taken today, rather than merely being short-term and profit orientated. That's right—a risk manager and decision maker with that rare commodity—a *conscience*. In this is the era of the collision between man-made and natural risk, agile risk managers are a new breed of organizational leaders who are part *psychologist* and part *financial guru*. To rise to the occasion and help organizations harness the unprecedented, yet unforgiving upside that can be created, we must move the discipline of risk management from being a *business prevention* function—a cost of doing business—to being a *catalyst for longevity*.

Agility implies a *deftness of movement*; whether the required pace given the circumstances is slow or fast—movement is the key. Standing still is usually not an option. We do not need to be taught that observing smoke rising from a few floors beneath us and standing still is clearly ill advised. While the data may suggest that everything is fine, silencing your instincts occurs all too often in organizational decision making. Agile risk managers therefore need to be as courageous in confronting authority as they are poised in the face of new challenges.

The conflation of CEO–chairman roles in large public firms, and the larger than life personalities in politics, means that critical risk signals will often fall on deaf ears. Who can confront an all-powerful charismatic CEO? As the *Economist* magazine says, *in a meeting of the minds, you would not be sent out for the biscuits* (Taylor 2007). Herein lies one of the principal organizational challenges to building agile risk management organizations: When decisions are made at the global level, risk managers are most often not in the room—and they need to be.

Changing this reality will require discipline. Professionals in the decision-making and risk management domains need to take a long look in the mirror. They need to *adapt personally* as much as the organizations they work for—particularly since organizations are beginning to leave boardroom and corner office doors ajar. For organizations to fully harness the power of risk agility, the competitive advantage of survivorship—long-range planning, entrepreneurial culture and bounded risk taking—those at the top need to be challenged. Many of them need a cold hard slap in the face, but not by providing doom and gloom tales of what can go wrong. That is unlikely to be effective. Rather, the challenge must come in the form of dialogue that fosters the ability to make sensible decisions under opacity. Sometimes the best long-range results are created when the playing field has been entirely abandoned, such as with investments in frontier markets.

Much as incalculable wealth was created by the pioneers and robber barons that led America's westward expansion, so too pioneering firms rely upon bold boardrooms and sound risk strategies. As we have seen with great frequency, the biggest cause of loss to a company's market value is not the risk of the "smoking crater," but, rather, a firm's inability to respond to and anticipate market change. Thus, the ideal of the entrepreneur is the basic building block of the agile risk manager; someone who is not afraid of making decisions, but who at all times is intimately, almost instinctively, aware of how dire the consequences can be. A master of no particular risk domain—for there is a risk in consulting a specialist when a generalist is needed—the agile risk manager has a healthy dose of fear of risk, but constantly tries to tame it in order to propel an organization forward.

2

Risk Management as a Process

Risk Is a Process Not an Event

Risk management is often treated as if it were a discrete time-defined event, rather than a process. Many commonly used methods to 'control' risk, such as annually renewable insurance policies, employee manuals, and other 'check-the-box' processes, may only be referred to infrequently over the course of a year (if at all). Such 'passive' methods produce a placebo effect, creating a sense of security, but often failing to identify and manage risks that may be lurking in the recesses of an organization's processes, and in the minds and through the actions of its people. When the external and unpredictable forces at play in our turbulent world are added to the mix, it is clear that new solutions must be included in the risk manager's arsenal in order to stay ahead of the risk curve.

This placebo effect is dangerous for many reasons. The first, as illustrated by the course of the financial regulation process in the U.S. since 2008, is that merely checking boxes does not necessarily mean a firm is safer, or that the systems protecting it are, in fact, effective. For example, Dodd–Frank (the Dodd–Frank Wall Street Reform and Consumer Protection Act) has been watered down consistently as a result of the efforts of lobbyists in Washington since it became law. The second, as seen with audit committees and in corporate governance, is that hitherto safe, thoughtful, and high-quality organizations have been brought to their knees by not effectively managing risk—which often

© The Author(s) 2016
D. Wagner, D. Disparte, *Global Risk Agility and Decision Making*,
DOI 10.1057/978-1-349-94860-4_2

could have been tamed had a more adaptive risk management approach (one that recognizes that risk is not an event, but rather, a process) been adopted.

As is the case with other processes, the notion of dynamism and change is as true for risk management as it is for other complex undertakings. A meteorologist, no matter how often he or she may make the right call, cannot meaningfully forecast the weather annually, monthly, or even necessarily weekly or daily with true precision and accuracy. The same is true for risk leaders and key decision makers in an organization. There is simply no way they are going to get it right every time. The management process should track in sync with anticipated changes in the risk landscape. One should not assume that conditions that were once acceptable or measured through observations will remain constant over time.

The Germanwings disaster, mentioned earlier, underscores the perils that can emerge when organizations treat risk as a constant variable. This is one of the principal drivers of the sense of repetition we see so often in the news headlines—the feeling of history repeating itself when calamities occur. That is what is produced when change, risk and other dynamic variables are treated as constants, never properly mitigated, and in most cases entirely misunderstood. From time to time, truly unprecedented risks emerge, requiring that the entire edifice of risk management and resilience be rebuilt. We are squarely in this era.

While many classical risk management frameworks labor under the challenge of being static approaches, they nonetheless offer useful heuristic frameworks for how to think about risk. If there is a constant variable in risk management, it is the notion of an exposure, and of being *exposed* to a peril. The exposure may vary over time, however, the notion that some aspect of a firm's global operations (or the people behind the profit and loss) are exposed should be thought of as a permanent condition. Even when an exposure is withdrawn, new exposures inevitably emerge.

Consider, for instance, companies that either retreat from or refrain from investing in countries with perceived high risk. While wanting to safely manage their exposures, these companies may inadvertently create exposure in their financial results by deliberately foregoing higher returns and market access in pursuit of the next wave of consumers. 'Safety' is a relative term. Just ask companies with the bulk of their investment in the U.S. and Europe prior to and during the Great Recession (which also serves to demonstrate just how inaccurate *perceived* risk can be). Purely based on limitations for growth, over time, remaining on the sidelines of emerging markets is a real challenge for global organizations, which are, nonetheless exposed to global supply chain shocks, financial contagion and other risks in a highly interconnected world where decoupling the fortunes of companies from countries is not possible.

Ironically, in protecting against a known risk, we can be exposed to entirely new forms of risk. As previously noted, fortified cockpit doors certainly made flying in commercial airplanes safer—or at least created the placebo effect of safety. However, in so doing, what seemed like an incredibly remote risk—being exposed to a suicidal airline pilot—manifested itself with great ease in the Germanwings case (even though there had been more than one previous incident in recent history where a pilot or first officer are believed to have deliberately crashed a commercial airplane—such as SilkAir 185 in 1997 in Indonesia, EgyptAir 990 in 1999 in the U.S., and the infamous Malaysia Airlines 370 in the Southern Indian Ocean). In attempting to allay this concern, however remote, risk managers will surely create new deterrents based on what we understand to be the risk at the time.

However, as noted above, these deterrents can create new risks. For example, the risk that terrorist groups would enroll devotees into flight school specifically to train pilots for suicide missions had not reached our collective consciousness until 9/11. Reinforced cockpit doors and adding cameras outside those doors cannot prevent an otherwise qualified pilot from making his (or her) way through the ranks to become an airline pilot. It does not prevent a passenger from accessing an unlocked aircraft's electronic equipment compartment.[1] It does not prevent a terrorist from shooting down a plane with a shoulder-held anti-aircraft missile. And it certainly does not prevent a shoe bomber from trying to bring down a plane (premium passengers in many countries no longer need to take off their shoes during the security screening process, and requiring passengers to remove their shoes during the process never became part of the protocol in many countries to begin with). In short, removing one form of exposure can create new forms of exposure that are amplified by the hardening of vulnerabilities in a system.

While no airline company is invulnerable to incredible odds against unforeseen events occurring—as was underscored by the dual inexplicable calamities visited upon Malaysia Airlines in 2014—the airline security system has in general been made safer by the occurrence of individual catastrophes. System-

---

[1] It is worth noting that in the case of MH370, Malaysia Airlines had not adopted the protocol of ensuring that no unauthorized individuals were able to visit the cockpit. Passengers (particularly attractive young ladies) were reported to have been occasionally invited inside. Malaysia Airlines also opted *not* to lock the electronic equipment compartments in their aircraft. One of the theories behind MH370s disappearance is that someone disabled the aircraft's transponders from this compartment, the entrance for which is located just outside the cockpit door. There was no requirement that the compartments be locked prior to MH370s disappearance.

wide exposure in the airline industry is not the mechanics of a single aircraft or even the risk of terror attacks involving multiple carriers. System-wide risk is in the end technological, as was highlighted in the August 2015 computer glitch in the air traffic control system along the northeastern corridor of the U.S. More than 400 flights were cancelled and the image of a 'black hole' emerging in the otherwise busy U.S. air traffic radars was an ominous portent of airline vulnerability.[2] Of course, plenty of other system-wide computer glitches have occurred in the U.S. and elsewhere over the years, but it seems it is only when an 'event' impacts thousands of people that it really gets attention.

Similarly Chris Roberts, an ethical hacker and cyber security expert, claimed to have hacked an airplane's flight control systems while in flight, bringing a new specter of risk from the so-called internet of things.[3] On a lesser scale, in 2015 General Motors (GM) experienced an embarrassing recall of its vehicles using the OnStar guidance system, over concerns that a vehicle's operations could be taken over electronically while being driven. The system-wide exposure to technological and cyber risks has never been greater, yet our ability—at all levels of risk management—to effectively address these challenges has rarely been weaker. With a low overall level of readiness, it is worth reviewing some classical risk management frameworks to understand how they can be adapted to suit the twenty-first century risk landscape.

Current Frameworks and Their Utility

Risk management as a profession is coming of age, buoyed in part by systemic failures during the global financial crisis, with some academic institutions trying to shed new light and vigor on this practice. Risk management as a discipline remains largely the domain of quants and insurance professionals who have far too many acronyms following their names. Too often, ordinary businesspeople have a hard time understanding how risk management works, and how it should be integrated into the work that they do. In order to more fully integrate risk management into business lexicon and bring it to its full potential, the function of risk management needs to move from being perceived as

---

[2] F. Kunkle, "FAA, Airlines Still Working to Resume Normal Air Traffic After Major Glitch," *Washington Post*, August 16, 2015, accessed November 24, 2015, https://www.washingtonpost.com/local/trafficand-commuting/faa-airlines-still-working-to-resume-normal-air-traffic-after-major-glitch/2015/08/16/2f97 3a48-442c-11e5-846d-02792f854297_story.html.

[3] K. Zetter, "Feds Say That Banned Researcher Commandeered a Plane," *Wired*, May 15, 2015, accessed November 24, 2015, http://www.wired.com/2015/05/feds-say-banned-researcher-commandeered-plane/.

being strictly the domain of risk prevention, or a regulatory function, to being understood to be a catalyst for effective decision making. While part of this transformation is underway at leading business schools, some of which have turned their legendary faculty to the task of modernizing these frameworks, lasting change requires adapting classical methodologies to a new era.

To begin with, risk management has two very different broad definitions, both of which are prevalent in the insurance and financial worlds. The insurance domain, where so much of this discipline has withstood the test of time and has in some ways become ossified, with well-established frameworks to address pure risk based on the principle of indemnification and diversification. Simply put, there is no upside in insurance for the recipient of coverage—if a smoking crater is all that remains of your business, the principle of indemnification would hold that you should not profit from such loss, but merely be made whole again.

Risk in the financial services world, particularly in banking, has a more volatile and speculative connotation, in which practitioners must contend with both a downside and an upside. It is through this concept of volatility, periods of rapid swings in market signals (e.g. share prices) where above-average profits can be made. As a result, banks have come the closest to using risk management as a source of competitive advantage. However, few risk managers in global financial institutions could pull the handbrake if their firm was careening over the edge. This much remains true despite the entire overhaul of the global macroprudential regulatory framework. Merely checking boxes, even if billions of dollars and entire economies are at stake, does not make for a sound risk management framework.

Rather than making the financial system safer, the relentless pursuit of more boxes to check and added layers of regulatory complexity to comply with have triggered a perilous game of regulatory capture. While the resulting system derived greater shock absorption capacity from the *Great Deleveraging*, inherent risk in the system has been *shifted*, but not entirely *removed*. The next financial crisis will emerge in an entirely new domain, for which all the added layers of complexity and regulatory capital will not suffice. It is likely these risks will emerge from the shadow banking space, with opaque hedge funds, which are arbitraging entire economies and high debt ratios. So, for all the efforts to make the general banking system safer, few safety nets are present in areas of the global economy that are overheating, and a raft of 'check-the-box' risk management approaches is causing well-resourced global banks to combat regulatory complexity with operating complexity, yielding a riskier system.

One of the chief factors that amplifies risk in the twenty-first century is speed. A greedy, data-rich global economy that is binging on instant gratification is a dangerous place to be—and yet so many people seem to want to be in the middle of it. To draw a parallel, think of all the safety features in an automobile: seatbelts, headrests, airbags, and a vehicular risk management checklist are ultimately of little use in a high-speed head-on collision. The same holds true in the modern economy where speed is the dominant force, and speed blindness is a dominant malady.

Perhaps one of the greatest examples of this speed trap is the case of Knight Capital, a high-frequency trading company. Algorithmic trading firms exploit information asymmetries in the market by executing buy and sell decisions in fractions of a second. Knight Capital—a paragon of its industry—lost $440 million and was closed and sold as parts to rival firms following the release of a so-called "rogue algorithm". This untested set of decision rules over the course of seconds in the trading day on August 1, 2012 bought up millions of market positions that Knight Capital could not afford. In the twenty-first century, companies worth hundreds of millions or even billions of dollars, and surely a best practitioner of check-the-box risk management, can be destroyed in seconds through this practice. As a result of these events, Knight Capital was left in a smoking crater nearly twice its market capitalization of $296 million.[4]

In an earlier, much slower trading day, one can imagine a code of honor enabling a firm like Knight Capital to survive following the activities of a rogue trader, a 'fat finger', or an errant 'buy' decision—especially one executed by an algorithm. In the era of Black Friday, Prime Day, Cyber Monday and flash crashes, the market does not tolerate a single misstep, yet the ability of rogue traders and rogue algorithms to continue to function, and flourish, continues, despite continuous lessons learned.

Classical Risk Management Frameworks

Risk managers have a number of established approaches in their arsenal, each seemingly more formulaic and involving additional layers of control, checks, and balances. Enterprise Risk Management (ERM) remains *en vogue*, and many decision leaders pay lip service to it as if it were the panacea for internal risk governance. ERM's basic building blocks of identifying, analyzing, measuring, and controlling often treat risk as if it were a static object. Devotees to ERM spend countless hours going through internal risk audits involving

---

[4] M. Philips, "Knight Capital Shows How to Lose $440 Million in 30 Minutes," *Bloomberg Business*, August, 2, 2012, accessed November 24, 2015, http://www.bloomberg.com/bw/articles/2012-08-02/knight-shows-how-to-lose-440-million-in-30-minutes.

all organizational levels and geographies. This annual ritual is usually a part of the planning process and serves to identify solutions for known risks, but often fairs poorly in the face of known unknowns. Needless to say, it does not bode well for an organization's risk agility that it must go through an annual bureaucratic process to respond and survive unknown dynamic forces.

The International Organization for Standardization (ISO), a global non-profit organization based in Switzerland, has made its core business the development and adoption of global standards. In the area of risk, ISO 31000 is widely considered to be the gold standard for organizational risk management. It is a useful framework in that it has helped codify common language and processes around risk management. However, much like its predecessor (ERM, from which it borrows much of its structure), ISO 31000 invokes too much rigidity to be the guidepost for a risk-agile organization.

Such standards give organizations the impression that they are risk-ready, when in reality they often overlook key structural blind spots. The first is paycheck persuasion. Which business line risk manager is equipped to halt an operation in the face of a known calamity? Risk governance as the reality of how corporate decisions are made and who ultimately remains in the room when the shots are called diminishes the value of these otherwise sound approaches. The key to moving risk management from being a *business prevention function* to being a *catalyst for growth* is to adapt both the talent pool and the methodologies to a new landscape.

ERM and ISO 31000 are the prevailing standards for risk management in non-financial firms. Financial institutions, particularly banks, rely on quantitative mathematical models as their principal risk scorecards. Chief among these is the value at risk (VaR) measure, which is supposed to give banks a forward-looking assessment of how much value is exposed to losses, based on prior trading periods or market positions. In addition to VaR, regulated financial institutions rely on capital buffers and solvency standards imposed on them by their supervisory authorities. These layers of liquidity are supposed to serve as shock absorbers for the global financial system by serving as the first line of defense in a financial crisis. If the 2008 financial crisis taught us anything, liquidity is indeed king in guarding the system, but much work remains to be done, particularly on systemic interconnections and the shadow banking arena. If liquidity is king, confidence to lend is queen in resuscitating an economy.

ERM, ISO 31000, and quantitative financial models all offer value in their own ways, with their distinct applications. The reason they labor to keep pace within a dynamic global environment is principally due to the preventable risk of poor risk governance, where most risk management departments are consigned to being business prevention specialists. While typically an adjunct to finance, risk managers struggle to find their place in non-financial and financial firms.

In the non-financial firm, a risk manager is often conflated with an insurance procurement officer or security specialist. In truth, a risk manager is a hybrid of the two, with the added function of unlocking *upside* from risk and organizational decision making. In financial institutions, risk managers are most assuredly in some kind of internal or external compliance function, keeping the organization's guardrails up, as line managers in profit centers frequently toe or transgress the regulatory line. In all global organizations, the fear that geographically remote risk managers go "native", or are entirely isolated, creates many organizational blind spots. Fewer still are able to sound the alarm warning about imminent danger, whether for paycheck persuasion, fear and other reasons.

Matching Process with Process

Risk is dynamic, yet most tools in the risk manager's arsenal are not only static, but may in fact be backward-looking, creating a fundamental mismatch between the object of control and the methods used. Take, for example, the average insurance policy, whose conditions are held constant for 12 months, with limited or no interaction, save for the occasional amendment or claim. The placebo effect on the average insured that everything is covered during the 12-month operating cycle often leaves blind spots and coverage gaps— Murphy's Law would have it that this is precisely the areas where losses will arise. The typical purchaser of any form of insurance fails to read the fine print until there is a claim, and typically does not consider how the policy can be improved at the outset. There is also the pernicious effect of taking and enhancing risks that would have otherwise not been incurred under the guise that they are covered or hedged. Similarly, there is 'inertia risk'—the risk of doing nothing. Often entire market opportunities and blue oceans are missed when firms stay on the sidelines.

As an example, Sony once possessed all of the ingredients to beat Apple at the mobile content and platform game, where Apple is now ravenously redefining a number of adjacent markets, including banking, with the introduction of Apple Pay. Not only did Sony control mobile music with its once-ubiquitous Walkman, but it also had in-house content creation, movie and record studios, solid state drives and all other elements of what is now Apple's principal revenue driver and a category defining segment. One can only imagine the cultural and organizational force of gravity that held Sony back from piecing together these elements. Professor Clayton Christensen famously surmised this organizational blindness as the *Innovator's Dilemma*, where for fear of displacing their own revenues and making parts of their business obsolete, firms overlook potentially disruptive innovations.[5] Sony's embrace of risk was too rigid and narrowly framed around downside protection for it to take controlled risks in a market where it used to be a dominant force. It paid a very steep price for its failure to do so.

Indeed, when looking at organizational mortality, the leading causes of failure are market-related and often times insidious and preventable. SkyMall and its parent company Xhibit Corporation is another example of a firm that quite literally had a captive audience, but failed to adapt to a changing market environment that included airline passengers enraptured by their mobile devices.[6] Surely, the market for gewgaws, curios, and weird things for your pet did not vanish suddenly—demand merely shifted to alternate channels where SkyMall was obviously not present in a competitive way, or at all. What the SkyMall example illustrates is that business continuity is threatened not only by risks of the smoking crater variety, but also by the gradual degradation of market positions hastened by a managerial inability to adapt. In risk management, there is no 'glory' in fending off slow decay, and it is likely most strategists view their loss averse risk counterparts as part of the so-called business prevention department, rather than as a source of value creation.

Firms that build enduring business models and thus enduring value, harness risk and are therefore bestowed with enviable staying power. These firms also have an innate ability to reinvent themselves, define new markets, build ecosystems, and exploit adjacencies. Perhaps most importantly, these risk-ready firms seem to promote bounded risk taking at all organizational levels—ringfenced by a thin 'line of control' between profit and loss, set against the typical forces that shape the market. Risk-ready firms create a balancing act between risk factors and mitigating factors; the trick is to find the right balance.

Risk factors do not live in isolation from one another; rather, they thrive in often discernible domains, such as economic, political and technological risks, among others (see Illustration 2.1). While idiosyncratic risks may emerge, confined to a single domain, their amorphous and contagious quality often enables risk to cut across boundaries, shattering misapplied mitigating factors. Consider Sony's now-famous cyber-attack. What initially began as a technological risk also caused millions of dollars of hardware losses in the

---

[5] C. Christensen, *The Innovator's Dilemma: Creating and Sustaining Successful Growth* (Boston, MA: Harvard Business School Press 1997).

[6] T. Corrigan, "SkyMall Files for Bankruptcy," *The Wall Street Journal*, January 23, 2015, accessed November 24, 2015, http://www.wsj.com/articles/in-flight-catalog-skymall-files-for-bankruptcy-1422025308.

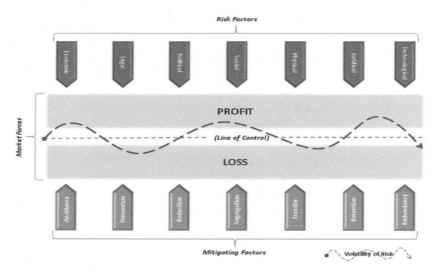

Illustration 2.1   Matching process with process (*Source*: Author Dante Disparte)

physical domain and millions more in litigation exposures in the legal and juridical domains. Although the financial exposures pale in comparison to the initial technological loss, like battling a wild fire, firms need to hold the line on many fronts in order to achieve dynamic risk control.

Evolving Over Time

In risk management there is an old adage that complex systems fail in complex ways, yet many risk management approaches deal with risk as if it were a discrete, time-defined event, rather than a dynamic and highly volatile process. Certain types of risk are acute in nature, but all risk lives along a continuum that is shaped by time and compounded by action (or inaction). In his books *Black Swan* and *Antifragile*, Nassim Taleb popularized the notion that fat-tailed (or leptokurtic) events are not as rare as once thought, calling for new frameworks for coping with their likelihood and, critically, surviving them altogether.

To begin with, a stylized probability distribution of risk can serve as a useful guide for how to think about risk, and how it evolves by severity and likelihood. The left-hand side of the diagram highlights high-probability low-impact risks that are traditionally mitigated or optimized in a firm's business model. These are acceptable attritional risks that many firms bear and have learned over many years of observation how to mitigate. Let's call this the

Fragile Domain, as organizations that do not adequately absorb or mitigate these risks tend to break easily.

The next category is the Robust Domain, the home of unexpected losses or, to borrow from Donald Rumsfeld, the 'known unknowns'. The priority in this category is to transfer risk and remain vigilant to emerging threats and exposures. By transferring this category of risk, firms gain a fixed price on uncertainty that can be built into their pricing models and overall strategy. They also gain robustness afforded by the liquidity, response of the private market (or risk pools), and by the process of diversifying away from potentially crippling losses, such as Sony's cyberterrorist attack. Judging by the upsurge in the purchase of cyber liability insurance, the market agrees that attempting to transfer these risks is the right approach. This category often produces systemic risk when risks are "transferred" (but not actually removed) from the system, much as during the Great Recession.

A brief etymology is needed to accurately explain the powerful concept of antifragility and the last domain in this model. Nassim Taleb coined the term 'antifragile' as he found there was no useful antonym to fragile in the English language (e.g., something that would benefit from shocks). If "fragile" means easily broken, and "robust" implies something that will remain in the same state, then antifragile implies something that by default becomes stronger with shocks. An apt classical example would imply that the Sword of Damocles is fragile (for it is held by a string), Phoenix is robust as she rises from the ashes in the same state, and Hydra is antifragile for the more you attack and cut off its head, the stronger Hydra becomes (see Illustration 2.2).[7]

Uber is a good an example of a firm that benefits from some antifragility in its business model. The more governments and angered taxi drivers protest Uber's arrival, the stronger demand becomes among prospective passengers and drivers. It is also difficult to "kill" a firm whose drivers and customers alike are part of a distributed on-demand business model. Yet, even asset-less firms are beginning to show cracks in their armor, as the backlash by Uber drivers against the company for its failure to provide a salary or benefits, attests.

Risk in the Asset-less Economy

As Uber continues to make headlines with both fawning stories of its hyper-valuation and tales of popular dissent, it is clear that risk is beginning to creep through the cracks of the asset-less economy. Economic risk always follows economic gain, and the so-called disrupters are particularly vulnerable to this

---

[7] N. Taleb, *Antifragile: Things That Gain from Disorder* (New York: Random House, 2012).

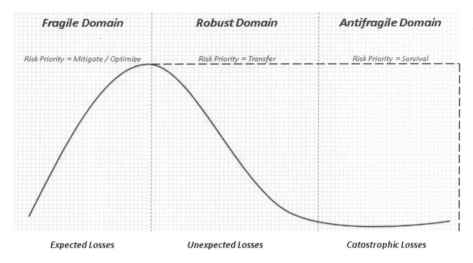

Fragile Domain   *Robust Domain*   *Antifragile Domain*

Risk Priority = Mitigate / Optimize   Risk Priority = Transfer   Risk Priority = Survival

Expected Losses   Unexpected Losses   Catastrophic Losses

Illustration 2.2   Stylized probability distribution (*Source*: Author Dante Disparte)

exposure. As the mounting case against Uber mutates from angered taxi drivers to now include state and national legislators, asset-less firms are not only fending off massive legal and regulatory complexity, the very essence of their business model, and thus their survival, is in the crosshairs.[8]

While Uber may be the world's largest fleet operator without owning a single vehicle, and Airbnb the largest hotelier without owning any property, their outsized returns are making them easy prey for economic risks. The chief culprit in most cases is not the angry people who have been disrupted (although they tend to be the loudest voices of dissent), it is regulatory bodies whose loss of oversight and revenues is triggering sustained attacks.

Counter-valence options against the state are few and far between, particularly as asset-less firms operate as extra-territorial web-based platforms. Clever arguments to combat legal impositions, such as "we have no employees" and "we are entirely amorphous and outside of your regulatory purview" are being shot down one by one in legal and geographical jurisdictions around the world. With growing legal precedence, asset-less firms will now need to acknowledge, price for, and ultimately own the reality that all the risks in their value chains are in fact theirs. To paraphrase Descartes, "you operate here, therefore you are".

---

[8] A. Breeden and A. Rubin, "Uber Suspends UberPop in France and Awaits Court Ruling," *New York Times*, July 3, 2015, accessed November 24, 2015, http://www.nytimes.com/2015/07/04/technology/uber-suspends-uberpop-in-france-and-awaits-court-ruling.html.

Against this backdrop, Uber's drivers may ultimately need to be treated like employees, as has been argued by California's legislators.[9] Uber will need to take an active view of vehicle-related risks, particularly abiding by third-party liability, personal injury and property damage provisions, which in most countries around the world are legally mandated. Airbnb may have to contend with multi-jurisdictional property insurance and a raft of legal and reputation risks, as numerous examples of rentals gone wrong attest.

Imagine the exposure of unionized Uber drivers, who can readily argue, particularly in socialist environments like France, that they represent a collective and thus need to be afforded a collective voice. Imagine the imposition of stringent carbon emissions and road safety standards on large fleet operators, which any state could readily argue includes firms like Uber. In the face of these changes, which a host of headlines suggest are very real threats, the competitive pendulum would swing back into the favor of firms with more direct control over their assets, human capital, and business models. Uber would struggle to abide by state-imposed emission standards, as it is merely the just-in-time aggregation of any driver and any vehicle available. A Pyrrhic victory in these cases would be to quit unfavorable markets or to cull parts of a business model, such as the decision to suspend UberPop, the firm's entry level service in France.[10]

An asset base is a source of both weakness and resilience. Asset-less firms are growing in all segments of the economy as consumers look for fractional access to everything from corporate jets to city bikes and vacation properties. The true asset of firms like Uber and Airbnb is their incredible marketplace for pairing supply and demand in consumer-controlled ways. Like all marketplaces around the world, the advent of regulatory burdens will not only erode profitability, it will create a source of friction that these firms are only beginning to experience in their rapid growth and relative youth.

While many asset-less firms are on a war footing, hiring people such as David Plouffe (a former White House advisor), they will need to take a more conciliatory tone if they are to remain in play in many countries.[11] Many of the financial and risk transfer mechanisms that are needed to keep the asset-less economy vibrant and global are themselves not keeping pace. For example, most auto insurance policies are designed as annualized contracts

[9] G. Miller, "California's Uber Ruling Could Erase Billions," *Wall Street Daily*, June 26, 2015, accessed November 24, 2015, http://www.wallstreetdaily.com/2015/06/26/uber-california-ruling/.

[10] Z. Sheftalovich, "UberPop Suspended in Paris," *Politico*, July 3, 2015, accessed November 24, 2015, http://www.politico.eu/article/uberpop-suspended-in-france/.

[11] N. Gass, "Obama's Damage Control Team Goes Corporate," *Politico*, June 9, 2015, accessed November 24, 2015, http://www.politico.com/story/2015/06/obama-white-house-staff-corporate-jobs-118786.

and not geared as a risk-per-kilometer traveled instrument. Similarly, financial regulation of the insurance industry makes it virtually impossible to offer the same global standards in terms of limits of liability and other policy provisions, which are often governed by arcane legal statutes.

Uber and Airbnb's customers, 'non-employee employees', and stakeholders rightfully have an expectation of the harmonization of the terms of trade, irrespective of borders. Herein lies one of the greatest existential risks to the asset-less economy. As Uber, Airbnb, and the like grow ravenously around the world, bringing incalculable value to freelancers, people will demand common treatment and benefits. So, if an Uber driver in California needs to be afforded certain company-sponsored protections, it is reasonable to assume a chorus for these demands will be raised elsewhere. Asset-less firms and their readiness to respond to and, indeed, anticipate this pressure will not only make these important businesses more enduring, they will also improve the conditions for freelancers who engage these models for their household income.[12]

Complex Systems Fail in Complex Ways

Notwithstanding its razor-thin profit margins, the airline industry is also considered antifragile, since each calamity does not cripple the entire system, but rather strengthens overall air safety. The concept of 'near-miss' management that has been codified in aviation has broad applications outside that industry. Even cyber risk management, for example, can benefit from this approach, if firms would only destigmatize the advent of a security breach and develop a central clearinghouse to report and record these losses. The risk priority in the antifragile domain is survival, and the general theme is that large losses (black swans) are not nearly as rare as once thought, and firms would be wise to prepare for their arrival.

When looking at the management and philosophical approaches to complexity, most tomes are themselves complex, leaving the reader more confused after reading them than they were when they began. Addressing complexity in our everyday lives, as well as in truly complex professional decisions, is more artistic or intrinsic than scientific. Yet, through social norms and codes of conduct, we have trained ourselves to ignore our instincts, which are often the best guides in truly complex and potentially perilous situations.

---

[12] D. Disparte, "Risk in the Asset-less Economy," *Huffington Post*, July 14, 2015, accessed November 24, 2015, http://www.huffingtonpost.com/dante-disparte/risk-in-the-assetless-eco_b_7789512.html.

Think of the number of times a fire alarm may go off in a downtown office building only to find that 99 out of 100 times it is a false alarm. Surely, every time, the instinct to run for the exit is suppressed by a sense of comfort while waiting for the fire marshal to announce the true reason for the alarm. In effect, we may put ourselves and others in danger in order to keep up appearances—particularly if we believe that there really is little reason to be concerned. Consider the true story of people who went back into the World Trade Center on 9/11, after the first tower had been hit, simply because someone in a position of authority said it was safe to go back in.[13] We have a tendency not to necessarily think about what we are doing if others are doing the same thing.

In many instances the ensuing chaos or paralysis can shatter even the most carefully mapped-out emergency response. When a hitherto unthinkable earthquake struck Washington, D.C. in August of 2011, there were a series of compounding "micro-failures" that culminated in the complete paralysis of the city's infrastructure. Traffic was at a standstill, firms were confused about whether to send people home or keep them in their offices, and the city's mobile and land lines were overloaded with traffic, causing many calls to drop. Amid the shutdown, which spared the District of major losses, the calm and instinctive were already home, while the scientific and socially correct may have been back at their desks or stuck in traffic.

While D.C.'s rumbling in 2011 was by no means a calamity, it provided a good study in how people respond to complexity and new information. The normative response of an individual to this situation may have been the 'wrong' one. Many individuals framing decisions as a collective often escalate their commitment to 'bad' choices, imperiling themselves and others. As in the case with the false fire alarms, it is often done with a sense of calm. These cultural norms that shape organizations are often the first and last lines of defense when it comes to complexity reduction, read here as a proxy for risk.

As an example, British Petroleum's (BP) 2010 Gulf oil spill is often seen as an isolated, 'black swan' event that could have happened to any of the large oil firms. Yet operational safety experts cited BP's previously lax culture of safety, habitual violations and routine 'micro-failures' as ominous signs that the 'big one' was coming.[14] Indeed, as later became clear in proceedings, there was a long and very complex string of failures that led to the disaster, for which the first and last lines of defense failed; namely, an individual culture of safety

---

[13] Daniel Wagner knew someone who died that day as a result of this.

[14] E. Hass-Edersheim, "The BP Culture's Role in the Gulf Oil Crisis," *Harvard Business Review*, June 8, 2010, accessed November 24, 2015, https://hbr.org/2010/06/the-bp-cultures-role-in-the-gu.

that superseded corporate pressures and the cutting of corners. As was the case with the 1984 Bhopal disaster that preceded it, BP taught us that complex systems fail in complex ways.

Complexity arises not only in large, headline-grabbing events. It also governs subtle relationships with equal mastery, and can be a source of both risk and reward. While risk can be measured, complexity or uncertainty cannot. This absence of measurable information causes the scientific firm or manager to freeze in the face of complexity. Firms and individuals whose culture allows for instinctive risk taking can often capitalize on the playing field (or market) being temporarily abandoned by firms trying to grapple with the absence of information.

In these instances, intrinsic leaders favor action, often gaining a personal mythology that they are 'naturals' at calling a market or making decisions under opacity. Steve Jobs offers a paramount example of these qualities. His ability to reinvent himself and countless industries was due in large part to his stubborn commitment to his instincts. While this approach did not make him very popular among his colleagues (he was once fired from Apple), his iconoclastic "people don't know what they want until you show them" approach gave Apple enduring value.

How many product managers would eschew focus groups and costly market analysis? How many firms would allow this "foolhardy" approach at all? Like most things in life, when confronting complexity in all its forms, there is no 'one size fits all' approach. A first order of business is to not attempt to rationalize complex information intellectually, but rather to respond to it, at first (or at least pay heed) instinctively. Often the instinctive decisions we suppress for social order are the right ones. In extreme examples, studies of authority have shown how quickly social pressure breaks down social good whereby a lab coat, clipboard and a clear directive will cause otherwise sound individuals to willingly carry out grotesque instructions. So, large or small, when confronted with complexity, the best tool is the simplicity of the 'voice inside' that so rarely speaks and is seldom heard.

Bank culture, particularly as it relates to risk taking and profit making, needs to be honed with simplicity, rather than complexity. As described by Andrew Haldane's seminal work *The Dog and the Frisbee*, imagine the staying power of the Volcker Rule if it read "thou shalt not play with other people's money" versus the current 1,077 page regulatory rodomontade.[15] Simplicity,

---

[15] A. Haldane, "The Dog and the Frisbee," Bank of England, August 31, 2012 (adapted from a speech given by Mr Haldane), accessed November 24, 2015, http://www.bankofengland.co.uk/publications/Pages/speeches/2012/596.aspx.

the 'via negativa' or subtractive path, is key in tackling adverse selection and moral hazard in banking culture. This is true internally as much as externally.

In a 2014 survey of risk managers, when asked about corporate value systems in the U.S.A.'s systemically important financial institutions (SIFIs), 64 percent likened SIFIs to either 'useful marketing' or 'meaningless words on a wall'. Undoubtedly, there are good people in banking, in fact they may very well be the majority of operators in an otherwise 'bad' system. Yet banking and modern finance more generally are not sectors in which one can reasonably expect safety and prudence to reign in an environment that "expects the best of people", yet turns a blind eye to the powerful structural and psychological inducements of moral hazard.[16]

Take, for instance, the case of Jamie Dimon, JP Morgan's all-powerful chairman and CEO. Despite a long series of obvious risk management failures that pointed right to the top, a pass by the bank's internal audit committee, and a fawning public hearing, Dimon's reward was a 74 percent pay raise in January of 2014, which raises all sorts of questions about the transparency of internal investigations, the 'Teflon' nature of people at the top, and a double standard vis-à-vis risk and reward inside corporations. Being the best of the worst does not make for sound corporate strategy—yet there are no signs that excessive risk-taking in banks will ever be abated. It will merely morph into different forms and appear in different segments of the financial system.

Hubris as company culture seems to be the predominant corporate value in the financial system. American International Group (AIG's) grotesque nationally televised "Thank You America" ads, while Hank Greenberg (AIG's iconic former CEO) was filing a lawsuit against the Federal government, are emblematic of the problem. In the end, a system of playing with other people's money with no real downside at best breeds indifference and at worst breeds callousness. Any attempt at stemming the excessive greed in the system is met with the threat—real or perceived—that there will be a flight of talent and capital to more permissive financial centers around the world. This is so because the modern bank, much like the modern multinational, flies flags of convenience and can unberth to "safer" harbors. Little thanks for the social safety net that kept the global banking system intact.[17]

The supposition then is that boom and bust cycles in banking cannot be stemmed. Much like the general economy, banks fall prey to a series of per-

---

[16] D. Disparte et al. *Moral Hazard Perceptions Survey*, New York University, Stern School of Business, 2014 (administered in 2014 and carried out with NYU Stern risk management students and practitioners in banking and finance).

[17] G. Turner, "HSBC Threatens to Leave London, Again," *Wall Street Journal*, April 24, 2015, accessed November 24, 2015, http://blogs.wsj.com/moneybeat/2015/04/24/hsbc-threatens-to-leave-london-again/.

sistently intrinsic risks that are a part of the economic creative–destructive cycle. It is reasonable to suspect that the public financial debacles and crises are merely the tip of the risk-taking iceberg. Near-misses and unreported or unseen misdeeds are likely to be far too numerous and would make regulators and the public alike blanch in the face of such excessive risk taking. Survivorship, rather than the illusion of control instilled by regulators, makes for a sounder macroprudential objective. Weaving moral fibers into banking culture can only commence when the public, regulators and bankers themselves (beginning at the top), demand stronger cultural norms. Until then, depositor beware.

Agile Risk Control

The sad case of the Rana Plaza building collapse in Bangladesh in 2013 is emblematic of risk in the twenty-first century. In the late 1980s and 1990s, despite the rise in the number of protests demanding improved workers' rights, it was hard to imagine Nike's share price being adversely affected by industrial risk in a subsupplier's operation, let alone the adoption of voluntary safety standards. In Thomas Friedman's *Hot, Flat and Crowded* world, however, the notion of vicarious liability thrives.[18] A vicarious liability is a risk associated with the actions of third parties. In Rana Plaza's case, identifying the tags of global clothing brands in the rubble was all it took for the entire industry to take a hit. While many of the leading clothing manufacturers put forward a voluntary code of conduct dedicated to improving workers' rights and conditions, the industry has been put on notice by instantaneous global information sharing, which can exacerbate even the most robust company's reputational exposure.[19]

Apple's symbiotic relationship with Foxconn, its systemically imperative Chinese supplier, highlights this exposure. Foxconn has made frequent headlines over the years for its low safety standards and poor operating conditions. At one point, the company resorted to putting safety nets around interior balconies to stop people from killing themselves by jumping from high places.[20] Needless to say, a more callous managerial approach by Apple would indicate that this risk is not their doing and is merely part of the operating

---

[18] T. Friedman, *Hot, Flat and Crowded* (New York: Farrar, Straus and Giroux, 2008).

[19] A. Moffat, "Have Supply Chains Changed Since the Rana Plaza Calamity?," *GreenBiz*, April 24, 2014, accessed November 24, 2015.

[20] T. Randall, "Inside Apple's Foxconn Factories," *Bloomberg*, March 30, 2012, accessed November 24, 2015.

reality of global supply chains. Rather than taking the "we're the best of the worst alternatives" approach to their own supply chain risks, Apple responded with an uncharacteristic level of transparency. Not only is Apple famous for being a 'black box' in terms of its corporate decisions, the broader technology industry is highly opaque for fear of patent infringements and the speed of the obsolescence cycle of new products.

Marred by persistent issues around Foxconn, Apple pioneered an annual report on operating standards throughout their global supply chain. With this report, which has been published since 2008, Apple was saying to its stakeholders that workers' rights and other standards matter, and it took steps to improve these standards in an otherwise imperfect world and highly secretive industry. So, a risk that may have harmed lesser technology companies became a source of competitive advantage for Apple. As China's 'Black Monday' erupted, and its broad financial condition began to deteriorate in 2015, even Apple's leading-edge supply chain transparency could not buttress it from the shocks on the supply and demand side of its business, as the entire Chinese economy experienced massive economic corrections.

Apple underscores the notion of Agile Risk Control (ARC). As a framework, ARC built upon on the steps of ISO 31000 and ERM before it, but added a necessary layer of agility to the equation. Being nimble and responsive in a hypertransparent and interconnected world is the key to both combating and harnessing risk. Looking at Apple's Foxconn response from the outside, it is clear that its culture of risk agility stems from the top of the organization and not from some remote silo in the supply chain management function. Indeed, Apple is famous for its centralized command and control structure, where all corporate decisions, large or small, are made in Cupertino. For some risks this centralization is dangerous, in that the field will remain paralyzed while awaiting for guidance from the head office. For others, like the Foxconn example, a centralized decision for supply chain transparency put the entire company's resources to bear in responding to a potentially crippling event.

In risk, as in life, there are maximizers and optimizers. The maximizers are concerned with securing as much as possible as fast as possible. In risk, these responders tend to suppress unfortunate news for fear of the short-term impact on share prices. Think of the obfuscation surrounding BP's response to the Gulf oil spill, or Toyota's muted response to product recalls. Optimizers, by contrast, aim for as much as possible for as long as possible. This framework aims for survivorship, as opposed to short-term gain.

An optimizer like Vincent Herbert, the founder and CEO of Pain Quotidien, a Belgian-styled farm-to-table food chain, aims to confront risk head-on and, as warranted, in a public fashion.[21] When a potentially business-killing field mouse was found in a customer's salad, Pain Quotidien publicly maintained that field mice were a business risk, albeit a remote one, stemming from using only fresh pesticide-free produce. Instead of being hobbled by this exceptional circumstance, or trying to hide from it behind public relations and crisis management firms, Herbert came forward and candidly accepted and apologized for the incident in a way that large company leaders can learn from. Consumers understood that Pain Quotidien's business model implied risks that other businesses in the restaurant space do not expose themselves to. Rather than punishing the company for something that happened outside its control, consumers applauded its candor and transparency.

Apple and Pain Quotidien's supply chain responses show that risk agility is not just the domain of small, nimble firms, but that large publicly traded companies can also harness risk in value creating ways. These cases also highlight the fact that the traditional speculative risk management approaches that rein in finance can be adapted to suit non-financial firms, like tech giants and food service companies, among others. Risk agility and ARC, therefore, are all about being nimble and, in some respects, counter-intuitive to responses espoused by traditional approaches. Hiding (or hiding from) bad news does not improve its impact. On the contrary, sweeping bad news under the corporate carpet, as is the conventional wisdom in large firms, causes these issues to fester and spread, because they will eventually be uncovered and the attempt to hide it from public view will come back to bite them. Such a practice also sends a dangerous message that transparency, trust and accountability do not hamper the ability for agile risk culture to emerge.

Gaining Agility and Speed

The best risk leaders and decision makers are, by default, oriented toward process, and not static thinking. Getting organizations to follow suit is complex, as the modern organization has evolved over many centuries to become a finely tuned, albeit bureaucratic, machine. The rigidity of these organizational structures and decision-making layers often blinds them to risk, and stymies the speed of their response—when speed and agility can create a competitive advantage, as much as it can create danger. The example of Apple's unusual response to supply chain pressures and Pain Quotidien's field mouse incident

---

[21] J. Altucher, "A Mouse in the Salad Means your Business is Growing," *Business Insider*, July 2011, accessed November 24, 2015, http://www.businessinsider.com/mouse-in-the-salad-2011-7.

give guidance that large or small firms can and should embrace risk agility. Risk management needs to apply as much to downside protection as it does to upside creation.

Speed and agility are not singularly the remit of well-resourced private sector players. Although examples of risk agility in the public sector are rare, the best places to see relative speed and agility at work are in the nimbler environments of state houses and regional legislative bodies. In the U.S., for example, individual states like California, a constant epicenter of risk and risk innovation, and Massachusetts, are pioneering processes that have eventually become national paradigms.

Flawed as its implementation may have been, the Obama Administration's Affordable Care Act was modelled closely after healthcare reforms first implemented in Massachusetts under Governor Romney[22] (who, as is widely known, was and remains a Republican). Surely, from a national risk management point of view, however it is implemented, bringing the uninsured and under-insured into the fold will have long-run benefits. To be truly agile and in order to gain the ability to speed up or slow down, decision makers need to embrace the thinking of an entrepreneur rather than a risk-averse corporate leader.

Entrepreneurs are often thought of as risk takers and visionaries who sometimes pursue lofty goals well beyond their financial resources. Revered for their ability to either enhance the value of existing businesses or create entirely new categories of economic value, entrepreneurs play an important role in national economic health and economic competitiveness. What is less recognized, and becoming increasingly important, is the role entrepreneurs also play in national security.

Just as the rate of global human immunodeficiency virus (HIV) infection was acknowledged to be a national security threat by the Bush Administration, the deficit of entrepreneurs in many parts of the world is contributing to economic and political instability. The world is facing a dangerous convergence of high urbanization (the highest rate in human history) and perilously high unemployment rates (particularly among young people and women, who in many countries are entirely excluded from the economic process). The deficit of risk capital and a cultural aversion to failure are in many countries sowing the seeds of increasing economic and political volatility that creates a Catch-22 environment and can eventually become a self-fulfilling prophecy.

---

[22] A. Roy, "How Mitt Romney's Health-Care Experts Helped Design Obamacare," *Forbes*, October 11, 2011, accessed November 24, 2015, http://www.forbes.com/sites/aroy/2011/10/11/how-mitt-romneys-health-care-experts-helped-design-obamacare/.

The self-proclaimed IS (Daesh) would have a hard time attracting young recruits in an environment favoring repeated economic risk taking. That Daesh has attracted adherents from all over the world suggests that marginalized, underemployed, and angry youth need something to be hopeful about in the here and now—not in the hereafter. However, entrepreneurship is not an export product that can be promoted as part of a grand bargain or a trade transaction. Rather, it is an attribute that thrives on indigenous resources, such as the ease of doing business, vibrant capital markets and, most importantly, a local talent pool of risk-takers.

While there is no index measuring the fear of failure, the inability of entrepreneurs and aspiring risk takers to reinvent themselves (or not) in many parts of the world can create a cultural impediment to building new businesses and business models. This challenge is particularly evident in Europe, where many of the preconditions for entrepreneurship are in place, yet cultural norms and strict educational and career paths make entrepreneurship the road less traveled. Ironically, the 'entrepreneur deficit' is even alive and well in a thriving economy such as Singapore, where the state has assumed an overwhelmingly entrepreneurial role and many locals have not been encouraged to be entrepreneurial.

While it is not feasible to measure the precise economic and policy conditions that spark entrepreneurship, successful entrepreneurs are masters at overcoming the odds and thinking outside of the box. It seems clear that creating the basic enabling conditions that foster entrepreneurialism—such as a strong regulatory environment, clean judiciary, and emphasis on good government—is surely a prerequisite to creating the conditions that encourage entrepreneurial behavior. That said, it is easier to measure and identify the impediments. Chief among them is a top-heavy economy with significant state involvement that crowds out new market entrants, slows the pace of innovation, and consumes the lion's share of available capital. State patronage is a breeding ground for cronyism, nepotism, and corruption, leading to a mistrust of business.

Such challenges are highly prevalent in emerging and developing markets, putting a heavy yoke around the neck of aspiring entrepreneurs. While a strong regulatory regime and rules for market conduct improve the state of entrepreneurship, in many advanced markets overregulation creates friction and high costs that only the largest organizations can overcome. Economic Darwinism already produces a high mortality rate for most start-ups. Adding regulatory complexity merely increases the risks start-ups and small businesses must overcome.

Many international efforts to spur entrepreneurship, particularly in developing countries, often labor under the aid and development apparatus, which has difficulty measuring an economic return on investment. Creating a cadre of risk takers and forming entrepreneurial clusters is a stabilizing force in any economy. People attempting to create these preconditions through international efforts would be wise to remember that entrepreneurship is not an export, but rather an idea.

Frameworks granting market access and removing barriers for firms of all sizes to trade their goods and services would be a great catalyst for growth. Similarly, structures that would increase the fluidity and patience of equity capital would draw more prospective entrepreneurs off the sidelines. Solutions that strip away risk from cross-border investments while protecting investor returns and the ability to repatriate earnings would lure more investors in the support of global entrepreneurship. Where entrepreneurship thrives, economies grow and the foundation for political stability strengthens. That is a national security imperative that should not be ignored, and should be a key lesson to established players.[23]

---

[23] D. Disparte, D. Wagner, "Entrepreneurialism and National Security," *Huffington Post*, March 25, 2015.

3

The Risk Continuum

In the previous chapter we outlined some ways in which risk is a process. It is important to have a risk management process that mirrors the stages in which risks are likely to become manifested. In this chapter we examine some of the long-range implications of risk, particularly how a single kernel of change or exposure can lead to often-calamitous results. Similarly, small improvements in anticipation of change can increase overall risk-readiness and long-range planning capabilities. Many of these risks are preventable and, as is the case with most of the themes throughout the book, managerial decision making is a central tenet that can make the difference between simply surviving, or thriving.

The word 'stochastic' derives from the Greek word *stochastikos,* meaning "skillful in aiming". In the financial world, it has come to denote random or probabilistic variables.[1] *Confirmation bias,* in which people use analytical methods to "prove" their point of view, is a common pitfall in organizational behavior. Yet, like other stochastic processes, risk is incredibly difficult to predict and understand. Indeed, mathematical attempts to model the universe often produce a distinct class of risk known as *model error,* where the model or the modeler miss key variables, resulting in the wrong conclusions.

While stochastic thinking is most often applied in the area of complex finance, particularly in pricing and analyzing stocks and fixed income securities, it is a very useful area of thought for risk management more generally.

---

[1] Merriam-Webster definition of stochastic.

© The Author(s) 2016
D. Wagner, D. Disparte, *Global Risk Agility and Decision Making,*
DOI 10.1057/978-1-349-94860-4_3

More important than analyzing the probability of a particular adverse risk occurring is determining what an organization should do and how its people and systems should respond in the face of adversity. A simpler and often more accurate approach to forecasting is to conduct a *Delphi analysis*, whereby a group of experts are asked the same question and the consensus among them is often directionally correct.[2]

Out of Sample Events: Hurricanes

The mathematical domain of risk modelling has become quite sophisticated over time, particularly in the area of analyzing natural catastrophes. Yet, even in this domain, which has (by statistical standards) a deep and 'noise-free' data set, *out of sample* events continue to occur. One such event, which nearly broke the Saffir–Simpson scale measuring hurricane intensity, is Hurricane Patricia, which struck the Pacific coast of Mexico in October 2015. Hurricane Patricia was a monster Category 5 storm, the strongest ever recorded, which largely spared Mexico's sparsely populated Pacific coast[3] because it had weakened somewhat by the time it struck. Had this same storm struck an urban center or a heavily populated coastal area, it would clearly have been incredibly destructive. The same would have been true for the next strong hurricane ever recorded, Haiyan, which hit a mostly rural area of the eastern Philippines in 2013, and left a wasteland in its wake. The closest proximate cause for both hurricanes was the most intense El Niño phenomenon ever recorded—another out of sample weather event, underscoring the continued acceleration of climate change.[4]

While Hurricane Katrina was a much weaker storm than either Patricia or Haiyan, the devastation it wrought on the Gulf Coast of the U.S., and particularly New Orleans, is worthy of study. Although Katrina made landfall in the Gulf as a weaker Category 3 storm, a confluence of factors made its punch much worse than it should have been. From a national risk-readiness point of view, Katrina's wrath was certainly a predictable risk, but city, state, and

---

[2] G. Rowe and B. Wright, "The Delphi Technique as a Forecasting Tool: Issues and Analysis," *International Journal of Forecasting* (1999): 353–75.

[3] W. Cornwall, "Hurricane Patricia, More Pacific Storms, and 4 Other Signs of El Niño," *National Geographic*, October 25, 2015, accessed November 24, 2015, http://news.nationalgeographic.com/2015/10/151023-hurricane-patricia-el-nino-extreme-weather-storms/.

[4] A. Fritz, "Forecast Models Are Now Calling for This El Niño to be the Strongest on Record," *Washington Post*, August 21, 2015, accessed November 24, 2015, https://www.washingtonpost.com/news/capital-weather-gang/wp/2015/08/21/forecast-models-are-now-calling-for-this-el-nino-to-be-the-strongest-on-record/.

federal planners never apparently imagined the storm would hit in exactly the way that it did, or that the bodies of water surrounding New Orleans would turn out to be its greatest source of risk. Coming, as the storm did, after the largest reorganization of the U.S. Federal Government in recent history, with the aim of improving homeland security, rather than demonstrate how prepared emergency services were to react effectively prior to, during, and following the storm, it served to show just how feeble they were.

In the aftermath of 9/11, report after report cited the lack of coordination between local, state, regional, and Federal first responders as one of the critical contributors to weakened national readiness. These challenges lingered and affected everything from the ability to coordinate and have inter-agency communication on portable radios, to more insidious jurisdictional matters that paralyzed the country as one of its most iconic cities faced what was ultimately a predictable homeland security threat.

Home to about 500,000 people pre-Katrina, New Orleans has since had an anemic population recovery, reaching less than 400,000 people a decade later, and leaving more than 100,000 internally displaced people.[5] While the rise and fall of once-great cities has many causes, New Orleans' decline was clearly exacerbated by poor disaster preparedness, shoddy frontline risk mitigation in the city's levee system, and dreadful response and recovery plans.

The Department of Homeland Securuty (DHS), formed three years before Katrina struck, brought the nation's security and first response apparatus under one federal umbrella. Yet, as Katrina approached New Orleans, even though the Army Corps of Engineers predicted catastrophic levee failure, there was no semblance of coordination at any level of government. Some critics contend that the effects of the aftermath were compounded by a so-called 'backdoor draft', wherein the otherwise ever-ready National Guard were deployed on long tours of duty in Iraq and Afghanistan, rather than protecting the homeland as "weekend warriors".[6] Whatever the cause, the U.S. scorecard on risk, response, and resilience post-Katrina may be little better today.

Unlike other natural disasters, hurricanes give people the true benefit of foresight, and predicting their paths is becoming increasingly accurate, with the benefit of combining differing prediction methods. Even so, it remains more of an art than a science, and we can expect future hurricanes to continue

---

[5] J. Adelson, "New Orleans Area Population Still Growing Post-Katrina, But Slowly," *The Advocate*, March 26, 2015, accessed November 24, 2015, http://theadvocate.com/news/neworleans/neworleans-news/11941581-123/new-orleans-area-population-still.

[6] E. Knickmeyer, "Troops Head Home to Another Crisis," *Washington Post*, September 2, 2005, accessed November 24, 2015, http://www.washingtonpost.com/wp-dyn/content/article/2005/09/01/AR2005090101661.html.

to remind us just how powerless we are in the face of climate change, and how much room there is for improvement in terms of predictive models, disaster preparedness, and disaster response. One day a Patricia or Haiyan will hit a major city. We would like to believe that the response to such a disaster would be better than was the case for Katrina, but the truth may be that in the face of such a mammoth calamity, and as was the case in New Orleans, there is only so much that can be done. All the planning in the world may not be enough to effectively manage risk for that type of event.

Macro Agility in Panama

Underscoring the need for national, regional, and international coordination on climate change readiness, Panama's energy matrix offers an instructive case study. Panama's heavy reliance on hydropower increases the country's vulnerability to climate-related risks such as those caused by droughts and the El Niño weather phenomenon. During the dry season, which runs from December through March, generation capacity drops significantly, further straining the energy utility and its ability to meet Panama's growing electricity demands. To alleviate this seasonal shortage, the country's leadership has prioritized energy diversification and increasing energy security with thermal generation, making it an issue of strategic national importance.[7]

Panama is the fastest-growing regional economy and the second-largest economy in Central America,[8] and is slated to continue its economic growth at a steady annualized rate of 6 percent through 2020.[9] To support this economic growth, increased and diversified generation capacity must keep pace. Being agile in its response is not only the key to meeting growing energy demands, but to quelling a restive public who see Panama's development as unequal.[10]

Panama is well positioned to benefit from a wave of energy investments reshaping the energy landscape throughout the Caribbean Basin and the world. The shale gas ('fracking') bonanza has now positioned the U.S. as the

---

[7] "Panama Orders Power Rationing as Drought Continues," BBC News, May 8, 2013, accessed November 24, 2015, http://www.bbc.com/news/world-latin-america-22449328.

[8] "A Singapore for Central America?," Economist.com, July 14, 2011, accessed November 24, 2015, http://www.economist.com/node/18959000.

[9] IMF forecasts.

[10] K. Schneider, *Panama's Hydropower Development Defined by Fierce Resistance and Tough Choices*, Circle of Blue, February 13, 2015, accessed November 24, 2015, http://www.circleofblue.org/waternews/2015/world/panamas-hydropower-development-defined-fierce-resistance-tough-choices/.

world's largest producer of oil and gas.[11] As gas producers seek to expand exports, Panama is well suited to operate as a logistics platform to trans-ship fuels and other gas derivatives, and the U.S. government has encouraged Caribbean and Central American countries to invest in energy infrastructure and shift to gas products as feedstock fuel.[12]

These developments are driven in part by the collapse of Venezuela's economy and its PetroCaribe program, in which artificially low oil pricing was maintained as a means of holding political influence over the Caribbean Basin.[13] As a result, power utilities invested relatively little in modernizing generation and distribution, while aging turbines rely on so-called "dirty fuels" to supply unreliable energy. The U.S. has pledged financial and technical support the wider Caribbean Basin filling in Venezuela's diplomatic energy void.

Looking ahead, Panama's large-scale investments in infrastructure—including a metro system, the enlargement of the Tocumen Airport, and the expansion of the Panama Canal—are expected to accelerate energy demands. The Canal's expansion will not only increase the size of vessels that can traverse from the Pacific to the Caribbean Sea beyond the Panamax class, but will also boost the country's revenue stream as a shipping and logistics hub—further straining energy capacity.[14]

Infrastructure investments and continued urban development, including in the Colon Free Trade Zone, will put a further strain on Panama's nearly 8 billion kilowatt hour (kWh) of energy production. Energy investments in Panama, as in the wider Caribbean Basin, present a compelling energy price arbitrage. On the one hand, much of the region has relied on the heavily subsidized oil from Venezuela to power its aging turbines. On the other hand, as a result of inadequate economies of scale and antiquated and disconnected energy grids, utility companies and independent power producers in the Caribbean Basin charge end users one of the highest energy costs per kWh in the world.[15]

---

[11] R. Katakey, "U.S. Ousts Russia as Top World Oil, Gas Producer in BP Data," *Bloomberg Business*, June 10, 2015, accessed November 24, 2015.

[12] "Caribbean Energy Security Summit – Joint Statement," White House Press Release, January 26, 2015.

[13] "PetroCaribe and the Caribbean: Single Point of Failure," Economist.com, October 4, 2014, accessed November 24, 2015, http://www.economist.com/news/americas/21621845-venezuelas-financing-programme-leaves-many-caribbean-countries-vulnerable-single-point.

[14] S. Black, "This Central American Nation is Spending Roughly 50 percent of its Entire Economy on Infrastructure," *Business Insider*, March 11, 2012, accessed November 25, 2015, http://www.businessinsider.com/infrastructure-projects-in-panama-equivalent-to-8-trillion-in-the-us-2012-3.

[15] C. Barton et al., *The Caribbean has Some of the World's Highest Energy Costs—Now is the Time to Transform the Region's Energy Market*, Inter-American Development Bank, November 14, 2015, accessed November 25, 2015, http://blogs.iadb.org/caribbean-dev-trends/2013/11/14/the-caribbean-has-some-of-the-worlds-highest-energy-costs-now-is-the-time-to-transform-the-regions-energy-market/.

Investments to help drive down costs are a national priority to stimulate economic development and improve energy security in the region. With regard to Panama, to alleviate the country's energy crunch in the short term, the government has permitted two private sector firms to install temporary generation capacity to supply electricity to the grid.[16] Panama's case underscores the perils of an undiversified 'green' energy strategy that relies largely on hydroelectric generation, in a time of unprecedented climate change. Having insufficient thermal energy capacity has placed Panama in the crosshairs of El Niño, and puts its development trajectory and energy needs on a collision course.

Anti-Risk Agility: Neglect, Bankruptcy, and Crime in the Caribbean

Being home to approximately 42 million people and 28 island nations and territories, the Caribbean has a hard time gaining attention on the global stage and competing for resources.[17] The Caribbean is a veritable microcosm of the many challenges and opportunities facing countries around the world, yet the region gains very little air time with such large and important neighbors nearby. No part of the region's economy highlights this more than the challenge of energy security.

Powering an island economy involves a complex, and in many ways antiquated, supply chain of diesel fuels and petroleum products to keep aging turbines whirling. The majority of Caribbean electric utilities are state-owned enterprises (SOEs), which increases the pressure to keep these public services as a part of the electoral apparatus, rather than reforming them to become more efficient and cost-effective for the public—an objective that is understandably hindered by the limited purchasing power individual islands enjoy, and the geographic inability to form energy blocs or build interconnected power grids.[18]

---

[16] *AES Awarded Panama's First Natural Gas-Fired Generation Plant*, Press Release, Business Wire, September 11, 2015, accessed November 25, 2015, http://www.businesswire.com/news/home/20150911005301/en/AES-Awarded-Panama%E2%80%99s-Natural-Gas-Fired-Generation-Plant.

[17] "Ranking Caribbean Countries by Population Density," *Caribbean Journal*, October 22, 2013, accessed November 25, 2015, http://caribjournal.com/2013/10/22/ranking-caribbean-countries-by-population-density/#.

[18] The American Security Project, *Energy Security in the Caribbean*, March 4, 2015, accessed November 25, 2015, http://www.americansecurityproject.org/energy-security-in-the-caribbean/.

When it comes to electricity supply, the Caribbean represents an ongoing battle between access and affordability. In Haiti, for example, just 28 percent of the population has access to electricity.[19] Across the border in the Dominican Republic, the number is closer to 96 percent, while in the U.S. territory of Puerto Rico, the number of residents with access to electricity is 100 percent. Access and affordability is a challenge across all three islands and the broader region. Even though Puerto Rico has been a de facto U.S. protectorate since the end of the Spanish–American War, the island's residents are saddled with energy costs three or four times higher than those on the mainland.[20] This would have a crippling effect on any economy, but in Puerto Rico, the high cost of energy has quite literally paralyzed households, corporate investments, and has cast a heavy yoke around the neck of the struggling economy.

While Puerto Ricans are comparatively well off by comparison with their Haitian and Dominican neighbors, if the island were a U.S. state it would actually be the poorest state in the union—imagine Mississippi, the poorest state in the U.S.,[21] with an energy bill three times higher than its present level. Puerto Rico's per capita GDP is less than half that of Mississippi (at about $15,000), and some pundits have referred to the island as America's "third world country".[22] Coupled with Puerto Rico's fiscal crisis (with $73 billion worth of public debt), the island has been likened to the 'Greece of the Caribbean'. Given this, reforming the island's ailing public utility, the Puerto Rico Electric Power Authority (PREPA), would appear to be essential. Doing so requires leadership, fortitude, and rapid decision making that simply do not exist. PREPA is itself laboring under $9 billion of debt. The country's transport authority was declared bankrupt in 2015, and a host of other government institutions are teetering on the verge of bankruptcy. Where were the agile risk managers and decision makers when Puerto Rico needed them most?

The consequences of lagging energy standards in the Caribbean, and Puerto Rico in particular, present a series of insidious threats to the region and to regional security. The Caribbean is a long-standing entrepôt for the drug trade. With a hamstrung regional economy and declining levels of corporate investment and tourism, the region's vibrant grey economy is sure to grow,

[19] *Energy Access Expansion in Haiti*, World Bank Energy Summary, April 4, 2014, accessed November 24, 2015, http://www.worldbank.org/en/results/2014/04/04/energy-access-expansion-in-haiti.

[20] E. Morales, "Puerto Rico's Soaring Cost of Living, From Giant Electric Bills to $5 Cornflakes," *The Guardian*, July 12, 2015, accessed November 25, 2015, http://www.theguardian.com/world/2015/jul/12/puerto-rico-cost-of-living.

[21] "The 10 Poorest States in America," CNBC, 2013, accessed November 25, 2015, http://www.cnbc.com/2013/09/27/the-10-poorest-states-in-america.html.

[22] C. Long, "Puerto Rico is America's Greece," Reuters, March 8, 2012, accessed November 25, 2015, http://blogs.reuters.com/muniland/2012/03/08/puerto-rico-is-americas-greece/.

and become darker still as people turn to the drug trade and criminal gangs to make ends meet. If Puerto Rico were officially part of the U.S., it would be the most violent state per capita by a wide margin.[23] This criminality is in large part driven by the political neglect that the island has suffered in the U.S., and the lack of political leadership and consensus among Puerto Rico's three political parties, each vying for statehood, independence and the ability to define the status quo.

Improving energy security in the Caribbean (and Puerto Rico in particular) involves the following key steps:

1. Weaning electric utilities off costly foreign oil and diesel, and looking to viable alternatives such as propane as a bridge to natural gas. This bridge is driven in part by the U.S. shale gas bonanza that is slated to make America a net exporter of energy by 2020.[24]
2. De-risking existing contracts with heavily indebted public utilities like PREPA through the global capital markets and insurance solutions that can attract viable commercial partners, giving them comfort in light of enfeebled state-owned counterparties.
3. Repealing crippling facets of the terribly dated and out-of-touch Merchant Marine (more commonly known as the Jones Act) which nearly doubles the cost of ship-borne trade in U.S. islands, such as Hawaii and Puerto Rico, where it is more cost-effective to receive shipments from China, than it is from the U.S. mainland.

The conservative Heritage Foundation has argued vociferously that the Jones Act is an outdated protectionist policy that not only hampers economic competitiveness, but also harms national security.[25] The crippling effects of the Jones Act and its narrow interests are felt more in Hawaii and Puerto Rico than in any other place in the world—two island territories that can ill-afford an unfair playing field. Failure to take concerted action will only lead

---

[23] A. O'Reilly, "Plagued by Violence, Bad Economy, Puerto Rico Rings in 2014 With Bang, 13 Murders in 5 Days," Fox News Latino, January 8, 2014, accessed November 25, 2015, http://latino.foxnews.com/latino/news/2014/01/08/plagued-by-violence-bad-economy-puerto-rico-rings-in-2014-with-bang-13-murders/.

[24] B. Chapel, "U.S. Predicted to be Net Energy Exporter in Next Decade: First Time Since 1950s," National Public Radio, April 15, 2015, accessed November 25, 2015, http://www.npr.org/sections/thetwo-way/2015/04/15/399843516/u-s-predicted-to-be-net-energy-exporter-in-next-decade-first-time-since-1950s.

[25] B. Slaterry, B. Riley, N. Loris, "Sink the Jones Act: Restoring America's Competitive Advantage in Maritime-Related Industries," The Heritage Foundation, May 22, 2014, accessed November 25, 2015, http://www.heritage.org/research/reports/2014/05/sink-the-jones-act-restoring-americas-competitive-advantage-in-maritime-related-industries.

to longer-term challenges. Further backsliding in the region will make the problem costlier and more complex to solve in the future. The case for energy security and diversification in the Caribbean Basin underscores the need for long range planning around risk and resilience.

How Kings Fall

The Kingdom of Saudi Arabia (KSA) is unaccustomed to being thought of as an enfeebled economy, but that has increasingly become the case in recent years. Much as the fallacy of perpetual property appreciation felled the U.S. economy during the Great Recession, KSA's failure to diversify its economy, combined with simmering social and sectarian tension, has endangered the country's relative stability. With 90 percent of Saudi revenues derived from the sale of oil, sustained oil price declines represent a grave risk to the country. The dramatic and sustained drop in oil prices in 2014 and 2015 prompted rating agencies to downgrade the country's sovereign debt with a negative long-term outlook.[26] By late in 2015, some pundits were even predicting the eventual disintegration of the House of Saud.[27]

For a country that has enjoyed budget surpluses for much of the last decade and that relies heavily on national paycheck persuasion to keep its often-restive population at bay, darkening economic clouds are an omen. According to the International Monetary Fund (IMF), for KSA to maintain a balanced budget, it needs oil to sell at a price in excess of $106 per barrel. If the average price of a barrel of oil were to remain at or below $50 a barrel for a prolonged period, the IMF forecasts that KSA will run out of cash by 2020.[28] That a country can have a marginal cost of production of $3 per barrel[29] yet base its national budget on $106 per barrel, calls into question some rather fundamental precepts of risk management and planning.

The scent of jasmine from the Arab Spring was, in many ways, quelled by the scent of money that KSA (as a rentier state) paid its citizens to calm their dissatisfaction with the country's heavy personal restrictions and high levels of

[26] A. Sabi, *Saudi Arabia's Deficit Problem*, Global Risk Insights, October 29, 2015, accessed November 25, 2015, http://globalriskinsights.com/2015/10/saudi-arabias-deficit-problem/.

[27] D. Goldman, "Cheap Oil Puts the House of Saud at Risk," *Asia Times*, October 22, 2015, accessed November 24, 2015, http://atimes.com/2015/10/cheap-oil-puts-the-house-of-saud-at-risk/.

[28] M. Egan, "Saudi Arabia to Run out of Cash in Less Than 5 Years," CNN Money, October 25, 2015, accessed November 24, 2015, http://money.cnn.com/2015/10/25/investing/oil-prices-saudi-arabia-cash-opec-middle-east/.

[29] "Cost of Oil by Country," Knoema.com, 2015, accessed November 24, 2015, http://knoema.com/vyronoe/cost-of-oil-production-by-country.

youth unemployment. In 2011, the population benefited from approximately $93 billion in public spending as an inducement not to take to the streets, as their cousins in other Arab nations had done by the millions.[30] This financial inducement, combined with a swift crackdown on public dissent, largely spared the Kingdom the fate that had been visited upon other countries in the region during the uprising. In future years an empty public purse will make it very difficult to restrain the public.

The strain of economic pressure, leadership transitions following the death of King Abdullah in 2015, and KSA's military misadventures in Yemen have all taken a toll on the Kingdom's usually cool diplomatic demeanor.[31] Building a nation or a business under the assumption of certain economic constants—such as a $100-dollar barrel of oil, steadily growing consumer spending and constantly rising home values—is a perilous approach. How KSA contends with years of pent-up internal frustration, royal internecine fighting, and waning influence in a tattered Middle East is sure to exacerbate regional instability and conflict, potentially signaling an end to the House of Saud. Long range planning is as important as economic diversification for the risk agile state.

Cities at Risk

While the individual who tipped the scale went entirely uncelebrated, at a specific time in the last decade, a new class of humanity emerged with the rise of *homo urbanus*.[32] This version of modern man is a city dweller, often living in one of the world's densely populated urban centers, in search of survival and a share of the economic prosperity that cities represent. Cities have always been at the center of human trade, and economic, cultural, and political history. In shaping the current century, cities are increasingly important as the nexus where humanity and risk intersect. Almost four billion people currently live in cities, representing more than 50 percent of the world's population, which is projected to grow by 66 percent by 2050.[33] Our collective detach-

---

[30] "Throwing Money at the Street," Economist.com, March 10, 2011, accessed November 24, 2015, http://www.economist.com/node/18332638.

[31] M. Bird, "Saudi Arabia's Game of Thrones Could Derail Global Oil Markets," *Business Insider*, October 30, 2015.

[32] "The World Goes to Town," Economist.com, May 3, 2007, accessed November 24, 2015, http://www.economist.com/node/9070726.

[33] United Nations, UN Department of Economic and Social Affairs, *World Urbanization Prospects*, 2014. Accessed November 25, 2015, http://esa.un.org/unpd/wup/highlights/wup2014-highlights.pdf.

ment from the natural world may be one of the more insidious threats of the city-dwelling era.

This concentration of human risk in cities represents a profound strain on the world's resources, but it also represents an opportunity—one that must be carefully managed, for cities can become profoundly volatile places if a development pathway is viewed as inequitable by its citizens. The violent scenes that have in recent history gripped Baltimore and the *banlieues* of Paris serve as reminders that cities can quickly turn into cauldrons of conflict anywhere in the world.

The promise of new hope and economic progress that cities represent requires a careful balancing act of economic development and public policy. In the era of mobile communication, the ability to fan flames of dissent is instantaneous. As we saw in Cairo during the protests in Tahrir Square, young organizers turned to social media to organize their revolt. The government of Hosni Mubarak went to great lengths to try to control the result, including a national shutdown of the internet and unleashing his henchmen to dissipate dissent.[34]

Even in the U.S., a bastion of civil liberty and democracy, police brutality is being kept in check by the proposition of body cameras on all police officers. That the protectors of the peace need a bigger brother watching them underscores a deep-seated public distrust of authority that can very quickly turn to civil commotion and unrest. This tinder box has been set alight in the U.S. with increasing frequency and alarm, from cities like Ferguson, Missouri and Baltimore, Maryland to many others. The advent of the mobile phone and the democratization of information are powerful tools in the hands of the citizenry of our urban world.

Beyond public unrest, cities are the central target for a range of threat categories, including man-made, natural, and emerging threats. From creaking infrastructure, market crashes and cyber risk to natural disasters and pandemics, the fact that humanity is so densely clustered in cities may come back to haunt us if we do not improve our resilience, collaboration, and agility.[35] In a compelling study produced in 2015 by the Cambridge Center for Risk Studies for Lloyd's of London, the notion of the GDP at risk in cities is explored.

---

[34] C. McGrath, "Mubarak Regime Shuts Down Internet in Futile Attempt to Stop Protests," *The Electronic Intifada*, January 28, 2011, accessed November 24, 2015, https://electronicintifada.net/content/mubarak-regime-shuts-down-internet-futile-attempt-stop-protests/9794.

[35] Lloyd's of London, *Lloyd's City Risk Index: 2015–2025DOUBLEHYPHENExecutive Summary*, 2015, accessed November 25, 2015, http://www.lloyds.com/cityriskindex/files/8771-city-risk-executive-summary-aw.pdf.

The City Risk Index captures both short-term and long-range GDP at risk in 300 of the world's cities, collectively representing approximately 50 percent of the world's economic output, and a large share of the world's urban population. The Index captures the economic impact of 18 individual threat categories, falling into either a man-made, natural, or emerging risk domains. The top 10 threat categories represent a potential GDP at risk of $4.14 trillion – or 91 percent of the total.[36] As noted in earlier chapters, risk neither emanates from nor conforms to a single risk domain, and therefore it is wise to remember the notion of contagion and the ability for things to unravel across multiple domains. The speed with which risk spreads is particularly evident in cities.

The Index offers a framework for the public and private sectors to improve their understanding of long-range risks and how to marshal collective resources and financial solutions, thereby improving recovery and resilience.

In advanced markets many aspects of insurance penetration are almost like a force of gravity, with everything from motor insurance to property and healthcare being mandated coverage areas. The same cannot be said for developing and emerging market cities, where there is a dangerously low level of insurance penetration. Research suggests that a one percent increase in insurance penetration translates into a 13 percent reduction in uninsured losses and a 22 percent reduction in the recovery burden taxpayers must shoulder following a loss. Improving the resilience of cities and, by extension, countries will challenge the financial services industry to continue innovating to adequately price for and absorb risks.[37]

Of the 18 threats listed in the Index (see Illustration 3.1), it is surprising that war and interstate conflict is not one of them. Despite all of the efforts of the international community to establish and maintain peace since the the two World Wars, the risk of conflict remains ever-present. From the constant saber rattling on the Korean Peninsula to the increasing militarization in the Pacific Rim (with China's construction of man-made islands in the South China Sea) and the Russian and U.S. military fighters bumping into each other in the Syrian theater, the possibility of war triggered by miscalculation seems greater than ever. Even pacifist Japan has transcended its post-war pacifist constitution so that it may increasingly exert its considerable military capabilities abroad.[38]

---

[36] Lloyd's, *City Risk*, Executive Summary.

[37] C. Edwards, *Lloyd's Global Underinsurance Report,* Center for Economic and Business Research/Lloyd's, 2012, accessed November 25, 2015, https://www.lloyds.com/~/media/files/news%20and%20 insight/360%20risk%20insight/global_underinsurance_report_311012.pdf.

[38] M. Ford, "Japan Curtails its Pacifist Stance," *The Atlantic*, September 19, 2015, accessed November 24, 2015, http://www.theatlantic.com/international/archive/2015/09/japan-pacifism-article-nine/406318/.

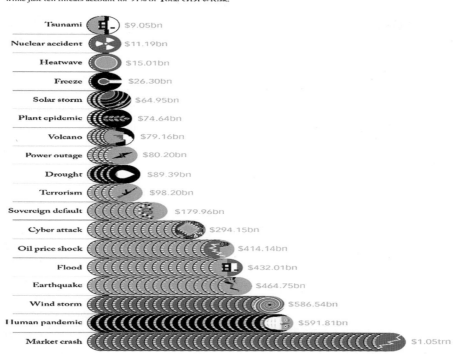

Total GDP@Risk: All cities
$4.56trn

Market crash represents nearly a quarter of all cities' potential losses, while just ten threats account for 91% of Total GDP@Risk:

Tsunami	$9.05bn
Nuclear accident	$11.19bn
Heatwave	$15.01bn
Freeze	$26.30bn
Solar storm	$64.95bn
Plant epidemic	$74.64bn
Volcano	$79.16bn
Power outage	$80.20bn
Drought	$89.39bn
Terrorism	$98.20bn
Sovereign default	$179.96bn
Cyber attack	$294.15bn
Oil price shock	$414.14bn
Flood	$432.01bn
Earthquake	$464.75bn
Wind storm	$586.54bn
Human pandemic	$591.81bn
Market crash	$1.05trn

Illustration 3.1   Lloyd's City Risk Index (18 Loss Categories). (*Lloyd's City Risk Index: 2015–2025--Executive Summary*, Lloyd's of London, 2015, accessed November 25, 2015, http://www.lloyds.com/cityriskindex/files/8771-city-risk-executive-summary-aw.pdf)

With the Middle East at the beginning of what may turn out to be a 30-year war, with arms sales and the militarization of countries around the globe at a higher level than ever before, and with the threat of global terrorism at unprecedented levels, the risk of conflict occurring anywhere and at any time will remain high. As previously noted, whereas risk managers and decision makers previously had the luxury of presuming that this type of risk was someone else's concern, maintaining that mindset bears potentially huge consequences in the future. What happens in Cairo or Baltimore, Syria or North Korea, now has potential implications that may stop at your doorstep.

Reputation Risk Has a Price

In another example of how long-range risks evolve and can endure, the emissions-rigging scandal that has wreaked havoc on Volkswagen (VW) offers clear guidance about the dangers of cutting regulatory corners, and how self-inflicted pain can become a self-fulfilling prophecy. As news of the scandal trickled out, and then failed to stop emerging (with new elements being revealed many weeks after it first emerged), questions were naturally raised about how, and indeed whether, the world's largest automaker (by volume) could regain its footing. What most shocked so many consumers and the business community was the brazenness of the emissions rigging, and that the largest manufacturer of pollution-producing automobiles in the world would so callously attempt to completely bypass pollution control guidelines specifically put in place by governments around the world to attempt to limit the environmental damage caused by cars. It may well prove to be the worst affront to environmentally conscious citizens and corporations in history.

Some surveys identify reputation risk as one of the principal threats to global firms, and yet it remains undervalued and amorphous.[39] With one large scandal after the other, from BP's Deep Water Horizon disaster to the scandals involving football's governing body FIFA and Sony Entertainment, corporate leaders are beginning to give reputation risk the deference it deserves.

VW's case sheds new light on how reputation risk can fell even the largest, best-known, and most robust enterprise. In 2014, VW was ranked number 3 in the automotive industry in Fortune's list of most admired companies, yet shortly afterward 40 percent of its market cap was eradicated in a matter of days. Given the volume of lawsuits, regulatory fines, and injunctions, VW's financial viability was thought to be at risk, and its reputation likely irreparably harmed,[40] and even though in time consumers may forget about the scandal (as they have a tendency to do), the corporate world and global regulators certainly will not.

The U.S. Justice Department declared its intention to prosecute individuals and corporate entities for serious corporate fraud cases, and this case involved an intent to deceive and circumvent pollution limits, not just a failure to report an unintended design defect, as was the case with previous suits

---

[39] Allianz SE and Allianz Global Corporate SE, *Allianz Risk Barometer: Top Business Risks 2015*, 2015, accessed November 25, 2015, http://www.agcs.allianz.com/assets/PDFs/Reports/Allianz-Risk-Barometer-2015_EN.pd.

[40] E. Henning and H. Varnholt, "Volkswagen Assesses Emissions Scandal's Impact on Its Finances," *The Wall Street Journal*, October 4, 2015, accessed November 25, 2015, http://www.wsj.com/articles/volkswagen-evaluating-emissions-scandals-impact-on-companys-finances-1443980626.

against Toyota and GM.[41] "The criminal charges at play were violations of the U.S. Clean Air Act, as well as mail and wire fraud (for deceiving consumers, regulators and car dealerships) and possibly racketeering charges. Given the scope and breadth of VW's scheme reported in the press, criminal penalties for the company would include fines as much as $18 billion under the Act. In addition to that and vehicle recalls, other "collateral costs" of debarment, civil fraud lawsuits, and shareholder lawsuits were possible. The penalties under the advisory federal sentencing guidelines for individuals prosecuted would include jail time and substantial fines."[42]

Consider the words "intentional intent to deceive" and think about what impact that can have on a company's reputation. Reputation risk now has a price attached to it and the entire automotive industry has been put on notice. Boardrooms across all industries should take heed. Until recently, corporate malfeasance rarely produced individual punishment, other than corporate leaders voluntarily stepping down and falling on their own swords. This in part is driven by the so-called "personhood" of a corporation, shielding directors and officers from certain types of personal liability. Moreover, deep financial pools covering corporate directors and officers, as well as product-related insurance, defray many of the economic costs of these types of cases. VW may have to go it alone, however, its insurers will likely claim, along with courts and regulators, that their allegedly tortious and willful acts invalidate any insurance coverage that might have been afforded.

Anything in the Name of Growth

VW, like Ford before it, has a large and disparate number of automotive holdings, ranging from supercars for the 'one percent' (including Bugatti, Bentley, Lamborghini, and Porsche) to 'the people's' car wearing the VW, Audi, Škoda, and Seat badges. In all, VW has 12 automotive brands in its holding structure – two more than GM.[43] VW's strategic impetus was to overtake Toyota as the world's largest automotive manufacturer, the difference being that Toyota largely operates as a monoline franchise, with spinoffs such as the high-end Lexus and the low-end Scion brands largely developed on core Toyota platforms. VW's growth strategy depended on cobbling together many non-core

---

[41] *Anticipating Criminal Charges Against Volkswagen*, Law 360, September 24, 2015, accessed November 24, 2015, http://www.law360.com/articles/706830/anticipating-criminal-charges-against-volkswagen.
[42] *Criminal Charges*, Law 360.
[43] Volkswagen AG, *Volkswagen AG global website*, accessed November 24, 2015, www.vw.com.

automotive holdings. Perhaps this is the reason VW lost sight of its core focus and embarked on the alleged effort to manipulate performance by making certain diesel engine vehicles appear more efficient than they really were. As is the case with many corporate risks, early warning signs were reportedly ignored dating back to 2007.[44] Even worse, history's lessons were ignored altogether, as VW was fined in 1973 in an eerily reminiscent case.[45]

In late 2006, Ford had an automotive holding structure similar to that of VW, including mid-market and high-end brands, such as Volvo, Jaguar, Land Rover and Aston Martin, among others.[46] When it started experiencing downward pressure, eroding market share and quality in its core (legacy) brand, the company embarked on a concerted divestiture process. By spinning off all non-core holdings, Ford emerged as a leaner and more agile automaker, creating a "One Ford" strategy that saw the adoption of automotive platforms for world markets, and a 'rightsizing' of the company's balance sheet.[47] A decidedly smaller Ford emerged, yet the company's survival was no longer in question (survival being the ultimate purpose of risk agility). In contrast to VW, which lost 39 percent of its share value within the first two weeks of the announcement of the scandal[48] in 2015, Ford had record third-quarter profit in the U.S. in 2015—with the first three quarters surpassing all of 2014 combined.[49]

This type of bold change agenda required a leader from outside the firm and outside of the industry. Alan Mulally, formerly Boeing's CEO, is largely credited with turning Ford around, making it the only U.S. automaker that did not require a direct government bailout during the Great Recession—although it was supplied with a government line of credit.[50] CEOs from Detroit's besmirched automakers were famously paraded in front of Congress

---

[44] "Volkswagen was Warned Years ago about Illegal Emissions Tricks: Report," *Chicago Tribune*, September 27, 2015, accessed November 25, 2015, lhttp://www.chicagotribune.com/business/ct-volkswagen-scandal-20150927-story.html.

[45] H, Mustoe, "VW and the Never-ending Cycle of Corporate Scandals," BBC News, October 26, 2015, accessed November 24, 2015, http://www.bbc.com/news/business-34572562.

[46] A. Wilson, "Ford Overhauls Way Forward Plan: Cuts Include 14,000 White Collar Jobs," *Autoweek*, September 14, 2006, accessed November 24, 2015, http://autoweek.com/article/car-news/ford-overhauls-way-forward-plan-cuts-include-14000-white-collar-jobs.

[47] Ford global website, accessed November 24, 2015, www.ford.com.

[48] Volkswagen AG, CNBC website, accessed November 24, 2015, http://data.cnbc.com/quotes/VLKAF/tab/2.

[49] "Ford Earnings: New Products Drive Record Third Quarter Profit," Fool.com, November 1, 2015, accessed November 24, 2015, http://www.fool.com/investing/general/2015/11/01/ford-earnings-new-products-drive-record-3rd-quarte.aspx.

[50] J. Muller, "Ford Looks Hypocritical In New Anti-Bailout Commercial," *Forbes*, September 19, 2011.

for a public mea culpa and asked whether they drove to Washington or flew in in their corporate jets.

Public grandstanding aside, since the Great Recession, the automotive industry has worked full tilt to keep its sprawling manufacturing base at capacity, often at the expense of safety and quality. While few global automotive firms are spared from the occasional product recall, VW's case takes this challenge to a whole new level, with more than 11 million vehicles affected, and many new models facing a potential embargo, which will have a serious knock-on effect on VW's dealer network.[51] This gave VW the dubious distinction of having the second largest automotive recall in history.[52]

Divesting non-core brands on the higher end of the spectrum will not only create the capital that VW will need to fend off litigation and other costs, it will also enable the firm to reinvent itself and gradually regain the confidence of consumers. VW has a very long road ahead. Whether it sees its way through this crisis depends as much on corporate focus as it does on whether regulators want to make an example of it. U.S. and global regulators will need to weigh the importance of sending the right message, making the punishment fit the crime with the risk to VW's thousands of direct employees and thousands more in the company's dealer and supplier network.[53] During the Great Recession, the U.S. government took some actions that were, at the time, judged to be inappropriate (such as bailing out GM and AIG). In retrospect, those turned out to be the right decisions. VW, its shareholders, employees, and suppliers, must be hoping for a similar result.

Municipal and Sovereign Risk

Some difficult choices had to be made by governments during the Great Recession. Many people still believe that governments saved banks' balance sheets at the expense of their own. From the hitherto unthinkable concept of U.S. cities like Detroit going bankrupt, to entire countries (such as Greece, Argentina and Spain) needing seemingly endless rounds of financial intervention, the bailout is one of the more pernicious aspects of modern economic risk, and the root of the concept of privatizing gains while socializing losses. In short, saving one branch of the world's economic tree means contagion will

---

[51] "The Volkswagen Scandal: A Mucky Business," Economist.com, September 26, 2015.

[52] G. Wallace, "Biggest Auto Recalls Ever," CNN Money, May 27, 2014, accessed November 24, 2015, http://money.cnn.com/2014/05/27/autos/biggest-auto-recalls/.

[53] *Volkswagen U.S. corporate website*, Volkswagen AG, 2015, accessed November 24, 2015 (www.vw.com).

affect the other branches, or, as is the case with the spread of sovereign and municipal debt, the entire tree may be affected.

We cast our gaze once again on the curious case of Puerto Rico, which highlights the game of economic brinksmanship that is being played, potentially imperiling wider economic stability. Puerto Rico serves as a poster child of the incessant link between financial and economic risk, but many outside the Americas may not have paid much attention to it. They should, because at least twice in recent memory the game of financial chicken being waged in Washington, D.C. has had dire economic consequences for the U.S. and the world.

The most recent example is the political brinksmanship over the debt ceiling and so-called fiscal cliff, which cost the U.S. its AAA credit rating and damaged the dollar's credibility as the world's reserve currency. The second example was in not extending the "too-big-to-fail" line to include Lehman Brothers, leading to the bankruptcy of what was not recognized at the time as a systemically important firm.[54] Lehman's failure triggered a massive financial panic that caused the credit markets to freeze, and prompted the U.S. government to acknowledge that if it failed to stop the bleeding, the entire financial system of the country, and the world, could collapse.

While many argue that Puerto Rico's debt crisis is largely decoupled from the broader U.S. economy, as an American territory and dollar-denominated economy, allowing the island to default would be tantamount to a Lehman moment, but this time on a national scale. As neither a state nor an independent nation, Puerto Rico's financial crisis is a hybrid between a sovereign debt crisis of an independent nation and the "safe" debt of a U.S. municipality, city, or state, which bondholders tend to flock to because of historically low default rates and the possibility of orderly debt restructuring, enshrined in law. For the agile risk manager, this 'hybrid' of a looming financial disaster has been on the radar as a warning sign of the potential danger of the intersection between financial and economic risk.

Financial crises generally take a long time to fester and have many moving parts. Puerto Rico's profligacy is definitely a big part of the problem, but creating a false deadline for when the crisis needs to be resolved, while at the same time not granting Puerto Rico debt restructuring rights afforded to the rest of the U.S., would not only hasten the island's economic collapse but may also bring down the municipal bond market and other parts of the economy with it. The question is whether or not the people who need to be thinking about what would happen if the economy collapsed are considering the wider impact.

---

[54] B. Hallman, "Four Years Since Lehman Brothers, 'Too Big to Fail' Banks, Now Even Bigger, Fight Reform," *Huffington Post*, September 15, 2012, accessed November 24, 2015, http://www.huffington-post.com/2012/09/15/lehman-brothers-collapse_n_1885489.html.

Some creditors were in agreement with Puerto Rican officials in their appeals that Washington grant the island the same rights as other municipalities and states, including Chapter 9 of the Bankruptcy Code. Rather than waiting until the 11th hour, as was the case with the national debt ceiling debate, U.S. political leadership has an opportunity to fend off an economic calamity. Will they seize the opportunity or do they prefer to dive into the unknown, and, given all the other fiscal calamities that have befallen the world since the Great Recession, does Puerto Rico even matter to U.S. law makers?

Painful economic choices lay ahead. Puerto Rico has a chance to emerge as a stronger and leaner economy. For debt restructuring to be serious and to not only address the interests of hedge funds, Puerto Rico must consider all options, however unpopular, including the privatization of many of its sclerotic state-owned enterprises. Its electric power utility, PREPA, would be a good place to start. PREPA contributes 13 percent ($9 billion) to the island's debt burden and delivers exorbitantly high energy costs that hurt households, businesses, and tourists alike.[55]

For Puerto Rico to remain a business and tourism destination able to compete in the region—particularly with 50 years of pent-up interest in Cuba being unleashed—deep reforms are needed once the debt crisis is resolved. There are promising green shoots, such as the idea of making Puerto Rico a regional financial hub for the Caribbean and Latin America. With billionaires like John Paulson calling Puerto Rico home, due to its favorable tax code on capital gains, the island's last chance may rest in the hands of financiers whose paycheck persuasion and lobbyists will make them heard in Washington.[56]

A New Host for Systemic Risk

MetLife's designation by the Financial Stability Oversight Council as a SIFI has sparked furor by the life insurer, and has resulted in a series of lawsuits contesting this label.[57] MetLife now joins Prudential as the second large life

---

[55] L. Lopez, "Puerto Rico Keeps the Lights on, but Debt Crisis Far From Over," Reuters, August 15, 2014, accessed November 24, 2015, http://www.reuters.com/article/2014/08/15/us-usa-puertorico-utility-insight-idUSKBN0GF0C320140815.

[56] K. Burton, "John Paulson Calls Puerto Rico Singapore of Caribbean," *Bloomberg Business*, April 24, 2014, accessed November 24, 2015, http://www.bloomberg.com/news/articles/2014-04-24/john-paulson-says-puerto-rico-to-become-singapore-of-caribbean.

[57] I. Katz et al. "MetLife to File First Lawsuit Over Systemic-Risk Label," *Bloomberg Business*, January 13, 2014, accessed November 24, 2015, http://www.bloomberg.com/news/articles/2015-01-13/metlife-sues-over-too-big-to-fail-designation-by-u-s-regulators.

and retirement firm with the dubious distinction, and the heightened regulatory scrutiny that goes along with being labelled a SIFI.

Looking beyond the headlines, there appears to be merit in the designation, and cause for concern that large life and retirement firms are steadily climbing the systemic risk rankings. According to market-based analysis available at Nobel Laureate Robert Engle's Volatility Lab, MetLife has been a top 10 contributor to U.S. systemic financial risk since 2010. Since 2013, MetLife, Prudential, and Lincoln National Corporation have shown surprising staying power among the top 10 systemic institutions in the U.S., contributing nearly $100 billion in expected capital shortfalls, or 25 percent of the total as of 2015.[58] Principal Financial Group also joined the top 10 as of the first quarter of 2015.

To be designated a SIFI a financial institution's health must be considered to have system-wide implications, and its relative fiscal health will be closely monitored for its ability to pull down an entire economy. These firms are more commonly referred to as 'too-big-to-fail'. SIFIs have special regulatory oversight, higher macroprudential standards, and they must file so-called living wills outlining their orderly resolution in the event of financial distress. This analysis applies an 8 percent regulatory capital requirement. By lowering the required capital buffer in 2013 from 8 percent to 6 percent,[59] Prudential overtakes JP Morgan as the top systemic financial institution in the U.S.

While these results are produced in a simulation, the intuition behind V-Lab is that it uses market-based metrics to replicate how the financial system would perform during periods of economic duress. The scrutiny of systemic financial institutions in the public interest is increasingly being coordinated among regulators around the world, as they try to keep pace with the sprawling enterprise of global financial services firms—many of which have thousands of subsidiaries registered all over the world.

MetLife is no different. With more than 100 million customers and operations in 60 countries, not only would MetLife's failure have broad regulatory interest, but it would also spell disaster for its 60,000 employees and its customers.[60] To be clear, designation as a SIFI does not telegraph the imminence of financial distress any more than it conveys safety. What it *should* signal,

---

[58] Financial system capital shortfalls are based on a simulated 40 percent drop in broad U.S. equities. Firms with high leverage will face particularly punishing conditions and financial distress. Under these conditions, the few available suitors will face similar distress, increasing the likelihood of public bailouts or firm failure.

[59] M. R. Crittenden, "Plan Reins in Biggest Banks," *Wall Street Journal*, July 9, 2013, accessed November 24, 2015, http://www.wsj.com/articles/SB10001424127887323336870457859554039783694.

[60] MetLife, *Annual Report 2014*, 2015, 243.

however, is a warning call and a deeper search of not only MetLife, but its peers in the global life and retirement market that appear to be the new hosts of systemic financial risk. Comparatively, in the United Kingdom, which has a mature life insurance and retirement market, three large life insurers rank among the top 10 systemic financial institutions. This suggests that the U.S. is not alone in formulating a regulatory response to perceived riskiness among large life and retirement firms.

Taking a closer look at the financial characteristics of these firms, part of the implied risk is explained by their comparatively high financial leverage. On average, the four U.S. life and retirement firms appearing in this analysis have a leverage ratio of 17.21 percent, compared to JP Morgan's leverage of 11.82 percent. Prudential and MetLife have leverage ratios of 21.27 percent and 15.86 percent, respectively. Part of this high leverage is explained by the long-term balance sheet obligations these firms have toward their policyholders and other claimants. Longevity risk (the risk that people outlive actuarial models) is an ever-present risk for large life and retirement firms that must constantly recalibrate this exposure as part of their asset–liability management. At the opposite end of the spectrum, by contrast, Warren Buffet's Berkshire Hathaway has a leverage ratio of under 2 percent and is the lowest ranked financial institution out of the 75 monitored in V-Lab.

While V-Lab's systemic risk rankings do not perfectly mirror data kept by regulators who apply different models, when back-tested to the Great Deleveraging, V-Lab's results are predictive of firm failure and distress.[61] So, what do V-Lab's results and the staying power of life and retirement firms in the top 10 say about the condition of these vast institutions? For one, these firms are substantial contributors to systemic financial risk in the U.S. and other advanced economies. Among the many lessons the Great Recession taught us is that contagion can spread with dazzling speed, collapsing the market for suitable buyers of distressed firms, and increasing the cost and chaos of a public bailout (recall the marathon negotiations to save Wachovia). Imagine how the entire life and retirement market would fare under forced merger negotiations if all of the largest suitors were exhibiting signs of financial duress. While it should bring little comfort that MetLife and Prudential officially rank among global investment banks in terms of systemic risk, perhaps shining the light on them will yield some future benefit (Table 3.1).

---

[61] V. Acharya, R. Engle, M. Richardson. *Capital Shortfall: A New Approach to Ranking and Regulating Systemic Risks*, AEA Meetings, NYU Stern School of Business, January 7, 2012, accessed November 25, 2015, http://pages.stern.nyu.edu/~jcarpen0/Chinaluncheon/Capital%20Shortfall.A%20New%20 Approach%20to%20Ranking%20and%20Regulating%20Systemic%20Risks.pdf.

Table 3.1 U.S. systemic risk rankings using V-Lab data (February 2015)

U.S. systemic risk ranking

As of 2.6.2015

Institution	% Contribution U.S. systemic risk	Rank	Est. capital shortfall (b. USD)	Leverage	% Contribution Top 10
JPMorgan Chase & Co.	21.04%	1	$89,442	11.82	22.82%
Bank America Corp	18.21%	2	$77,414	11.73	19.75%
Citigroup	14.78%	3	$62,803	11.98	16.02%
Prudential Financial	10.37%	4	$44,081	21.27	11.25%
MetLife	9.75%	5	$41,423	15.86	10.57%
Morgan Stanley	5.78%	6	$24,585	11.36	6.27%
Goldman Sachs	5.30%	7	$22,532	10.56	5.75%
Lincoln National Crop	2.84%	8	$12,055	16.59	3.08%
Principal Financial Group	2.27%	9	$9,632	15.12	2.46%
Bank of New York Mellon	1.88%	10	$8,008	8.9	2.04%
	92.22%		$391,975		100.00%

Data derived from the Volatility Laboratory of the NYU Stern Volatility Institute, accessed November 25, 2015, http://vlab.stern.nyu.edu/en/. Produced by author Dante Disparte.

A National CRO

With the reauthorization of the Terrorism Risk Insurance Act (TRIA) of 2002, a question emerges about whether a siloed risk transfer approach is the most effective strategy for national risks borne by taxpayers. TRIA was initially designed to shore up the insurance industry, which suffered large property, casualty, and life losses due to 9/11.[62] Additional losses were borne by the airline industry and many other sectors, which had never anticipated such a large-scale event. Thus, without the federal backstop, which is ultimately borne by taxpayers, risk appetite for these exposures would have dwindled, especially in light of the massive litigation that followed 9/11 and continues to this day.[63]

There are inherent strengths and weaknesses associated with federated political systems like the U.S. In part, strength is gained through the principle that the whole is greater than the sum of its parts. At the same time, when the 'whole' cannot operate as a single unified unit because of disagreement among its constituent parts, cracks begin to emerge in the edifice. Such cracks are revealed by the absence of a national enterprise risk management strategy.

A "we'll deal with it when it happens" approach does not make for sound policy or preparedness in the face of countless risks, from cyber warfare to terrorism and natural catastrophes. There is no cabinet-level CRO in the U.S.; the occupant of the Oval Office is the closest proxy. A national CRO would be the person who bears overall accountability for the financial well-being and resilience of the nation following acute or attritional losses, wherever they arise and regardless of the cause. This function would serve as the risk champion of the federal government, coordinating and aggregating existing risk resources, such as TRIA, the Federal Emergency Management Agency (FEMA), and natural catastrophe insurance pools (e.g. the National Flood Insurance Program) among others. In effect, the CRO would have oversight over a national risk-bearing pool cutting across all risk domains that are paid for ultimately by taxpayers, thereby reducing the average cost.

The vital role of coordination among response agencies, such as DHS, oversight agencies such as the General Accounting Office and the Office of Management and Budget, and financial agencies such as the Treasury

---

[62] R. Hartwig, *Terrorism Risk and Insurance Program: Renewed and Restructured,* Insurance Information Institute, April 2015, accessed November 24, 2015, http://www.iii.org/sites/default/files/docs/pdf/paper_triastructure_2015_final.pdf.

[63] C. Simmons, "Firms' Discovery Fee Request Is Trimmed in 9/11 Litigation," *New York Law Journal,* November 2, 2015, accessed November 24, 2015, http://www.newyorklawjournal.com/id=1202741208985/Firms-Discovery-Fee-Request-Is-Trimmed-in-911-Litigation?slreturn=20151024090609.

Department and the Federal Deposit Insurance Corporation (FDIC), would lead to the better alignment and stewardship of taxpayer risk bearing. The premise is that rather than dealing with national risks ad hoc (as TRIA does), long-term risk-bearing capabilities would be established. Removing this national risk strategy from electoral politics, which held TRIA's re-authorization hostage and saw special interests being baked into the legislation, would create a more robust and lower cost risk financing solution.

Over time, coaxing the participation of the capital markets through insurance-linked securities and government-backed catastrophe bonds covering man-made and natural perils, will improve overall resilience and liquidity. Institutional investors are attracted to catastrophe bonds as their returns are uncorrelated to the business cycle. A U.S. government-backed program of this nature would not only have the allure of a 'flight to quality', it would benefit from the implied backstop afforded by the Treasury Department across systemic institutions and programs.

Following large-scale losses or economic duress, the U.S. has traditionally followed the Keynesian school of thought, whereby deficit or reconstruction spending helps fill the economic void and restart the engine. Thus far the record has favored Keynes, however, largely at the expense of the nation's debt, which has reached perilously high levels, stoking the potential for a sovereign debt crisis. An aggregate national risk pool, by contrast, would lower the average cost of risk to the taxpayer and break the pattern of treating risk financing like a surprise event. Risk often *is* a surprise event that the U.S. and all other nations must address. Therefore, a national risk pool that cuts across all risk domains and shores up private sector risk appetite will enhance national security.

In the wake of TRIA's reauthorization, which carries the Act to the year 2020, it may be prudent to pose the question: will terrorism risk will somehow diminish by that date? Given its proliferation as a global perennial phenomenon (further explored in Chapter 6), financial solutions to absorb these risks should have similar staying power, adaptability, and reach. Under the oversight of a cabinet-level CRO who can harness the power of the capital markets to absorb risk and coordinate a government response, any country will be better prepared to absorb foreseen or unforeseen shocks.

In Search of Causality While Ignoring Change

Organizations that fail to adequately anticipate long-range risk—the risk continuum—often search for causality while ignoring change. Far too often certain "truths" are held as constants, such as $100 for a barrel of oil or perpetually rising home values, and the entire edifice of economies and companies can be, and have been, constructed on what is essentially a weak foundation. Stress-testing these long-term risks is often impossible because they tend to emerge after many years of 'normal' performance, during which time multiple rounds of leadership change have likely occurred and collective amnesia sets in. Additionally, the confluence of emerging sources of influence that were not contemplated when 'normality' was established often disrupt entire business models and national foundations.

The clarion call of Global Warming that first emerged in the 1980s and was largely ignored, has finally gained a place in our collective consensus. Herein lies the importance of multi-disciplinary, multi-generational risk governance. The feeling of history repeating itself is indeed a very real sentiment that we should not ignore. For example, the EU experiment, which for many years proffered a vision of *Pax Europaea*, now seems to be fraying at the periphery under the weight of tremendous debt burdens, incredibly high levels of unemployment, humanitarian refugee flows, and under the specter of a new Cold War.[64] Peripheral economies, such as Portugal, Ireland, Italy, Greece and Spain (disdainfully referred to as the PIIGS), have for many years now been trying the nerves of less spendthrift Germans and the French.

Similarly, the prospect of European Union (EU) expansion seems to have run aground, along with the security canopy afforded by the North Atlantic Treaty Organization (NATO) and its ultimate peace guarantor, the U.S. Both have been severely challenged by Russia's actions in the Ukraine, which not only saw the annexation of Crimea, but continues to grind away at Ukraine's economy and the détente that reigned for decades in post-War Europe.[65] In short, we have been in, and will remain in, dangerous times.

The right-wing political resurgence in Europe that has accompanied its rolling economic recession and the waves of immigrants are a reminder that extremist political forces lurk beneath the surface, and that they may reappear quickly when times get tough. The term Islamophobia was coined to denote the deep-seated fear of Islam and Muslims, which comprise a large population in Europe. Nowhere is this so-called struggle of 'East meets West' more apparent than in France, where 7.5 percent of the population are Muslim.[66]

---

[64] A. Merelli, "The 30-year-old Agreement that Symbolizes European Unification is Fraying at the Edges," *Quartz*, September 13, 2015, accessed November 24, 2015, http://qz.com/501064/the-30-year-old-agreement-that-symbolizes-european-unification-is-fraying-at-the-edges/.

[65] A. Sytas and A. Croft, "Ukraine Crisis will be Game Changer for NATO," Reuters, May 18, 2014, accessed November 24, 2015, http://www.reuters.com/article/2014/05/18/us-ukraine-crisis-nato-insight-idUSBREA4H01V20140518#RfbjufyjAAHIZeri.97.

[66] C. Hackett, *5 Facts About the Muslim Population in Europe*, Pew Research Center, November 17, 2015, accessed November 24, 2015, http://www.pewresearch.org/fact-tank/2015/11/17/5-facts-about-the-muslim-population-in-europe/.

France's colonial misadventures in North Africa are now posing new, predictable challenges to secular French society. France's erstwhile colony in Algeria, considered a French overseas department for 130 years, profoundly changed French demographics and national character.[67] Under Charles de Gaulle, Algerians and the *pieds-noirs* (the North Africans of European descent) were granted the right of return to mainland France.[68]

To many French Muslims, secularism flies in the face of their religion, sowing the seeds of dissent and what they would argue is a pervasive double-standard in their country and, indeed, in Europe as a whole. Institutions like the EU, NATO and other international organizations are supposed to redouble their efforts when it is least convenient. Instead, Europe has been racked by rising nationalism and a return of "beggar-thy-neighbor" policies that are pitting one country against the other, and threatening the very viability of EU integration.

Throughout this exploration of the risk continuum we have discussed how a range of man-made risks have collided with themselves, whether it is in response to a natural disaster, a financial crisis, a government's approach to indigenous risk, or refugee flows resulting from conflicts. Their common thread is that the ability to thoughtfully anticipate and respond to these risks makes the difference between surviving and thriving—a theme we will be returning to throughout the rest of the book. The risk continuum is the realization that risk resides in every direction we may look, and that the solution to managing that risk may be sitting right in front of us. The ability to recognize this is one of the keys to becoming an agile risk manager and decision maker.

---

[67] A. Hussey, "Algiers: A City Where France is the Promised Land – and Still the Enemy," *The Guardian*, January 26, 2013, accessed November 24, 2015, http://www.theguardian.com/world/2013/jan/27/algeria-france-colonial-past-islam.

[68] M. Kimmelman, "Footprints of Pieds-Noirs Reach Deep into France," *The New York Times*, March 5, 2009, accessed November 24, 2015, http://www.nytimes.com/2009/03/05/world/europe/05iht-kimmel.4.20622745.html?_r=0.

4

Complexity Reduction

Despite the plethora of information and signals driving one choice over another, today the task of making decisions is more complicated than ever before. Among the plethora of reasons is that organizations have more and more moving parts. For example, an entirely new category of jobs exist, such as Chief Gender Equality Officers and Chief Happiness Officers (a Silicon Valley invention, known officially as the "Jolly Good Fellow" at Google), in which individuals busy themselves diagnosing the emotional well-being of their workers as well as adjusting workplace policy and culture in order to create the conditions for corporate happiness.[1] This naturally leads to asking the question: How can anyone make a decision when those involved now have to take account of so many amorphous variables which had never previously needed to be taken into account in the process?

For data scientists, who shepherd organizations through the information maze (or perhaps despite the abundance of information), one of the key risks of our times is the depth of complexity. While many aspects of organizational decision making have been made routine by following algorithmic patterns (e.g. decision trees, layers, and protocols), this is the very antithesis of risk agility. The preponderance of information is stultifying decision makers, rather than lighting up bright paths where choices can lead to desired organizational outcomes. In a word, we have so much information at our fingertips that we end up having no information at all.

---

[1] J. Kovensky, "Chief Happiness Officer is the Latest, Creepiest Job in Corporate America," *New Republic*, July 23, 2014, accessed November 24, 2015, http://www.newrepublic.com/article/118804/happiness-officers-are-spreading-across-america-why-its-bad.

© The Author(s) 2016
D. Wagner, D. Disparte, *Global Risk Agility and Decision Making*,
DOI 10.1057/978-1-349-94860-4_4

Another inhibiting force is how large organizations have ossified into slow-moving, top-heavy, unresponsive structures. This implies that all organizational decisions, especially those of any import, are made by the few people left in the room, while the risk manager and others are "sent out for the biscuits." This too is a profound source of internal distrust and external disdain of the modern global enterprise. Stakeholders feel left out, both literally and figuratively. Even shareholders in public corporations are feeling the bite of obfuscation and short-termism. In short, the global system, whether economic or geopolitical, has itself become far too complex and opaque to effectively respond to emerging risks. Arguably, this complexity is a key driving force behind increased systemic risk.

Complex systems fail in complex ways. Everyone from so-called ethical hackers to self-styled cyber Robin Hoods (like Edward Snowden) to state-sponsored cyber terrorists (like the North Korean-backed Guardians of the Peace) are taking every effort to shine an unwanted light on complex global organizations.[2] Neither governments nor private enterprise are spared from the wrath of disclosure that is being visited upon one organization after another. While cyber risk will be dealt with at length later in the book, it is worth noting—with some trepidation—that large, value-driven organizations and governments hate transparency.

Sunlight Is the Greatest Disinfectant

This aversion to transparency is partly driven by the somewhat dated zero-sum notion of competitive strategy, wherein all gain in a competitive market comes at someone's expense. While the right to privacy is indeed a pillar of a modern democracy, fear of transparency should not be one of the foundation stones. This is the very reason the house of cards of the modern enterprise may be tumbling down. Nothing is sacrosanct anymore. The few glimpses the public may see inside even the most venerable institution's inner workings are cause for deep concern. From the systemic child abuse scandal that racked the Vatican to one round after another of corporate scandal to FIFA's mockery of 'the world's sport', rare revelations of how some organizations behave behind closed doors is doing the modern enterprise no favors.

---

[2] D. Auerbach, "The Sony Hackers are Terrorists," *Slate*, December 14, 2014, accessed November 24, 2015, http://www.slate.com/articles/technology/bitwise/2014/12/sony_pictures_hack_why_its_perpetrators_should_be_called_cyberterrorists.html.

This is the reason why agile risk managers need as much emotional intelligence as they have IQ. Staying ahead of risks that lie in the dark recesses of people's minds, or between the keyboard and chair, requires sensitivity to human motives and behavior—particularly focusing on how they behave when no one is watching. Yet the dilemma in tackling these behavioral risks is that trust is the other cornerstone of the global economy. Trust of government and global institutions is at an all-time low. This is evident not only in the persistent 'public' tension and crises around the world, but also in the persistence of the scandals that are gradually dulling our nerves to what can occur under the cover of darkness.

Edelman, the world's largest public relations firm, produces an annual trust index of both private enterprise and government. In its 2015 report, unsurprisingly, trust was at an all-time low for government, business, media and NGOs. Table 4.1 highlights 27 countries and how they rank according to the Trust Barometer. Some surprising countries are on the list below the 50 percent mark, including Sweden, whose firms like Ikea and Volvo have been paragons of corporate citizenship. Equally surprising is the fact that trust in Indian companies and government, as well as in the United Arab Emirates (UAE) and China, is so high, raising question about how trust is measured in autocratic regimes, and serving to emphasize how difficult it can be to accurately interpret the findings of global surveys. Survey methodology aside, Edelman's report sheds light on how trust in the four pillars of modern society—government, business, the media and NGOs—is waning around the world.[3]

Naturally, practicality calls for balancing transparency on the one hand with the need for discretion on the other. Where the natural line will fall depends on how organizations respond to the thrust of globalization and the speed of sharing in the information age. For the world to regain trust in decisions that are made in darkness, risk managers and, above all, boards and leaders, need to combat internal malfeasance—where even the perception of wrongdoing erodes confidence.

For the better part of the last decade, Barclay's, a leading global bank, persistently manipulated the London Inter-bank Offered Rate (LIBOR), the interest rate that sets the benchmark for exchanges between banks, consumer loans, and savings. Manipulating LIBOR enabled banks to gain outsized profits. That this occurred at all is bad enough, but that it occurred following the global bailout of the banking industry and a wholesale overhaul of banking

---

[3] Edelman, *2015 Edelman Trust Barometer: Global Results*, 2015, accessed November 25, 2015, http://www.edelman.com/insights/intellectual-property/2015-edelman-trust-barometer/.

Table 4.1 The Edelman 2015 Trust Barometer

	2014		2015	
	Global	56	Global	55
Trusters	China	79	UAE	84
	UAE	79	India	79
	Singapore	73	Indonesia	78
	Indonesia	72	China	75
	India	69	Singapore	65
	Malaysia	65	Netherlands	64
	Canada	60	Brazil	59
	Netherlands	60	Mexico	59
Neutral	Hong Kong	59	Malaysia	56
	Mexico	59	Canada	53
	Australia	58	Australia	52
	Brazil	57	France	52
	Germany	57	U.S.	52
	Argentina	53	Germany	50
	U.K.	52	Italy	48
	S. Korea	51	S. Africa	48
	Sweden	51	Hong Kong	47
	S. Africa	50	S. Korea	47
Distrusters	U.S.	49	U.K.	46
	France	46	Argentina	45
	Japan	44	Poland	45
	Italy	43	Russia	45
	Turkey	41	Spain	45
	Ireland	39	Sweden	45
	Spain	39	Turkey	40
	Russia	37	Ireland	37
	Poland	35	Japan	37

2015 Edelman Trust Barometer: Global Results, Edelman, 2015, accessed November 25, 2015, http://www.edelman.com/insights/intellectual-property/2015-edelman-trust-barometer/

regulations speaks volumes about how risks can fester under opaque and complex conditions.

The Economist observed in 2012 that "what may still seem to many to be a parochial affair involving Barclays, a 300-year-old British bank, rigging an obscure number, is beginning to assume global significance. The number that the traders were toying with determines the prices that people and corporations around the world pay for loans or receive for their savings. It is used as a benchmark to set payments on about $800 trillion-worth of financial instruments, ranging from complex interest-rate derivatives to simple mortgages. The number determines the global flow of billions of dollars each year."[4]

---

[4] "The LIBOR Scandal: The Rotten Heart of Finance," Economist.com, July 7, 2013, accessed November 25, 2015, http://www.economist.com/node/21558281.

Complexity, combined with obscurity and greed, are a combustible mix that can fell companies and economies with equal ease. Indeed, when vast personal riches are to be made, the agile risk manager should be on alert.

One of the principal sources of risk in the financial and other sectors is moral hazard—risk taking without bearing the consequences of said risk. A moral hazard is most likely to emerge where there is a combination of factors,[5] which include:

1. *Organizational remoteness or opacity*: Think of the number of cases where remote traders—such Jérôme Kerviel, the day trader who nearly brought down Société Générale, or JP Morgan's 'London Whale'[6]—nearly brought down a bank. The risk of doing so is the reason why certain professionals in the banking sector must take an annual two-week hiatus when they are generally not permitted to use company systems, and why a number of financial institutions forbid the use of any personal electronics inside a trading floor. The two-week policy does little, however, in the face of corrupted teams, or the top-down pressure to perform to close a quarter or bridge a budget shortfall.
2. *Financial condition*: The financial condition of an organization says a lot about the likely incidence of moral hazard. Of particular importance is the time mismatch between firm indebtedness (e.g. high long-term debt) and short-term performance incentives. High leverage is a breeding ground for the preconditions of moral hazard because debt implies 'playing with other people's money' and leverage implies risk.
3. *Performance incentives*: Financial incentives are of particular importance when combating moral hazard. Excessive pay packages, golden parachutes, golden handshakes, and soft landings all imperil the system because they effectively remove the direst of consequences from certain classes of managers and professionals, incentivizing them to push and transgress boundaries. Not getting caught emboldens them to make toeing the line their modus operandi.
4. *Competitive environment*: A lot of otherwise good people may operate in an immoral way in a bad system. There is no assumption of morality in the global economy. Highly competitive, high-stakes environments or teams are likely to push operating boundaries. Again, think of the stakes for FIFA and prospective World Cup host countries. The Herfindahl–Hirschman Index (HHI) is a useful measure in determining the relative competition

---

[5] D. Disparte et al., "Moral Hazard Index: Detecting the Preconditions of Moral Hazard," unpublished master's degree research at New York University, 2014.
[6] M. O'Brien, "Meet the Most Indebted Man in the World," *The Atlantic* (November 2, 2012), accessed November 24, 2015, http://www.theatlantic.com/business/archive/2012/11/meet-the-most-indebted-man-in-the-world/264413/.

and density of play in an industry segment.[7] It shows that the higher the concentration of power and wealth in a few hands, the higher the propensity to rig markets, collude and engage in monopolistic practices. While they have been defanged in the age of cheap oil, the Organization of Petroleum Exporting Countries and other price cartels have this property.

5. *Governance*: The separation of powers, transparency, composition and structure of boards, managerial layers, and command and control functions are all vital. What is most often seen in global organizations is a very fine point at the top. At the tip of this point you are most likely to find CEO and board chairs wearing the same hat, or a charismatic public figure that is above the law. Think, for instance, of Italy's former prime minister, Silvio Berlusconi—at once the country's political head and one of its wealthiest citizens—controlling a vast media empire. What systems keeps this type of leader in check when he was the second and fourth estate all at once? Far too often, where CROs exist in organizations, they most often report to the very CEOs they must keep in check. Fewer still report to truly independent board-level risk management committees. The potential conflicts of interest are noteworthy.

Beyond this, organizational culture, national norms, and differences in what constitutes acceptable behavior in different countries complicates the task of both measuring and controlling trust and ethical boundaries. For U.S. companies, the Foreign Corrupt Practices Act governs U.S. corporate actions overseas vis-à-vis accounting transparency and bribery. While being legally forbidden in the U.S., in a host of other countries, such as France, bribes may be seen as a cost of doing business and can be deducted as a business expense in a firm's annual taxes.[8]

The First and Last Line of Defense

"Let's hope we are all wealthy and retired by the time this house of cards falters."[9]

Values sometimes matter most when they are least convenient. As shown in the global decline of trust of government and business, paying lip service

---

[7] The U.S. Department of Justice uses the index to measure market concentration and power as one of the benchmarks informing its positions on mergers.

[8] "Writing Off Tax Deductibility, *OECD Observer*, no. 220, April 2000, accessed November 24, 2015, http://www.oecdobserver.org/news/archivestory.php/aid/245/Writing_off_tax_deductibility_.html.

[9] D. Kopecki and L. Woellert, "Moody's, S&P Employees Doubted Ratings, E-Mails Say (Update2)," *Bloomberg*, October 22, 2008, accessed November 24, 2015, http://www.bloomberg.com/apps/news?pid=newsarchive&sid=a2EMlP5s7iM0.

to organizational value systems, while not believing in and upholding them, presents a problem and seriously hinders both decision making and risk management. Failure to adhere to values can also severely harm brand equity and market capitalization. Corporate and national values—and the people that rally around them—are the first and last line of defense against many of the man-made risks affecting our world. From climate change to grinding cyber risk and terrorism, a values-based response is often not only appropriate, but strengthens organizational resilience, combats complexity and improves overall decision making.

Values transcend all organizational levels and inform internal stakeholders, as well as the market and broader community, about what an organization stands for. As the old adage goes, you either stand for something or you stand for nothing. Over the last decade the number of examples where values have meant nothing trumps the number of cases where values were upheld. This is troubling for a number of reasons, although some promising green shoots are cropping up, signaling a rise in corporate activism for more than short-term shareholder gain or political lobbying.[10]

An organization's value system is not comprised of the nice words emblazoned on walls and in marketing materials, such as Google's former motto "Don't be Evil".[11] To truly understand organizational values one must look at the qualities of the people who thrive the most in an organization (the so-called '*tone at the top*') and not the often-gimmicky slogans, such as "Responsibility and Sustainability", the values that the beleaguered VW shares with prospective recruits—a far cry from the systematic cheating that has tarnished the firm.[12] The 'truer' values that pre-crisis VW espoused are captured in its 2014 annual report, wherein former chairman Martin Winterkorn declared "… Our pursuit of innovation and perfection and our responsible approach will help to make us the world's leading automaker by 2018—both economically and ecologically."[13] Clearly, the pursuit of being number one trumped the other pursuits Mr Winterkorn referenced.

In the fast-paced, high-stakes banking landscape, commonly conveyed organizational values such as integrity, trust, and meritocracy appear entirely

---

[10] D. Disparte and T. Gentry, "The Rise of Corporate Activism: From Shareholder Value to Social Value," *CSR Journal*, June 30, 2015, accessed November 24, 2015, http://csrjournal.org/the-rise-of-corporate-activism-from-shareholder-value-to-social-value/.

[11] C. Thompson, "Does 'Don't Be Evil' Still Apply to Google?," CNBC, August 19, 2014, accessed November 24, 2015, http://www.cnbc.com/2014/08/19/does-dont-be-evil-still-apply-to-google.html.

[12] VW AG website on corporate values, accessed November 24, 2015, http://www.volkswagenag.com/content/vwcorp/content/en/human_resources.html.

[13] Volkswagen AG, *2014 Annual Report*, 2014, accessed November 24, 2015, http://annualreport2014.volkswagenag.com/.

absent. When, for example, was the last time you read a headline about something purely altruistic done by a bank (i.e. with no direct or indirect benefit to itself)? In their place you are far more likely to find the values of competitiveness, wealth creation, and greed, as if the previous half-dozen banking crises, or the Great Recession, had never happened or meant anything. There have been too many instances where values ought to matter and serve as clear decision frameworks, yet they go entirely ignored. Hence the notion that values have the most meaning when it is least convenient. This is the source of much of the cynicism accompanying the banking sectors, and why corporate and institutional influence is waning with so many groups of people around the world.

The same can be said on a broader scale about the erosion of values. For example, following 9/11, the Geneva Convention and the Law of Land Warfare—the best-known international standards against torture and for the ethical treatment of prisoners of war—appears to mean far less to individuals and governments than it did pre-9/11. Since then, the notion that any means can and should be used to extract information from prisoners has gained acceptance from a wide array of individuals who consider such guidelines passé. Before 9/11, the western world had largely been spared the need to give these societal values much thought, or put them to the test. The U.S. fared particularly poorly, among Western nations, on this matter. The fact that so many prisoners of the War on Terror, and detainees from countries who were tortured using so-called 'enhanced interrogation tactics', had a serious adverse effect on America's credibility and hampered efforts to win over hearts and minds in the War on Terror, and in the protracted wars in Iraq and Afghanistan.[14] To win a war for hearts and minds, we must lead with our hearts and minds.

The word courage is derived from the French word *coeur*, meaning heart.[15] Extraordinary renditions of suspected terrorists, complotters and sympathizers to countries with lax rules on torture further diminished the standing of U.S. institutions, even though the actual blood may have been on another's hands. That parts of the U.S. Federal apparatus that are meant to uphold and enforce national values, such as the Justice Department, deliberately redrew boundaries around presidential war powers and torture is a national blemish akin to the internment of Japanese-Americans during World War II. Many Americans, and others around the world, never forgave the U.S. government

---

[14] C. Savage, "Election to Decide Future Interrogation Methods in Terrorism Cases," *The New York Times*, September 27, 2012, accessed November 24, 2015, http://www.nytimes.com/2012/09/28/us/politics/election-will-decide-future-of-interrogation-methods-for-terrorism-suspects.html?_r=0.

[15] R. Staub, "The Heart of Leadership: 12 Practices of Courageous Leader," North Carolina: *Staub Leadership Consultants,* January 2007.

for its imprisonment of Japanese-Americans, and it appears likely that the same will be true for the rendition of terrorism suspects. There is a price to be paid for moving goalposts, misleading the public, and refusing to take responsibility for misdeeds when they are uncovered. Our mothers taught us the difference between right and wrong. How did so many of us suddenly forget that once we became adults?

The U.S. is certainly not alone as a national example of where values are sometimes considered optional. History is littered with such examples. For example, in Brazil, a country that has received much attention over the last decade as a premier investment destination,[16] the country's largest firm Petrobras (a quasi-state-owned enterprise) was saddled with a massive corruption scandal pointing right to the top of the government.[17] Indeed, Dilma Rousseff, Brazil's president at the time the scandal erupted, was the chair of Petrobras. The company's robust corporate value system—including ten individual values, a code of conduct, ethics and anti-corruption policies—clearly meant little to those implicated in this case. Worse still, with the clear signs of government collusion at the highest levels of the Workers Party—Brazil's ruling party at the time—the people of Brazil were right to question the country's value system of *Ordem e Progresso* (order and progress),[18] which is emblazoned on the country's national flag. With approximately $800 million in bribes paid out over a decade, the Petrobras case is one of the largest uncovered corruption scandals in history, with far-reaching consequences for Brazil's standing as an investment destination.[19] Such a simple matter, yet so much economic value can be eradicated when values are not upheld.

Europe's slow and in some ways inhumane response to the Middle Eastern and African refugee crisis in 2015 flew in the face of its obligations under Article 18 of the EU Charter of Fundamental Rights, and the value system espoused by the broader union.[20] Some countries (such as Germany and

---

[16] Even though the statistics show that Brazil should not actually be considered a premier investment destination, for reasons ranging from an inability to sustain a GDP growth rate higher than 2 percent, to a failure to invest sufficiently in its own infrastructure, to a worker efficiency rates well below many of its competitors around the world.

[17] J. Leahy, "Petrobras Scandal Lays Bare Brazil's Political Fragilities," *Financial Times*, March 11, 2015, accessed November 24, 2015, http://www.ft.com/cms/s/0/0fdb4796-c6f8-11e4-9e34-00144feab7de.html#axzz3sQhRNpeW.

[18] J. Watts, "Brazil in Crisis Mode as Ruling Party Sees Public Trust Rapidly Dissolving," *The Guardian*, March 17, 2015, accessed November 24, 2015, http://www.theguardian.com/world/2015/mar/17/brazil-crisis-petrobas-scandal-dilma-rousseff-protests.

[19] Fortune and Reuters Editors, "Petrobras Takes $17 Billion hit on Scandal, Promises 'Normality'," *Fortune*, April 23, 2015, accessed November 24, 2015, http://fortune.com/2015/04/23/petrobras-takes-17-billion-hit-on-scandal-promises-normality.

[20] European Council for Refugees and Exiles website, accessed November 24, 2015, www.ecre.org.

Sweden) threw their doors open to the refugees (and have for many years) while others (such as Hungary) slammed the door shut. National interest and xenophobia have taken the place of solidarity and global obligations with respect to some of the world's millions of refugees from conflict areas. While the U.S. has for many years been a leader in welcoming refugees from around the world, it has failed miserably with respect to Syrian refugees, accepting only 1,500 such refugees per year in the first years of the Syria conflict, and raising that figure to only 10,000 in 2016[21] (compared with Germany's pledge to accept 800,000 in 2015 alone[22]).

Eroding confidence in global institutions is partly driven by the lack of operational alignment to the lofty goals, visions, and value statements these bodies espouse. From the UN's underrepresentation of developing and emerging countries in its most important groups, to the fact that the IMF and World Bank perpetually recycle European and American leadership, emerging markets are justified in their move to create parallel and, in their eyes, more representative alternatives. In the eyes of the underrepresented countries, this is not just something that would be nice to have; rather, it is essential.

To have legitimacy, the world's most highly populous and largest economies *must* be more appropriately represented in global institutions. It is silly (and unfair) that the 'five great powers' who ruled the roost in 1945 remain the 'great powers' at the UN today. Of these five (China, France, Russia, the UK and the U.S.), only two were among the eight most populous in 2015 (with China in the number one spot with 19 percent of the global population, and the U.S. in the number three spot, with less than 5 percent of the world's population[23]). All five did remain among the largest economies (based on nominal GDP) in 2015, but only the U.S. was among the top four (the others being China, Japan and Germany).[24] Unless the ownership percentages of these organizations are re-examined, their leaders will remain perceived as relics of World War II. This splintering of the global system is a dangerous portent and does not bode well for our collective ability to coordinate and act in the face of unprecedented man-made risks—all requiring consensus, shared sacrifice and a concerted global response.

---

[21] "U.S. Plans to Accept 10,000 Syrian Refugees Next Year," BBC.com, September 10, 2015, accessed November 24, 2015, http://www.bbc.com/news/world-us-canada-34215920.

[22] H. Horn, "The Staggering Scale of Germany's Refugee Project," *The Atlantic*, September 12, 2015, accessed November 24, 2015, http://www.theatlantic.com/international/archive/2015/09/germany-merkel-refugee-asylum/405058/.

[23] *List of Countries and Dependencies by Population*, Wikipedia, accessed November 24, 2015, https://en.wikipedia.org/wiki/List_of_countries_and_dependencies_by_population.

[24] *List of Countries by GDP (nominal)*, Wikipedia, accessed November 24, 2015, https://en.wikipedia.org/wiki/List_of_countries_by_GDP_(nominal).

The notion of sacrifice underpins values at work and creates an enduring example of risk agility. Values are upheld when organizations are prepared to accept responsibility and endure pain when their values are confronted. As a famous example, Johnson and Johnson, well known for its credo, led a global withdrawal of tainted Tylenol that posed a risk to the public in the early 1980s.[25] The Peanut Corporation of America, by contrast, led a blatant cover-up while its salmonella tainted peanut products worked their way into the food supply chain, killing 9 people and sickening hundreds of others. Stuart Parnell, its then president, and other executives, were given harsh sentences in what was intended to be a message that corporate wrongdoing will no longer consign a corporate officer to a golden parachute or a blue-collar summer camp.[26]

History has subsequently revealed numerous instances demonstrating that the message has not sunk in, but other examples prove that collective conscience, combined with creativity, is having an impact. Ikea, for example, facing pressure to bribe Russian officials in order to maintain its energy supply there, got off the Russian energy grid temporarily and installed its own power generation equipment.[27] Rather than foregoing the market entirely, ceding market share, and access to its competitors, Ikea's example shows that—contrary to popular belief—there is a place for values in action in opaque emerging markets.

Values in Action

Around the world there is a growing chorus of voices illustrating the institutional power of values in action. In the U.S., the national enactment of marriage equality by the Supreme Court in 2015 was an example of upholding the notion that we are all created equal.[28] While such fights are hard-won,

---

[25] M. Markel, "How the Tylenol Murders of 1982 Changed the Way We Consume Medication," PBS Newshour, September 29, 2014, accessed November 24, 2015, http://www.pbs.org/newshour/updates/tylenol-murders-1982/.

[26] L. Bever, "Former Peanut Plant Executive Faces Life Sentence for Lethal Salmonella Coverup," *The Washington Post*, July 24, 2015, accessed November 24, 2015, https://www.washingtonpost.com/news/morning-mix/wp/2015/07/24/former-peanut-plant-executive-faces-life-sentence-for-selling-salmonella-tainted-food/.

[27] A. Kramer, "Ikea Tries to Build Public Case Against Russian Corruption," *The New York Times*, September 11, 2009, accessed November 24, 2015, http://www.nytimes.com/2009/09/12/business/global/12ikea.html.

[28] B. Lowder, "Obama on Marriage Equality: America Should Be Very Proud," *Slate*, June 26, 2015, accessed November 24, 2015, http://www.slate.com/blogs/outward/2015/06/26/obama_on_supreme_court_gay_marriage_decision_america_should_be_very_proud.html\.

when they do occur they mark a profound sea change in the course of history and serve as examples around the world (although not everywhere, of course). Just as in combating apartheid South Africa, countries that stand for something when it is least convenient gain enduring leadership and authenticity. This lesson is beginning to translate to corporate board rooms as well, although not fast enough.

Apple serves as a great example of values in action.[29] Led by Tim Cook, the first openly gay CEO of a Fortune 500 company in the U.S. (and the highest-valued firm in the world by market capitalization as of 2015[30]) led a public charge in fighting for marriage equality, not as a personal crusade, but rather using his considerable gravitas in the CEO-suite arguing that the fight against gay rights was anathema to the company's value system.[31] Apple was not alone in its principled stance. Salesforce, led by CEO Marc Benioff, went as far as offering relocation support for an employee based in Indiana.[32] Contrary to the maximizer, zero-controversy tendency among CEO's in publicly-listed firms, Tim Cook and Marc Benioff risked harming shareholder value by quitting large markets like Indiana, potentially angering religious conservatives. In order to defend their corporate value systems for employees, partners and aligned external stakeholders, Apple and Salesforce took a stance for something they believed in, and guess what—it did not hurt their bottom line. Apple's uncharacteristically transparent supply chain report, highlighted earlier in the book, is another example of values in action, and the often counter-intuitive advantage that can be gained by standing for something when it is inconvenient.

Another example of emerging good practice in values-based decisions is Starbucks. Under the leadership of its founder and long-time CEO, Howard Schultz, Starbucks famously waded into the politically toxic waters of U.S. campaign finance and race relations—at a time when other firms were jockeying for influence through paycheck persuasion and lobbyists, or remaining muted as the embers of America's racist past were reignited.[33] Instead of

---

[29] T. Cook, "Tim Cook Speaks Up," *Bloomberg Business*, October 30, 2014, accessed November 24, 2015, http://www.bloomberg.com/news/articles/2014-10-30/tim-cook-speaks-up.

[30] V. Kopytoff, "Apple: The First $700 Billion Company," *Fortune*, February 10, 2015, accessed November 24, 2015, http://fortune.com/2015/02/10/apple-the-first-700-billion-company/.

[31] T. Cook, "Tim Cook: Pro-Discrimination 'Religious Freedom' Laws are Dangerous," *The Washington Post*, March 29, 2015, accessed November 24, 2015, https://www.washingtonpost.com/opinions/pro-discrimination-religious-freedom-laws-are-dangerous-to-america/2015/03/29/bdb4ce9e-d66d-11e4-ba28-f2a685dc7f89_story.html.

[32] L. Stampler, "Salesforce CEO Gave an Employee $50,000 to Help Leave Indiana," *Time*, April 2, 2015, accessed November 24, 2015, http://time.com/3768955/salesforce-boycott-indiana-religious-freedom/.

[33] A. Chatterji, "Starbucks' "Race Together" Campaign and the Upside of CEO Activism," *Harvard Business Review*, March 24, 2015, accessed November 24, 2015, accessed November 25, 2015, https://hbr.org/2015/03/starbucks-race-together-campaign-and-the-upside-of-ceo-activism.

standing on the sidelines on these highly divisive matters, Starbucks led a bold and controversial charge in defense of its value system, employees and aligned external stakeholders. Mr. Shultz drove a public campaign to get large corporations to stop financial contributions to both the Democrats and Republicans.[34] His argument, at the height of partisan politics paralyzing the U.S., was simply that neither party was serving the country well, and citizens of all political persuasions were suffering. In 2015, following racially divisive riots in Ferguson and Baltimore, and increased animus nationally, Starbucks launched the Race Together campaign.

While few people wanted to talk to their barista about race relations in America while waiting for their Venti Caramel Macchiato (with low-fat milk, even though it may only have been a teaspoonful), despite its controversy, the campaign underscored Starbuck's consistency in upholding its values. While institutional values can naturally gravitate toward either side of the political spectrum, their popularity does not matter nearly as much as the fact that they are upheld when it is least convenient. Consider President Vladimir Putin's support for the besieged Syrian President Bashir Assad. While irksome in the halls of power in the West, Mr Putin's steadfast support of the status quo in Damascus proved that he was not only a shrewd world leader, but also one who can be counted on by his friends—however loathsome others may perceive them to be. From holding a UN Security Council veto over the heads of western powers to kinetic military intervention in Syria, Russia proved its ability and value as an ally—it also proved that when it draws bright red lines, they mean something. These are among the reasons Forbes named President Putin the most powerful person in 2015.[35]

The key point to remember is that values only take on meaning when they shape actions, no matter the impact. In the face of risks that are reshaping the world's definition of normality, steadfast adherence to value systems may be the only certain thing we can hang on to in the midst of so much uncertainty. Values may be the only space in our world where absolutes are permissible. Therefore, if a nation says that it does not torture prisoners of war, it does not torture them ever—even when there is information that must be extracted for national security purposes. That was one of the reasons why the U.S. lost respect from so much of the world during the George W. Bush era. It said one thing, did another, and was revealed to be lying about it.

---

[34] Z. Miller, "Starbucks CEO Takes 'No Campaign Donations' Pledge to the Public with Full Page NYT Ad," *Business Insider*, September 4, 2011, accessed November 24, 2015, http://www.businessinsider.com/starbucks-ceo-takes-no-campaign-donations-pledge-to-the-public-2011-9.

[35] *The World's Most Powerful People*, Forbes.com, accessed November 24, 2015, http://www.forbes.com/powerful-people/list/.

If your firm stands for safe food products, make it your singular focus to provide safe food all of the time, to all of your consumers, in all markets. Sending tainted food to the developing world is not upholding values, it is skirting them, rewriting the rules, and putting the pursuit of profit ahead of doing the right thing. If your firm is a gun manufacturer and a state passes legislation that is anathema to its value system, you should have the right to move your operations, as Beretta did from the state of Maryland.[36] Equally, consumers, voters, and shareholders have a right to punish what they deem to be errant behavior.

If an organization truly stands for transparency, a cyber breach—although costly and inconvenient—would not faze them. Part of the negative impact of Sony Entertainment's breach over the film *The Interview* is the fact that quite a lot of rot was revealed to be festering in the framework of their value system, and in the recesses of key executives' minds, including that of Amy Pascal (Sony Entertainment's co-chair).[37] From the unconscionable thought of a black James Bond, a role considered for Idris Elba (a black British actor), to tawdry communiqués about Hollywood stars and President Obama's taste in film, Sony's breach was made all the worse for illustrating behavior unbefitting of a (supposedly) values-based organization.[38] Rather than improving risk management and loss mitigation, low internal adherence to Sony's stated corporate values ended up causing the firm more damage and lingering reputational harm. Again, living by values is free, and the values that were truly prized in the entertainment group were clearly those embodied by its leadership, rather than those stated on advertisements and in annual reports.

Primordial Lesson

In looking for examples of organizational and individual responses to risk and decision making, our closest cousins in the natural world offer some telling insights. In the 1960s researchers conducted a study involving behavioral patterns and the collective memory of monkeys. This study involved a group of rhesus monkeys (a proxy for a team or organizations), a ladder (a proxy for direction or risk taking), a bunch of bananas dangling above the ladder (a proxy for reward, purpose, or profits) and, finally, the risk of being sprayed

[36] M. Rosenwald, "Beretta, Moving Production out of Maryland, Joins Gunmakers Heading to Friendlier States," *The Washington Post*, July 27, 2014, accessed November 24, 2015.

[37] C. Rosen, "Amy Pascal Leaving Sony Pictures Role To Launch Own Production Company at Studio," *The Huffington Post*, February 2, 2015, accessed November 24, 2015, http://www.huffingtonpost.com/2015/02/05/amy-pascal-sony_n_6622920.html.

[38] A. Zurcher, "Rush Limbaugh and his 'Black Bond' Outrage," BBC News, December 29, 2014, accessed November 24, 2015, http://www.bbc.com/news/blogs-echochambers-30594460.

with cold water (a proxy for risk or an externality, since the ladder climber was not getting wet).[39] In the experiment, which was conducted over a series of rounds, the researchers wanted to understand the behavioral patterns of risk-takers, those that would climb the ladder and the learned response of the group to getting doused with cold water each time the bunch of bananas was taken.

Predictably, round after round of cold water angered the group of monkeys yielding a behavior in which the payoff of a bunch of bananas gave way to risk aversion; the group would stop any would-be ladder climbers from going up, despite the fact that members of the group were being exchanged for new subjects over the course of the experiment. The real surprise, and a powerful example of practicing risk agility and following instincts, is that even after an entirely new generation of test subjects was brought into the experiment, the ladder and banana had become an ill omen that they avoided almost superstitiously. In effect, the new generation almost had the lessons of their forbearers ingrained in their experiences, and the collective wisdom of not getting doused with water superseded the temporary reward of a few bananas. This case also highlights how learned behavior translates over time in organizations.[40]

This type of learned individual and collective behavior are also evident in humans; however, the pressure of modern society and keeping up appearances has taught us to silence our instincts. How tribes and traditional societies respond to the highly dangerous natural world is telling. Even in early childhood, members of the Yanomami tribe in remote Venezuela pass on survival guides as heuristics. Knowledge of poison dart frogs, edible and potentially dangerous plants, and other instincts are passed on generationally in simple, yet enduring ways.[41] Modern organizations, by contrast, labor to apply simple ways of improving their resilience. "Do Not Enter" signs almost beckon transgression, just as big red buttons beckon to be pressed.

Tempering Values with Risk Taking

Just as firms may have check-signing authorities clearly established, and clearly delineated lines of control, tempering risk-taking levels (levels of tolerance) with organizational values and hierarchy can go a long way toward improving

---

[39] N. Merchant, "How to Invent the Future," *Harvard Business Review*, October 17, 2014, accessed November 24, 2015, https://hbr.org/2014/10/how-to-invent-the-future.

[40] R. Martin, "Rethinking the Decision Factory," *Harvard Business Review*, October, 2013, accessed November 24, 2015, https://hbr.org/2013/10/rethinking-the-decision-factory.

[41] K. Milliken, *The Yanomami are Great Observers of Nature*, Survival International, 2015, accessed November 24, 2015, http://www.survivalinternational.org/articles/3162-yanomami-botanical-knowledge.

agility. It is healthy for line managers and those at lower levels in an organization to know how much risk to take. This not only spurs innovation and bounded risk taking, it also establishes a lower threshold for an early warning system—something that is almost non-existent in large, bureaucratic organizations. As is often the case in large enterprises, by the time the alarm bells sound in the captain's quarters, those in lower decks have either drowned, jumped overboard or failed to signal the impending loss. This is almost entirely preventable, costs next to nothing, and not only makes for better risk preparedness, but better competitive strategy and decision making under opacity.

Those closest to the frontline of where risks will emerge have been organizationally trained—like the primates and the ladder—to not convey bad news and disrupt order, lest they get doused by the cold water of organizational reproach. Similarly, stratified organizations send a clear signal to frontline staff that those higher up are not to be disturbed, but, if so, it should pass through the chain of command—even if one or more individuals in the chain of the command is part of the problem. People concerned with their own survival will rarely expend much effort overcoming organizational hurdles in order to convey potentially critical news. This increases the signal-to-noise ratio, and leaves many organizational leaders perilously out of touch with their enterprise and world. Global leaders are generally well versed in long range thinking by virtue of their roles, yet they are often entirely disconnected from the here and now, where disruptions are most likely to emerge.

Absence of Evidence

As Carl Sagan famously stated, the absence of evidence is not the evidence of absence. Another way to view this cryptic aphorism is that a path that may appear to be clear of threats may in fact be dangerous, as much as a path that appears fraught with peril may be safe. The risk-ready are not afraid of going down roads less travelled. On the contrary, these organizations are in many ways trailblazers carving pathways into hitherto dangerous places. These include the early investors in emerging markets, the pioneers behind breakthrough innovations, and those who drive political change in the face of fierce opposition. Paralysis, a product of uncertainty, which cannot be measured, is as dangerous a risk as any in our times. Of the many risks shaping our world, market failure stemming from bad strategic choices or inaction have felled many once-great enterprises. The risk ready realize the powerful influence risk has on our perceptions, and how to marshal our will to move forward by delicately balancing the risk–reward tradeoff.

JC Penney, a pioneer in the retail department store industry, was founded in 1902.[42] It once commanded sizeable market share along with longtime rival Sears, and is a well-recognized brand with more than 1,000 physical department stores, catalogue sales, white goods, and a wide variety of products.[43] Failing to anticipate changing market demands following the advent of online marketplaces such as Amazon, JC Penney foundered by responding too late to emerging trends. Those at the top proved to be tone deaf to the market's changing needs and made the mistake of hiring Ron Johnson, Apple's former retail head, to try and spruce up their stores and brand appeal. Without consulting with their typically conservative client base, who looked at JC Penney as a space for traditional shopping, JC Penney's failure to anticipate proved to be a costly misstep from which the firm is still recovering.[44] There is much speculation that retailers like JC Penney and Sears may go the way of RadioShack, an iconic retail chain that sold all variety of tech gewgaws and curios, which declared bankruptcy and folded in 2015.[45]

JC Penney's struggles, and those of many more companies around the world, underscore that the greatest potential killer of global enterprises are often not risks of the unforeseen "smoking crater" variety, but rather a slow attritional demise—one that in most cases is preventable with a greater degree of anticipation and a healthy dose of self-sacrifice. Companies and countries that ignore mega trends are consigned to this type of slow death. Ignoring Africa's emergence as a long-term investment destination, for example, is a decision that will consign corporate strategists to fierce battles for market share in the red oceans of saturated advanced economies, as much as diplomats will continue fighting for influence on the emerging continent.

Analysis Paralysis

As home to 1.2 billion people, a burgeoning middle class, with abundant natural resources and 54 countries, Africa may be the last place on earth where most businesses have long-term financial upside.[46] Clearly harnessing this

---

[42] *Forbes JC Penney Profile*, Forbes.com, accessed on the Forbes website, November 24, 2015.

[43] *JC Penney Company, Inc. corporate website*, JC Penney Company, Inc. www.jcpenney.com, accessed November 24, 2015.

[44] N. Tichy, "JC Penney and the Terrible Cost of Hiring an Outsider CEO," *Fortune*, November 13, 2014, accessed November 24, 2015, http://fortune.com/2014/11/13/jc-penney-ron-johnson-ceo-succession/.

[45] P. La Monica, "Can JC Penney and Sears avoid RadioShack's Fate?," CNN Money, February 10, 2015, accessed November 24, 2015, http://money.cnn.com/2015/02/10/investing/radioshack-jcpenney-sears/.

[46] United Nations, *World Population ProspectsDOUBLEHYPHEN2015 Revision*, 2015, accessed November 24, 2015, http://esa.un.org/unpd/wpp/publications/files/key_findings_wpp_2015.pdf.

global opportunity requires organizational deftness and risk agility the likes of which established players have not seen. Firms such as the Abraaj Group (an emerging market investment fund), AFIG Capital (a Dakar-based private equity firm), and large diversified firms like CFAO (based in France) have viewed Africa as an investment destination for many years. In CFAO's case, its trading history on the continent dates back to 1855.[47] Part of how these firms get their Africa strategies right is that they are not naive about the operating conditions. They also have a values-based in their approach. CFAO employs more than 12,000 people across the continent and has invested heavily in its corporate social responsibility programs.[48]

The continent is certainly not risk-free, however, and this explains why Abraaj has delivered outsized returns for its investors, with an impressive 17 percent annual rate of return since its founding.[49] The risk–reward tradeoff for a pan-African investment strategy, especially one that takes downside risks seriously, has the potential to beat 'safer', more liquid markets. Yet the continent is not for the faint of heart. Like the 'robber barons' who pushed America's westward expansion and developed vast empires in rail, steel, and all the foundations of the industrial age, Africa's corporate and investment pioneers will need a long investment horizon. Bolloré Africa Logistics and Blackstone's recently formed BlackRhino Group, for example, are focusing on logistics and infrastructure investments, respectively, for which the continent, with more than 30 million square miles of land mass, has a tiny fraction of the world's paved roads.[50] Some see this as an obstacle to growth, while others see it as an incredible untapped market—one larger than building America's transport infrastructure in the 1960s.

Entrepreneurialism

Remaining on the sidelines as a spectator can be more dangerous than getting on the playing field. Nowhere is this truer than in Africa. The continent is beginning to shake 50 years of post-colonial aid and development, which

---

[47] CFAO Group corporate website, accessed November 24, 2015, http://www.cfaogroup.com.

[48] CFAO Group corporate profile—2015, CFAOGroup.com, accessed November 24, 2015, http://www.cfaogroup.com/en/market-positioning.

[49] E. MacBride, "The Story Behind Abraaj Group's Stunning Rise in Global Private Equity," *Forbes*, November 4, 2015, accessed November 24, 2015, http://www.forbes.com/sites/elizabethmacbride/2015/11/04/the-story-behind-abraajs-stunning-rise/.

[50] "Cartography: The True Size of Africa," November 10, 2010, Economist.com, accessed November 24, 2015, http://www.economist.com/blogs/dailychart/2010/11/cartography.

has in many ways propped up kleptocratic regimes with little meaningful economic or institutional impact.[51] Pan-African leaders are emerging in business, governance and entrepreneurship (among others), shaping a uniquely African response to the twenty-first century.

Tony Elumelu, a Nigerian billionaire who made his money by creating UBA, a leading pan-African bank, serves as a good example of institution building. Through the work of his foundation, Mr. Elumelu is investing $100 million over ten years to raise a generation of African entrepreneurs. His aim is to invest in training a cadre of 10,000 pan-African entrepreneurs (many from the continent's well-educated diaspora) as a way of building market-based solutions to many of the problems plaguing the continent. Judging by the first year of this program, which had 1,000 participants from 51 African countries, Mr. Elumelu may be paving a way for an African century.[52]

After the sun, Africa's most abundant resource is its people—more than 960 million in Sub-Saharan Africa alone.[53] While some liken a 'youth bulge' to a curse and a source of social unrest, others see a youthful and dynamic population as a massive untapped opportunity. In the disruptive era of hyper-valued Unicorns like Uber, the ambitious goal of igniting Africa's entrepreneurial potential is entirely possible; it only takes one of the 10,000 startups to be a breakaway success to achieve these goals. Guided by the experience of successful African entrepreneurs and mentors, this new entrepreneurial movement underscores the need for continued investment to flow into African firms at all stages of growth. In effect, Mr. Elumelu has triggered a necessary standards war on which a model of entrepreneurship will set the African paradigm.

To be sure, importing the algorithmic decision making from advanced market banks, angel investors, and many private equity firms that confound entrepreneurs in the U.S. and around the world, would be a setback for their African peers. Prospective early-stage investors are often too formulaic in their approach. The formula of historical financials, market sizing, business plans and perfect five-year exit strategies means that even the most seasoned early stage investors approach African opportunities with trepidation and, far too often, the wrong yard stick.

---

[51] D. Moyo, *Dead Aid: Why Aid is Not Working and How There is a Better Way for Africa*, Farrar, Straus and Giroux: New York, 2010.

[52] D. Disparte, "President Obama and the Era of Afri-Preneurship," *The Huffington Post*, August, 3, 2015, accessed November 24, 2015, http://www.huffingtonpost.com/dante-disparte/president-obama-and-the-e_2_b_7920518.html.

[53] *World Population Prospects*, United Nations, 2015, accessed November 25, 2015, http://esa.un.org/unpd/wpp/publications/files/key_findings_wpp_2015.pdf.

Apart from the risk aversion that sinks in when dealing with multiple currencies and languages in Africa, opaque markets and the very real challenges of underdeveloped infrastructure and talent pools may place unreasonable financial demands on the risk–reward tradeoff, driving up the cost of capital and evaporating its already scarce availability in such a challenging environment. The other pernicious effect of this funding reality is that entrepreneurs are lured by the siren call of building businesses they want to fund, as opposed to businesses they want to run. In pitch decks and spreadsheets it is very easy to make a perfect business. The market is much less forgiving, in reality. Yet, herein lies the opportunity for new funding models and approaches to de-risked emerging market capital flows to emerge.

Just as banking in Africa was not constrained by the brick, mortar, and tellers of international banks, so too, funding for entrepreneurs need not conform to global norms. In fact, we would be wise to remember that more than 90 percent of startups in the U.S. go unfunded, and eventually fail.[54] Is the failure rate actually that high or is it that the funding model is too risk-averse and formulaic, driving up the failure rate? The truth is somewhere in the middle and it speaks to an entrepreneur's most valuable asset—namely, their culture and resilience. Even when you fall, most of the time you are still in the process of moving forward. If you fall backwards, get back up, dust yourself off, rinse, and repeat. These are qualities shared by entrepreneurs and agile enterprises alike.

There is a reason Silicon Valley is such a thriving place—even though women are woefully underrepresented.[55] While all of the preconditions for entrepreneurship are not export products, the most important ingredient is the easiest to translate—namely, attitudes towards failure. Entrepreneurship is all about attempts—repeated, foolhardy, whole-hearted attempts at something, no matter how fanciful or mundane, whether for profit or purpose. The best entrepreneurs think of failure as an iterative process, rather than a conclusive one. To paraphrase Winston Churchill, each failure is merely the end of the beginning.[56] The true end of an endeavor is an individual choice to give up and not one that can be imposed on an entrepreneur by societal

---

[54] D. Gage, "The Venture Capital Secret: 3 out of 4 Start-Ups Fail," *The Wall Street Journal*, September 20, 2012, accessed November 24, 2015, http://www.wsj.com/articles/SB10000872396390443720204578004980476429190.

[55] J. Yarrow, "Silicon Valley is 'Incredibly White and Male' and There's a 'Sort of Pride' About that Fact, Says Silicon Valley Culture Reporter," *Business Insider*, April 4, 2015, accessed November 24, 2015, http://www.businessinsider.com/silicon-valley-is-incredibly-white-and-male-2015-4.

[56] W. Churchill, *This is Not the End (Part 2)*, November 1942, accessed November 24, 2015, https://www.youtube.com/watch?v=pdRH5wzCQQw.

pressure. Many aspects of managing risk and confronting complex challenges have a similar property.

Pragmatic entrepreneurs see their choices not as a hobby, but rather as a means to an end—an imperative to survive. Nowhere is this truer than the vibrant entrepreneurial scene taking root across Africa. From Nairobi's tech entrepreneurs (with breakthroughs like Ushahidi and BRCK, a venture-backed mobile internet platform) to the continent's growing array of incuba-tors and hubs (such as AfriLabs, led by Tayo Akinyemi), Africa's burgeoning entrepreneurial ecosystem is rooted in age-old maternal survival instincts. Critically, these responses are increasingly equitable across gender lines, coun-tries and industries. In traditional societies, particularly matriarchal societies, women are the strongest thread in the economic safety net. Promoting the ideals of failing forward, equality and entrepreneurship can unleash the other half of the world's potential.

For all the research on the state of entrepreneurship in Africa (and there have been many high-quality studies like the collaborative project between Omidyar Network and Monitor Group), there is a common narrative that risk culture is alive and well. Despite the confounding reality of a very real and persistent tragedy of the commons grinding most African economies to an average GDP growth of 5 percent (enviable by advanced economy standards), there is an unstoppable optimism taking root across the continent.[57] So while the lights may go out for a while, and there are fewer paved roads across Sub-Saharan Africa than in the UK, countries can buy or barter their way out of an infrastructure deficit, but they cannot put a price on entrepreneurial culture. Harnessing this competitive advantage will take courage and the will to leap over a few creaking institutions that have dogged Africa's progress.

Complexity, Capital, and Supply Chains

The strategic imperative for Africa's continued development is for policies, funding and networks to catch up with the African entrepreneur's limited fear of failure. The Omidyar study found that Africa's budding entrepre-neurs increasingly favor this road less traveled compared to traditional career options, however limited. Of course, so much African output is in the infor-mal economy, where microentrepreneurs control their own fate through hawking goods and services. Armed with the ubiquitous mobile phone, these

---

[57] International Monetary Fund, *GDP Forecasts 2015*, 2015, accessed November 24, 2015, http://www.imf.org/external/pubs/ft/weo/2015/update/02/.

microentrepreneurs no longer suffer from the information asymmetry that spoiled crops, or the daily catch in the pre-information era. Now an entire ecosystem and supply chain has cropped up to support all facets of the communication value chain. From airtime recharging cards sold by human 'stations', to airtime ATMs, where minutes are exchanged for cash, Africa's mobile boom is occluding green shoots in other sectors.

One obvious place to start is for more Africans to move up the value chain across the continent's wide spectrum of natural resources. For a country like Nigeria, with one of the world's largest oil reserves, to import energy (as it does) speaks to the scope of this opportunity. Bringing more local refining capacity—and with it jobs—is a key pathway toward stabilizing restive regions of the country and improving income distribution. Adding value in agricultural products is also a massive untapped opportunity. As the Ghanaian-American entrepreneur Rahama Wright has shown with Shea Yaleen, a cooperative shea butter firm cherished by Oprah, the savvy entrepreneur is as much a supply chain master as an intercultural communicator.

Importing finished chocolate to West African nations, the world's largest producers of cocoa, is the equivalent of importing olive oil to Italy. While the higher value-added production facilities are largely absent in West African countries, the time to revisit the neoclassical concept 'economic comparative advantage' is upon us. Hinging an economy on commodities alone, without moving up the value chain, subjects countries to a wild rollercoaster ride that exacerbates income inequality and leaves them at the mercy of global markets, far from their control.

In the case of cocoa, the worldwide revenues in the market for finished chocolate exceeded $117 billion in 2014, according to KPMG, while a mere $9 billion was captured by West African originators, whose output was severely affected by the Ebola crisis. Global firms like Mars are constantly reminded to protect their supply chains. Coaxing them 'onshore' through shared models, special economic-zones for cocoa production, and refinement, can be a win–win, and an important pathway for skills transfer. The key, however, is to move demonstrably up the value chain.

Money is naturally the bloodline for every economy and entrepreneur. With 500 billion U.S. dollars of interest-free aid poured into Africa over the last 50 years, many of the continent's financial institutions are only now beginning to grapple with a massive gap in financial inclusion. This gap is partly driven by Africa's expansiveness, as much as by its rural–urban divide. What capital is available through traditional banks and investors often demands exorbitantly high rates of return, and forms of security available to the narrowest segment of the population.

Some estimates show that a mere 10 percent of African land is held privately. This negates traditional forms of collateral and drives capital up market to established firms that can more readily check all the underwriting boxes. While the need for broad based financial reform is dire, here too lies an opportunity for many African countries to leapfrog traditional financial models. Applications of peer-to-peer approaches are encouraging where 'social collateral' helps securitize investments. The business case for this model of security has clearly worked in micro-finance. Perhaps it is time to mutualize these approaches among entrepreneurial clusters with risk capital in exchange for equity pools.

Being an entrepreneur is difficult everywhere, of course, but in Africa, importing too many of the confines of established norms may prevent the emergence of truly unique funding and entrepreneurial paradigms. The conversations, ideas and networks brought together at the Global Entrepreneurship Summit in Kenya (in 2015) and catalysed in the market by Mr. Elumelu and other innovative programs building human capital are sure to trigger Africa's inexorable rise. The book *The Next Africa,* by Aubrey Hruby and Jake Bright (2015), chronicles the continent's inevitable ascension as a global powerhouse. While many obstacles remain, including grinding corruption, market opacity and a persistent lack of capital and market access, the die has been cast, and investors would do well to go long on Africa.

Education and ideas do not get stuck in customs. As President Obama remarked in Ghana in 2009 during his first visit to Africa as president, *"Africa doesn't need strongmen, it needs strong institutions".* Reinforcing the continent's educational institutions and connecting to the highly educated diaspora can unleash one of Africa's untapped advantages, namely, its entrepreneurial spirit. As the Arab Awakening reminded us, a country's greatest strength—its people—can also serve as its Achilles heel. Many of the long-term risks in Africa will be borne by how well its young, restive, entrepreneurial population is absorbed into the economy. Failure to meet this challenge, and to make the response equitable, may very well be Africa's greatest risk. Other man-made and natural risks—such as the Ebola outbreak, failed states, terrorism, war, and piracy—will be greatly mitigated, if not halted altogether, by how well Africa's leaders and the global community rise to the occasion.

Simplicity Is the Key

When combatting complexity and charting a course in the face of uncertainty, simplicity is the key. Just as Viking mariners used their fabled sun-stones to navigate in dark and treacherous waters, true north for modern organizations should not be occluded by so many layers, obfuscation, and dithering.[58] Simple navigational instruments can serve as a guide in good times and in bad. Simple solutions to complex problems are often the most robust and time-tested. Similar to the sunstone-guided Viking sailors, organizational value systems and clearly articulated decision-making frameworks and guidelines on risk tolerance will help organizations gain agility and resilience. Survival is the ultimate goal of any well-developed risk management framework. The best survival instincts are not theoretically rehearsed, they are honed through real world experiences and a survival imperative at all organizational levels. Put simply, in today's unpredictable risk landscape and the era of man-made risk, it is everybody's business to stay in business.

Boardrooms, CEOs, CROs, and political leaders should not be the only ones losing sleep over risk. Indeed, many of the risks in this age are so pronounced they affect the entire world. Given how widespread risk has become, it is essential for the broadest base of an organization and society to understand how their individual choices affect systemic change. Cynical attitudes in China or the U.S. toward climate change will have a profound impact on the entire world. Similarly, unabated consumerism and a carbon-driven economy in the U.S. and Europe will preclude the emergence of more sustainable alternatives. Attitudes towards risk that are shaped by simple guidelines will be the most robust. The enduring value of the Ten Commandments, for example, is partly due to their framing as negative options. 'Thou shalt not' is a clear declarative statement of what falls outside of acceptable norms.[59] Therefore, the presumption of negative options, establishes more clearly what can in fact be done.

Framing risk tolerance in a *thou shalt not* approach establishes the outermost limits of what an organization, team, or individual can risk, by clearly delineating what they cannot. The enduring simplicity of this approach will not only make for faster, nearly instinctive decision making, it will also engage people with a greater sense of participation and shared risk. In modern organizations, risk and risk-taking is all too often an amorphous thing that is most often felt in a highly impersonal way, if at all. With the removal of many of the downside risks leaders face amounting to systemic moral hazard, it is not surprising that a certain callousness has formed about the consequences of organizational decisions. Think of the recorded voices of Enron traders when they chanted "Burn baby burn" during severe wildfires in California

---

[58] J. Cohen, "Evidence of Fabled Viking Navigational Tool Found," *History*, March 7, 2013, accessed November 24, 2015, http://www.history.com/news/evidence-of-fabled-viking-navigational-tool-found.

[59] See Nicholas Nassim Taleb's book *Antifragile* for an in-depth discussion on the via *negativa* (or subtractive path), as well discourse on the enduring quality of negative options. *Antifragile: Things That Gain from Disorder*, Random House, N.Y., 2014.

that drove up energy prices.[60] Risk tolerance should not only be framed in terms of absolute negative consequences, it should also be framed by reward, especially if they are guided by adherence to value systems.

Facing a large number of layoffs, for example, corporate leaders are often confronted with the dilemma of prematurely telling affected employees about the impending downsizing at the jeopardy of shareholders or investors. This type of 'false choice' can potentially be reframed, not as one constituency winning at the expense of another, but as all those who may be on a foundering ship working to salvage the vessel and therefore their careers. If risk agility holds that it is everyone's business to stay in business, the agile leader would make employees a part of the solution rather than a part of the problem.

Counter-intuitively to traditional corporate norms, a number of firms are getting this powerful incentive structure right. Sir Richard Branson's corporate staff at Virgin, for example, benefit from unlimited vacations, managed in a delicate balancing act of "As much as you want, but not more than you need."[61] In another example, Gravity Payments, an entrepreneurial payment management firm, resorted to giving all of its staff a base salary of $70,000, the presumption being a true living wage at this level will get employees focused on growth and keeping the firm around, as opposed to merely focusing on making a living.[62]

If there is callousness at the top, ambition and greed in the middle, the bottom of the organizational pyramid is all too often indifferent. Broad organizational outcomes rarely translate meaningfully for people in lower organizational layers, who are most often the first to bear the brunt of any risk—the first to be fired, the first to be affected by a cyber breach, the soft targets in a terror attack—the most vulnerable are often indifferent about how they fit in the bigger picture. Tackling this indifference requires a combination of incentives, adaptive communication, and accountability.

Above all, transparency and putting stated values to action is key. For example, when the UN compound was bombed by Boko Haram in Abuja, Nigeria

---

[60] R. Oppel, "Word for Word/Energy Hogs; Enron Traders on Grandma Millie and Making Out Like Bandits," *The New York Times*, June 13, 2004, accessed November 24, 2015, http://www.nytimes.com/2004/06/13/weekinreview/word-for-word-energy-hogs-enron-traders-grandma-millie-making-like-bandits.html.

[61] M. Tabaka, *Why Richard Branson Thinks Unlimited Vacation Time Is AwesomeDOUBLEHYPHENand You Should Too*, Inc., October 6, 2014, accessed November 24, 2015, http://www.inc.com/marla-tabaka/richard-branson-s-unlimited-vacation-policy-will-it-work-for-your-business.html.

[62] P. Keegan, "Here's What Really Happened at That Company That Set a $70,000 Minimum Wage, Inc.," *Slate*, October 23, 2015, accessed November 24, 2015, http://www.slate.com/blogs/moneybox/2015/10/23/remember_dan_price_of_gravity_payments_who_gave_his_employees_a_70_000_minimum.html.

in 2011 there was no global life insurance available covering all UN staff in an indiscriminate way.[63] This made the most vulnerable local national and third-country national staff 'soft targets'. In Abuja, as in the rest of the world, the loss of a primary breadwinner will subject households to economic hardship or collapse. It took many years before drivers hired into the UN's vast vehicle fleet were offered driver training, increasing the impact of road safety and related risks.[64] Both of these examples smack of hypocrisy and increase the indifference toward those on the UN's frontlines.

Our indifference tends to wane with more acute risks, although we generally remain dangerously detached about all man-made risk. Long-term attritional risks, like road safety, are difficult to secure capture support for, even though the human toll on the world's roads is 1.3 million deaths and 20–50 million injuries per year. Left unchecked, preventable road traffic injuries, already the 9th leading cause of death, will leap to the 5th spot by 2030.[65] Risks of the slowburn variety, like climate change, do not galvanize action. By contrast, following one week of Icelandic ash and acute economic consequences to airlines in 2014, adversely affected business travelers and bored holiday makers, Europe's airlines quickly were back in the skies. This, of course, took some airborne bravado from Willie Walsh, British Airways former CEO, as he defied safety warnings and took flight in a 747 to 'prove' Europe's skies were safe. It was almost as if you could see the risk–reward tradeoff of the Icelandic ash event taking place in real time. As the economic costs reached into the billions, economics trumped air traffic safety and Europe was jetting once again.[66]

Another example of our collective response to perceptually acute risk is how air travelers the world over submit to often absurd, costly, placebo-level screening although air travel remains the safest mode of transport. In 2014, there were 904 commercial air traffic fatalities, a 7 percent increase over the prior year, corresponding to 98 accidents.[67] Yet, the often high-profile, high-anxiety nature of airline crashes has raised our sense of acuity, perception and

[63] I. Mshelizza, "Islamist Sect Boko Haram Claims Nigerian U.N. Bombing," Reuters, August 29, 2011, accessed November 24, 2015, http://www.reuters.com/article/2011/08/29/us-nigeria-bombing-claim-idUSTRE77S3ZO20110829#kp6w3qhjYCJBMEfe.97.

[64] D. Disparte, "The Weakest Link: The State of Humanitarian Fleet Management in Africa," *Monday Developments*, December, 2008, accessed November 24, 2015, http://www.reuters.com/article/2011/08/29/us-nigeria-bombing-claim-idUSTRE77S3ZO20110829#kp6w3qhjYCJBMEfe.97.

[65] *Decade of Action for Road Safety: 2011–2020*, 2011, accessed November 24, 2015, http://www.who.int/roadsafety/decade_of_action/plan/en/.

[66] M. Evans et al., "Volcanic Ash Cloud: British Airways Fly in the Face of Ban," *Telegraph*, April 18, 2010, accessed November 24, 2015, http://www.telegraph.co.uk/travel/travelnews/7605305/Volcanic-ash-cloud-British-Airways-fly-in-the-face-of-ban.html.

[67] International Civil Aviation Organization, *Safety Report 2015 Edition*, 2015, accessed November 24, 2015, http://www.icao.int/safety/Documents/ICAO_Safety_Report_2015_Web.pdf.

loss, despite the infinitesimally small likelihood of being involved in an airline calamity. While any loss of life is tragic, 908 air traffic fatalities is a paltry figure when compared to the 1.3 million road-related deaths each year. It is even smaller considering that 3.6 billion people are expected to fly in 2016.[68]

The difference is perception, and how acutely flying plays to our fears. Imagine the absurdity of everyone taking off their shoes at airports based on a single failed attempt at a shoe bombing.[69] Yet if a single life were to be lost as a result of a shoe bomber being successful, most people would probably say that all the hassle and time spent complying with regulations requiring us to remove our shoes during security screening was worth it. Then again, there are so many places in the world where removing shoes is not required that one does have to ask why we are bothering at all. Preventable gun violence has claimed many more lives than terrorism ever has, or ever will. However, because terrorism strikes a different nerve, people are prepared to change, sacrifice military lives and their national treasure. In 2013, 120 times more people were kills by guns in the U.S. than all those killed by acts of terrorism globally that year.[70]

Knowing the difference in how people respond to either acute or attritional risks will help risk managers craft adaptive approaches to controlling risk and wading through complexity. Part of the reason airline risk management is so robust in producing a safer system is because risk management goals are incredibly simple to comprehend and each loss provides system-wide benefits—risk management in all aspects of running a modern airline is a *zero-failure* mission. From the engine manufacturer, to onboard redundancy, pre-flight checklists and near-miss management, the airline industry's often simple approach to risk can be readily translated to other segments of the economy.

While idealistic, the goal of all risk management should be zero losses. Much like six-sigma approaches to manufacturing consistently aim for the absolute removal of imperfections in a system (e.g. 1 deviation in 6 million), there is no reason why agile risk managers cannot aspire to zero failure and

---

[68] International Airline Passenger Association, *Airlines to Welcome 3.6 Billion Passengers in 2016*, Press Release Number 50, December 6, 2012, accessed November 24, 2015, http://www.iata.org/pressroom/pr/pages/2012-12-06-01.aspx.

[69] K. Dutta, "Shoe Bomber Richard Reid Shows no Remorse After a Decade in Prison for Failed Terror Atrocity," *Independent*, February 3, 2015, accessed November 24, 2015, http://www.independent.co.uk/news/world/americas/shoe-bomber-richard-reid-shows-no-remorse-after-a-decade-in-prison-for-failed-terror-atrocity-10022074.html.

[70] E. Bower and J. Jones, "American Deaths in Terrorism vs. Gun Violence in One Graph," CNN, October 2, 2015, accessed November 24, 2015, http://www.cnn.com/2015/10/02/us/oregon-shooting-terrorism-gun-violence/.

zero loss. The goal is translatable at all organizational levels and is certainly one that is shared by the broadest cross-section of humanity. The most poignant example of this is the life or death choices faced by the military, whose commanders clearly aspire to a 100 percent survival rate among their soldiers. If that noble objective can be strived for in the theater of military battle, it should certainly be an objective in the theater of organizational battle.

5

Three-Dimensional Risk Management

Risk management is sometimes thought of as a necessary evil, increasingly consigned to being an adjunct to compliance, finance, and other so-called 'business prevention' functions. In too many organizations, risk management is considered little more than a 'tick the box' task that sales teams must maneuver around on their way to board-level approval. Such an approach is a prescription for disaster, but that does not stop many individuals within organizations—and even the organizations themselves—from simply going through the motions of risk compliance.

Non-financial firms traditionally address risk through a series of transfer mechanisms, such as insurance and self-funded vehicles, or they merely absorb unforeseen losses with their earnings. The financial sector, on the other hand, applies sophisticated statistical methods in a form of speculative risk management that captures the upside and the downside of risk taking. These approaches are used to calculate VaR, regulatory capital and other internal and external risk measures. Both types of organizations utilize employee directives such as value systems, codes of conduct and employee manuals to contain human-borne risks. Many of these methods, however, are based on backward-looking accounting book values and a permissive 'fox watching the chicken coop' environment, wherein financial institutions often develop their own internal risk metrics with loose guidance from regulators.

The global macroprudential structure often lags far behind the firms they are supposed to supervise because of a lack of global coordination in the financial system, the comparative sophistication of industries that regulators frankly too often do not understand, and the vast resources they can marshal to either

© The Author(s) 2016
D. Wagner, D. Disparte, *Global Risk Agility and Decision Making*,
DOI 10.1057/978-1-349-94860-4_5

ward off prying eyes or absorb regulatory fines. Injunctions and financial penalties may be seen as little more than rounding errors, given the vast profits financial firms tend to generate. The direst consequences for regulatory noncompliance are often spared for executives who let standards lapse in a pervasive, system-wide moral hazard.

All too often, the risk measures imposed by regulators to address 'the next' financial crisis were designed with last year's financial crisis in mind. Many of the more recent regulatory impositions have enacted stricter solvency standards and have attempted to limit proprietary trading activities with depositors' money, among other restrictions. Nonetheless, the system does not appear safer, and many of the Frankenstein products that accelerated economic collapse in the Great Recession are once again rearing their ugly heads. Additionally, the literal nature of regulatory impositions translates into many financial firms merely adhering to minimally required risk management standards (compliance), as opposed to embracing a more strategic and broadly based survival-oriented posture.

So called 'living wills', or resolution documents for SIFIs, fall into the category of 'nice to have' and not 'need to have' in guarding the global financial system.[1] What the living wills negate is that in conditions of broad financial distress (which could fell a SIFI) it is highly probable that prospective suitors wanting to acquire a distressed firm's assets would themselves be in distress. This contagion, as the Great Recession taught us, is amplified by the fact that a large amount of a financial institution's value is derived from highly volatile market-based valuations.[2]

Any fluctuations in a financial institution's market value can easily erode its intrinsic worth, wiping out its depositors, investors, and shareholders with equal ease. 'Too-big-to-fail' risks managed through living wills are inadequate and fail to address the correlations among systemic institutions, and the fact that any one of them can depress an entire market or economy. Too-big-to-fail institutions should not be too-big-to-dissect, however. Carving vast firms into constituent less risk-prone parts would not only make for a safer global financial system but would lower the public burden of privatizing gains while socializing losses. In complex financial firms, which are often sprawling global enterprises, truly understanding organizational risks is a confounding process—made all the more difficult by the fact that financial risk is intangible.

---

[1] N. Pakin, "The Case Against Dodd–Frank Act's Living Wills: Contingency Planning Following the Financial Crisis," *Berkeley Business Law Journal* (2013) accessed November 24, 2015, http://scholarship.law.berkeley.edu/cgi/viewcontent.cgi?article=1093&context=bblj.

[2] V. Acharya, R. Engle, M. Richardson, *Capital Shortfall: A New Approach to Ranking and Regulating Systemic Risks*, AEA Meetings, January 7, 2012, New York.

Much like combating a virus, by the time you see a patient's temperature rising (the only visible signal of a risk emerging) it may very well be too late and the risk has spread.

The same holds true in financial firms where, with great speed and ease, market risk can quickly morph into liquidity and operational risk—all before any alarm bells have sounded, due to the backward orientation of financial risk management. Part of the challenge stems from the fact that financial risk management largely relies on accounting information, which is backward-looking and aggregated across multiple lines of business from all over the world, while, in many cases, financial firms have exploited regulatory arbitrage opportunities, creating a mismatch in capital buffers and interpretations of acceptable risk. This makes the effort of staying on top of risk and remaining agile in the process very difficult for even the best of breed in the financial sector. It is easy to see how regulators can get caught in a perpetual game of catch up.

One-Dimensional Risk Management and the Placebo Effect

The frequency of potentially preventable losses, along with the calamitous effects of 'black swans' or rare events, suggests that quants not only need qualitative tools in their arsenal, they also need structural alternatives to one-dimensional risk management. Such a unidimensional structure often misses the mark and can suffer from confirmation bias in that risk managers who look for trouble may in fact find it by chasing misleading risk signals, or merely confirming their assumptions.[3]

As noted elsewhere in the book, JP Morgan—long considered one of the best practitioners of financial risk management in the banking arena—missed the London Whale's transgressions, despite some fairly obvious warning signs.[4] Using one-dimensional risk management, even though many firms like JP Morgan install 'native' CROs in their business lines, these individuals are often marginalized and kept on a "need-to-know" basis. This has the placebo effect of creating a false sense of comfort that risks are being managed, when in reality excessive risk taking behavior may be being carried out in the CRO's line of sight. The latency and backward orientation of traditional risk

---

[3] D. Disparte, *3D Risk Management: A Survivorship Framework*, Risk Intelligence News Center, NYU Stern, April 24, 2014, accessed November 24, 2015, http://blogs.stern.nyu.edu/riskintelligence/?p=1765.

[4] R. Kaplan et al., *JP Morgan's Loss: Bigger than "Risk Management*, HBR Blog Network, May 23, 2012, accessed November 24, 2015, https://hbr.org/2012/05/jp-morgans-loss-bigger-than-ri.

measures often negates proactive controls, and when the smoke begins to rise, it is often too late. At times, financial risk management is tantamount to standing on the third rail while trying to calculate the speed and likelihood of an oncoming train. New perspectives are needed to create a safer financial system and improve survivorship among systemic institutions.

This first dimension is mostly quantitative in nature, although some financial institutions are beginning to pay heed to operational and behavioral risk management. The placebo effect that everything is covered and that risks are well understood and managed is one of the perils of adopting a singularly mathematical approach to risk. As outlined earlier in the book, this also introduces the real and frequent danger of model error and a high degree of complexity stemming from often-arcane financial methods that only the supernumeraries understand. Another challenge with this approach is that it is hierarchical in nature and runs the risk of breeding internal loss aversion and the occlusion of potentially bad information that is difficult to quantify. There may also be the potential of developing so-called 'risk myopia', or the narrowed sight of only appreciating those risks that can be measured, as opposed to a more instinctive, probing approach.

One-dimensional risk management follows a top-down command-and-control structure. This not only creates isolation compounded by the vast remoteness of global firms, but is also commonly the case that CROs report directly to the excessively optimistic CEOs they are supposed to manage. This was the case with JP Morgan, whose all-powerful chairman and CEO (Jamie Dimon) pushed the boundaries of the firm's internal hedging activities carried out by its Chief Investment Officer (CIO) prior to the Great Recession. In order for a one-dimensional approach to be effective, studies have found that the one variable firms cannot compromise on is to have an independent board-level risk management regime.[5]

In cases where CRO's report directly to the CEO and not the board, paycheck persuasion is a powerful inducement not to ring the alarms too loudly or to turn a blind eye to potentially egregious behavior. As former Citigroup CEO, Chuck Prince, blithely said in 2007, "When the music stops, in terms of liquidity, things will be complicated. But as long as the music is playing, you've got to get up and dance. We're still dancing."[6] This often-callous herd mentality seen in large financial institutions fuels populist rhetoric about bailing out

---

[5] C. Levin, Senator, Chairman, et al., *JP Morgan Chase Whale Trade: A Case History of Derivatives Risks and Abuses*, Permanent Subcommittee on Investigations, U.S. Senate Hearing, March 15, 2013.
[6] T. Durden, "Citi Warns of 'Dancing', 'Music' and 'Complicated Things' for the Second Time," Zero Hedge, February 21, 2015, accessed November 24, 2015, http://www.zerohedge.com/news/2015-02-21/citi-warns-dancing-music-and-complicated-things-second-time.

Wall Street at Main Street's expense. Knowing, as he did that "things would get complicated" in a liquidity crisis, why did Mr Prince ignore clear warning signs and not prepare Citigroup for survival? Only a year later, in 2008, Citigroup would receive one of the largest government bailouts in history, topping $476 billion.[7]

The robustness of risk management in the financial sector has a spotty record due to the wholesale collapse and subsequent public bailout of the system during the Great Recession, and numerous other crises prior to that. This is compounded by the implicit or explicit mandate given to risk managers to merely toe the regulatory line and do what is minimally required in terms of solvency, capital buffers, and other standards. Indeed, many analysts suggest that JP Morgan's London Whale debacle was driven by persistent regulatory arbitrage to reduce the risk weighting of its capital buffers.[8] In short, meeting externally imposed compliance standards does not bode well for survivorship.

In fact, under persistent conditions of economic duress such as those seen during the Great Recession (defined as a 40 percent collapse of U.S. equities), even the mightiest institutions would fall. Failure of the top 10 systemic financial institutions in the U.S. in 2015 would have resulted in a loss in excess of $300 billion, and nearly $600 billion for their top 10 counterparts in Europe to fill their capital shortfall.[9] In short, doing what is minimally required by regulators under a one-dimensional risk management structure is risky business. And assuming that politicians will continue to bail out too-big-to-fail financial institutions is similarly risky. Citizens all over the world have grown tired of 'privatizing gains while socializing losses'.

Foresight in financial risk management is most often granted for risks that are commonplace, have a rich history of observations, and have been well documented. The advantages of the first dimension are that it is undeniably rigorous and many of the statistical methods that are applied have a predictive quality, albeit in the short term. When made accessible and deterministic of managerial choices, such quantitative rigor can be a powerful tool in the hands of an agile risk manager and decision maker. All too often, however, the insights coming from the first dimension get fed into decision teams and boards that may eternally debate their course of action, while the numbers

[7] E. Javers, "Citigroup Tops List of Banks Who Received Federal Aid," CNBC, March 16, 2011, accessed November 24, 2015, http://www.cnbc.com/id/42099554.

[8] J. Macaskill, "Rivals Hover over JP Morgan as Farce Threatens to Turn into Tragedy," *Euromoney* (June 2012), accessed November 24, 2015, http://www.euromoney.com/Article/3039413/Rivals-hover-over-JPMorgan-as-farce-threatens-to-turn-into-tragedy.html.

[9] *Volatility Lab Systemic Rankings*, 2015, New York University, accessed November 24, 2015, http://vlab.stern.nyu.edu/welcome/risk/.

tell a clear tale of which way to go. So-called 'out of sample events' or unprecedented risks, which have no deep historical dataset consistently shatter financial risk management norms. In addition, given that the underlying actor in financial risks are people, their behavior and their expectations, more attention needs to be paid to behavioral economics than purely statistical approaches.

While the issue of risk in financial firms makes for headline news and attracts broad public interest (since banks are increasingly viewed as utilities), preventable losses, and missed signals are not confined to Wall Street and the other financial capitals around the world. Toyota's ascendance as a global automotive powerhouse hit a few speed bumps following a series of worldwide product recalls, sticking accelerators, and an embarrassing mea culpa by Akio Toyoda, Toyota's CEO.[10] Long heralded as the paragon of *kaizen* (total quality management), Toyota's assembly line follows a horizontal risk management structure. The aim is to reduce the probabilistic incidence of errors in the manufacturing process to as many standard deviations as possible, for example six sigma. This two-dimensional structure famously empowers workers on the assembly line to 'pull the cord', in effect stopping production, in the event of an error or pattern of errors.[11]

Although many aspects of this flat structure are appealing, it often labors under a potentially high 'signal-to-noise' ratio that may lead to false positives and may only be applicable to firms or markets with slower velocity. For example, it is doubtful that a two-dimensional risk management approach would have saved the high-velocity trading firm Knight Capital, which crashed and burned with spectacular speed due to a rogue trading algorithm.[12] Besides, tracking defects is a slightly more gratifying task when you have a tangible product.

In financial services, however, defects are occluded by correlations and their formless nature. Much like a virus, financial risk is often identified through contagion and rising temperatures, neither of which are sound preventive measures. Despite all the passive and active safety features in a modern car, externally imposed speed limits and driving conditions, there is no risk management substitute for a well-trained driver. Similarly, most risks in financial

---

[10] J. Kanter, M. Maynard, H. Tabuchi, "Toyota Has a Pattern of Slow Responses on Safety Issues," *New York Times* (February 6, 2010), accessed November 24, 2015, http://www.nytimes.com/2010/02/07/business/global/07toyota.html?pagewanted=all&_r=0.

[11] D. Fowler, *The Toyota Way*, Simplicable, August 14, 2013, accessed November 24, 2015, http://management.simplicable.com/management/new/toyota-way.

[12] H. Touryalai, "Knight Capital: The Ideal Way to Screw up on Wall Street," *Forbes*, August 6, 2012, accessed November 24, 2015, http://www.forbes.com/sites/halahtouryalai/2012/08/06/knight-capital-the-ideal-way-to-screw-up-on-wall-street/.

and non-financial firms emanate from the behavior of people inside the firm and in the market.

One-dimensional risk management approaches were once regarded as the Ultima Thule, particularly among manufacturers and total quality proponents. These models have suffered persistent setbacks, however, as global firms continue to topple over a mix of deliberate and inadvertent errors in their once-vaunted low-error processes. Takata Corporation, a Japanese automotive original equipment manufacturer, triggered a massive ongoing air bag recall that affected most of the leading players in the automotive industry between 2013 and 2015. Takata's airbags suffered a serious defect that could trigger an explosive deployment, potentially injuring or killing passengers.[13] Approximately 40 million cars were affected, spanning 18 automotive manufacturers, and the financial consequences for Takata and its automotive customers continue to mount.[14] Takata's misfortune and the knock-on effect it created in the supply chains of major manufacturers underscores the importance of controlling quality in all facets of a complex system. The vicarious risk passed on by third parties almost always balances out the gains in cost savings or efficiency. In effect, your subsuppliers' risk is your own. There is no such thing as an arm's-length transaction in the twenty-first century.

Illustration 5.1 outlines the general structural differences between the three dimensions of organizational risk management—vertical, horizontal, and hybrid approaches. Agile risk managers espouse an equal measure of rigorous quantitative methods (used in finance) as they do the qualitative methods espoused by total quality management. This 'blended' approach is a hybrid of the first and second dimensions and a far more effective means of managing risk than either the vertical (first) or horizontal (second) dimension approaches in isolation.

Against this backdrop, three dimensional risk management offers a new framework for an increasingly punishing and interconnected world. In this approach, risk management is not merely a quantitative preventive feature, but rather an embedded firm-wide decision-making framework that makes using risk management to thrive in business an organizational imperative. Leveraging the best of the first and second dimensions filters the signal-to-noise ratio and

---

[13] H. Tabuchi, "Takata Saw and Hid Risk in Airbags in 2004 Former Workers Say," *New York Times* (November 6, 2014), accessed November 24, 2015, http://www.nytimes.com/2014/11/07/business/air-bag-maker-takata-is-said-to-have-conducted-secret-tests.html.

[14] C. Aityeh and R. Blackwell, "Massive Takata Airbag Recall: Everything You Need to Know, Including Full List of Affected Vehicles," CarandDriver.com, November 23, 2015, accessed November 24, 2015, http://blog.caranddriver.com/massive-takata-airbag-recall-everything-you-need-to-know-including-full-list-of-affected-vehicles/#list.

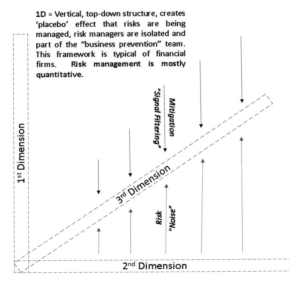

1D = Vertical, top-down structure, creates 'placebo' effect that risks are being managed, risk managers are isolated and part of the "business prevention" team. This framework is typical of financial firms. Risk management is mostly quantitative.

3D = Diagonal hybrid structure borrows from flat manufacturing process risk management, but filters signal to noise ratio – RMs embedded at all organizational levels as part of decision making framework, not prevention framework. Risk management is 50:50.

2D = Horizontal structure mirroring assembly line 'defects' management. Typically seen in the manufacturing context – the entire line is empowered to halt the system. The signal to noise ratio is a potential downside. May lead to "McCarthyism" and false positives. Risk management is mostly qualitative.

Illustration 5.1 Three-dimensional risk management (*Source*: Author Dante Disparte)

encourages bounded risk taking at all organizational levels. The power of this approach lies in replacing *complexity* with *simplicity* and removing the stigma of raising the alarm when things go wrong. Imagine the power, clarity, and resilience of the Volcker Rule, which, over the course of 1,105 pages, precludes proprietary trading, if it was simply stated as 'thou shalt not speculate with other people's money'—the very egregious behavior that got JP Morgan into trouble.[15]

This 'third way' provides the best of both worlds and brings risk management closer to the here and now, rather than merely laboring over last quarter's financial results in an attempt to gauge signals for the next 90 days, and remain compliant. Moreover, organizations that achieve some semblance of this standard will gradually move risk management from being a 'business prevention function' to becoming a part of the value-driving frontline—a business enabling function. The idea of embedding regional and business line CROs makes sense, however, these CROs only begin to add enterprise value when they are seen as a part of a local or regional team and not as a part of a corporate espionage regime.

---

[15] Federal Deposit Insurance Corporation, *Volcker Rule*, December 10, 2013, accessed November 24, 2015, https://www.fdic.gov/regulations/reform/volcker/rule.html.

Agile Risk Culture

In the era of man-made risks, culture, attitudes, and our general behavior—whether alone or in groups—may be our first and last line of defense. Shaping culture and modifying behavior are among the most complex challenges organizations face, yet doing so remains critically important for all decisions and overall alignment to the mission, vision, and values espoused by an organization. Different from the analysis on governance found later in the book, risk culture and tolerance are more subtle, yet equally important components of building risk agility and organizational resilience. The typical large and complex organization has so many decision rules shaping personnel behavior that, in many ways, the modern worker has become a feckless automaton.[16] Many more have been banished altogether by the rise of outsourcing and automation. Those who remain may be so highly detached or overpaid, so as to preclude the formation of a distinctive risk culture. Risk *aversion*, a more common phenomenon, is not the same as risk *culture*.

Risk culture is defined by unwritten rules, permissiveness around bounded risk taking at all organizational levels, and, above all, a sense of ownership—a feeling that an organizational loss is a shared loss. The aviation industry and the military provide vivid examples of risk culture, for the consequences of failure in both areas are so stark, both embark on zero-failure missions. In order to survive and thrive in the era of man-made risk, more organizations should adopt this approach, even if the apparent 'objective' is simply to make the most money.

A risk culture of zero failure goes far beyond six sigma, which allows for errors in a system, albeit rare. It goes further than other models and truly establishes cultural norms towards risk taking, readiness and, above all else, how to respond and decide under opacity. Where an organization's value system is the written and unwritten behaviors embodied by top performers, risk culture is the subtler expression of what actions would make those at the top of the management pyramid and in other parts of an organization cringe, such as asking if you can wear casual clothing to the office, or whether flip-flops are permissible, or bringing a bicycle into your office space so it is not stolen. Such requests do not negatively impact the ability of a firm to function, but they say a lot about whether employees feel empowered enough to ask such a question in the first place, and whether they feel comfortable that

---

[16] "Wealth Without Workers, Workers Without Wealth," Economist.com (October 4, 2014), accessed November 24, 2014, http://www.economist.com/news/leaders/21621800-digital-revolution-bringing-sweeping-change-labour-markets-both-rich-and-poor.

they will not be penalized for having done so. The normative reaction of peer groups, subordinates and superiors to individual behavior are the best determinants of risk culture.

Risk culture is set by the tone at the top and the ability to lead by example. So, by definition, an organization that does not have casual Fridays or allow someone to work from home one day per week is less likely to embrace such individual behaviors. Rather, that individual is likely to be perceived as a 'problem', 'obviously' outside the 'circle of trust', or the thumb that is sticking up and needs to be hammered down. By contrast, the leader that goes home at 5:00 or declares every Wednesday 'jeans' day is much more likely to be perceived to be open to embracing new ideas and adopting new ways of thinking that an organization ultimately stands to benefit from.

Another important attribute of healthy risk culture is its attitude toward failure. If people are fired the first time they make mistakes or even for honest errors (barring egregious examples in zero-failure settings), the organization's risk culture will be intolerant, and when employees identify or encounter potential enterprise risks, they are more likely to remain silent and 'not make waves'. Encouraging risk culture by not punishing failure, but rather by looking at failure as part of the creative–destructive cycle that drives enterprises forward, will remove the stigma of silence and improve the speed and quality of information sharing.

Assessing Risk Culture

In 2014 the Financial Stability Board produced important guidance for financial institutions to use to create a framework for assessing risk culture.[17] Its intention was to apply the guidance to SIFIs, but its recommendations would be of benefit to a range of organizations. The FSB identified common elements that support a sound risk culture, including effective governance, effective risk appetite frameworks and compensation practices that promote appropriate risk-taking behavior. It also noted a number of organizational practices that are indicative of a sound risk culture, which are mutually reinforcing:

1. *The tone at the top*: Senior management is where core values are established, and where behavior reflects those values. Doing the right thing and promptly reporting wrongdoing are two behaviors that can set the right tone at the top. An organization's leadership should be shown to be moni-

---

[17] Financial Stability Board, *Guidance on Supervisory Interaction with Financial Institutions on Risk Culture*, April 7, 2014, accessed on November 24, 2015, http://www.financialstabilityboard.org/2014/04/140407/.

toring and assessing risk culture and making changes when necessary, promptly.

2. *Accountability*: Employees at all levels of an organization understand what the risk culture is, agree to abide by its precepts, and are held accountable for their actions.

3. *Effective Communication*: The decision-making process encourages a range of views, challenges current practices, encourages critical thinking, and promotes an environment of open and constructive engagement.

4. *Incentives*: A system of incentives is put in place to reward good behavior and the promotion of the organization's core values.

There is no organization that would not benefit from adopting and abiding by these simple, common sense principles, yet many do not. The simple message is to put guidelines in place, abide by them, reward those who do, and punish those who do not. Doing so increases the likelihood of organizational success, and reduces the chance of failure.

Lessons from Entrepreneurs

The fear of failure is as dangerous for entrepreneurs and early-stage firms as it is for vast multinationals and countries. The term failure is often treated like a four-letter word—one that would make your grandparents blush. No one wants their children to be failures when they grow up, any more than a mentor wants an apprentice to crash and burn. The world over there is a clear predetermined pathway for what success looks like that has informed cultural norms, societal goals, and individual aspirations. In some countries, this means acing state exams and taking a preordained path to becoming an engineer, doctor, or lawyer. In others, it means taking the mantle of a family business and perpetuating a legacy. In far too many countries it means confining women to non-economic 'traditional' roles, leaving a world economy that can only fulfill half its potential. If there is any true failure on a global scale, it is the economic underrepresentation, underutilization and the unequal treatment of women in comparison with their male peers. If an organization wants to demonstrate twenty-first-century risk culture, the first place to start would be the empowerment of women in their midst.

In business and politics, failure most often means the end of the line and an ignominious defeat.[18] In other spheres, especially in the world's vibrant entre-

---

[18] D. Disparte, "Half Our Potential: Failing Forward, Women and Entrepreneurship," *The Huffington Post*, November 6, 2015, accessed on November 24, 2015, http://www.huffingtonpost.com/dante-disparte/half-our-potential-failin_b_8493312.html.

preneurial community, failure is not only an ideal to aspire to, but a badge of honor—the war wounds that separate serial innovators from mere tinkerers. Part of the reason the U.S. has a vibrant entrepreneurial culture is that it is a national pastime to reinvent oneself and to *fail forward*.[19] While the concept of failing forward has been popularized in recent years, a subtler and perhaps deeper point of reflection is needed.[20] Failure is derived from the French word *faillir*, meaning "almost," denoting a certain trial and error—as in almost making it. It shows an embrace of risk taking and action rather than defeat.

Truly successful entrepreneurs (and, yes, there is a French word for that!) are not naturally endowed with panglossian success. Rather they are 'repeat failures'—a proxy for people who repeat the prior attempt while making small course corrections. The best of breed are almost stubborn to a fault at this process—and it is a process akin to the scientific method. It is hardwired in their DNA. It is worth recalling again that Steve Jobs, now deified around the world as a transcendental business leader, was once fired from Apple. He was also a well-documented as an interpersonal "failure" known for his often-intolerant demeanor and for not suffering fools lightly.[21]

This unbending commitment to progress—even through repeated failure—is what sets risk takers apart and is a bedrock of risk agility. On the road to success, failures are the guardrails that keep us on track. This is a particularly important metaphor for early-stage firms and entrepreneurs to embrace—for the car they are building may not have the usual safety features of the fancier vehicles being driven by their corporate peers. This subtle aspect of the culture of entrepreneurship is perhaps its most important.

In contrast to the U.S., in many countries around the world, the 'F' of failure is like wearing a scarlet letter, or worse, like being branded with the hot iron of public and familial rebuke. This casts a heavy societal yoke on aspiring entrepreneurs, leaving incalculable economic opportunity un-ventured and therefore un-gained. Success through failure is an indiscriminate process that recognizes neither gender nor resources. It is not idealistic to believe that anyone, anywhere can achieve their dreams and become truly transformational. Just look at Malala Yousafzai's incredible rise from the ashes of hatred and fear of educated women the Taliban visited upon her in Pakistan. From her

---

[19] J. Maxwell, *Failing Forward* (New York: Nelson Business, 2000).

[20] M. Maddock, "If You Have to Fail—And You Do—Fail Forward," *Forbes* (October 10, 2012), accessed on November 24, 2015, http://www.forbes.com/sites/mikemaddock/2012/10/10/if-you-have-to-fail-and-you-do-fail-forward/.

[21] T. McNichol, "Be a Jerk: The Worst Business Lesson from the Steve Jobs Biography," *The Atlantic* (November 28, 2011), accessed on November 24, 2015, http://www.theatlantic.com/business/archive/2011/11/be-a-jerk-the-worst-business-lesson-from-the-steve-jobs-biography/249136/.

improbable beginnings to a global platform and brand, she reminds us all that adversity, like failure, is often a figment of our imagination. Countries with preordained bright paths of 'success' are often the least likely to embrace entrepreneurship and economic participation for women, resulting in waning competitiveness and growth. Yet the men in these societies have no concept of the collective price they all pay for adhering to such an approach.

The connection between entrepreneurship, national security and economic competitiveness is becoming increasingly clear as many countries grapple with the combustible mix of increasing urbanization, youth unemployment and shrinking commodity-based revenues. Against this backdrop, selling a national 'bill of goods' that promotes entrepreneurship and risk-taking is one way to satiate growing public discontent. The true believers in this narrative are not merely paying lip service to it—they are investing capital, establishing entrepreneurial clusters, and building a regulatory framework that encourages ease of doing business. As outlined earlier, entrepreneurship is not an export product, although it does thrive on global linkages among like-minded people. Getting the recipe right to galvanize latent entrepreneurial desires is not only key to spurring economic development at the national level, it is necessary to ensure that firms remain nimble and fully prepared for 21st century challenges.

Unequal Resilience

Given the low level of infrastructure development, political stability, and absence of emergency preparedness that is so prevalent in so many places around the world, organizations that seek to trade, invest, lend, or operate globally must constantly adapt to fast-changing environments. When comparing the risks typically faced in advanced versus emerging markets, the tone and tenor of risk naturally changes dramatically. The chief concerns that keep risk managers awake at night in advanced markets seldom involve a shortage of electricity or a war that displaces thousands—or even millions—of people. The communication infrastructure and the knowledge labor force that predominates high value-added activities in advanced markets is such that long-term business continuity is highly likely even in catastrophic scenarios.

The economic impacts of catastrophic losses in advanced economies are often quickly abated, as government reconstruction, stimulus funds, and insured losses fill the void. Along these lines, the 2004 Indian Ocean tsunami—which is believed to have killed more than 250,000 people, displaced millions, and affected 12 countries—caused less than $14 billion in economic losses.[22] Contrast this with the Sendai earthquake and tsunami of 2011, which claimed approximately 23,000 lives and caused an estimated $360 billion in economic losses.[23] The Japanese automotive industry, whose supply chains, plants, and people were severely disrupted following the dreadful trifecta of an earthquake, tsunami, and the ensuing nuclear fallout, quickly regained its footing to operate at pre-calamity levels. In a developing or emerging market context, however, the risk picture changes dramatically.

Risk Agility Meets Mobility

Firms are not only scrambling to compete in BRICS countries, they are also looking beyond the horizon for growth in frontier markets in Sub-Saharan Africa, Indochina, the Middle East, and other regions. Despite the inherent risks of expatriation to emerging and developing economies, international personnel assignments are slated to grow by 50 percent by 2020. This is so because of the dueling interests of identifying talent in frontier markets and the seemingly boundless economic growth rates, which are redefining company strategy and market expansion priorities. This shift is unsurprising, as emerging and developing economies will account for six in seven people on the planet by 2020 and are no longer solely the domain for extractive industries, but are now being treated as growth markets in virtually every sector of the economy. C.K. Prahalad,[24] in his seminal work, codified the notion that there is a fortune at the base of the pyramid.

When they choose to expand into new markets, firms are exposed to a new array of challenges, in places where such a mundane activity as driving down the road may be one of the leading causes of death and injury (as in many parts of Africa). Corruption, bribery, and fraud are often the bywords of doing business around the developing world and are as pervasive as taxation in advanced markets. So, too, are the ascendant risks of piracy, unlawful detention, creeping expropriation, and a myriad of terms that risk managers may be adept at handling when it comes to insurance placement, but often fail to manage at the individual level. Operating abroad can bring new meaning

[22] "Natural Disasters: Counting the Costs," Economist.com (March 21, 2011), accessed November 24, 2015, http://www.economist.com/blogs/dailychart/2011/03/natural_disasters.

[23] E. Ferris, and M. Solís, *Earthquake, Tsunami, Meltdown—The Triple Disaster's Impact on Japan, Impact on the World*, Brookings Institution, March 11, 2013, accessed November 24, 2015, http://www.brookings.edu/blogs/up-front/posts/2013/03/11-japan-earthquake-ferris-solis.

[24] C.K. Prahalad, *The Fortune at the Bottom of the Pyramid*, Pearson, N.Y., 2010.

to the importance of getting the '5 R's' (Recruiting, Relocating, Rewarding, Retaining and Returning) right.

Traditionally an adjunct to human resources, the global mobility profession is dominated by crucially important tasks, such as assimilation, language training, benefits packages, and the general transaction of relocating employees to international posts. This begs the question; how many mobility practitioners can look the CFO in the eyes, or are even on the same floor as the CFO, and can challenge the company's threshold for dealing with a turbulent market? Financial results are carefully looked after and there are countless risk management solutions readily applied by global firms to prevent and dissipate economic losses. However, many of the "people behind the P&L" in expatriate assignments are entirely exposed.

Who, for example, is liable for an expat's personal property left behind because of the effects of the Arab Spring (abandoned property is typically excluded from insurance policies)? How is duty of care extended to include the issuance of or access to a company car in emerging or developing markets, which account for around 85 percent of all road traffic deaths and injuries? In fact, left unchecked, World Health Organization (WHO) forecasts show that road traffic accidents will be the third most common cause of premature death by 2020. How many mobility practitioners arrange 'drive to survive' courses for their international assignees? What matters most in the mobility discipline is the pace of adaptation to the dynamic changes unfolding in the emerging world, and the dichotomy of sending the best talent to risky environments in the pursuit of growth.

In order to oversee this shift and better understand the growing trends affecting expatriation in developing markets, firms must elevate the role of the mobility practitioner to a c-suite executive responsible for all facets of internationally mobile human capital. Contrasted with the current practice, which tends to be transactional in nature, several layers removed from the top and focused on so-called 'soft' elements, the Chief Mobility Officer (CMO) will be a peer to other executives and reports directly to the CEO, from whom they are conferred firm-wide oversight and accountability. Strategic in nature, the CMO's priorities are the fulfillment of company strategy through the deployment and ongoing management of expatriates to the four corners of the world. Firms that adopt this functional role will gain an edge in competing for the great rewards in emerging markets, while mitigating the many risks of operating in the world's frontiers.[25]

---

[25] D. Disparte, *Market Expansion Risk and Global Mobility*, American Security Project, December 3, 2013, accessed November 24, 2015, http://www.americansecurityproject.org/market-expansion-risk-and-global-mobility/.

While international business assignments seldom confer diplomatic benefits, by definition, they can have political connotations, and many international assignees often find their freedom held in the balance by kidnappings, unlawful detention, and passport seizures pending charges. In many of these examples, nationality, mission, and the implied backing of hard currency can exacerbate the hardships borne by assignees, leading to inextricable connections between statecraft and the pursuit of overseas growth. In most cases, the opportunity for a quick profit sees expatriates routinely shaken down at traffic stops or perhaps at the airport. In more extreme examples, there is little hope of a quick or easy release from detention, as Western nationals become unwitting soft targets in the battle against terrorism. Humanitarians and journalists are especially vulnerable as they often carry out their work in remote and unstable regions. Their counterparts in business—especially at the upper echelons—manage to avoid many of these risks by traversing urban mazes in helicopters or armored convoys provided by private security firms.

Mobility Curtailed

One high-profile example is the seizure of passports by the Brazilian government of several foreign executives of Chevron Brazil's top management team related to the 2014 oil spill. The seepage of 2,400 barrels of offshore oil—for which criminal charges worked their way through the Brazilian legal system— struck a chord with the Brazilian government,[26] particularly at a time when its largest state-owned enterprise, Petrobras, was mired in a mammoth corruption scandal. The passport seizure had as much to do with simply making a point (that foreign executives are not above the law in Brazil) as it did with establishing the boundaries for future operation in the oil sector by foreign firms and, of course, demonstrating how "tuned in" Brazil was to issues of environmental concern.

By contrast, BP's foreign executives were never barred from leaving the U.S. following the Macondo explosion in 2010, and the calamity that followed. This highlights the increasing risk that even the most senior, well-protected expatriates can find themselves pawns on the chessboard of international business and politics. One noteworthy example of this was the detention and subsequent prosecution of a senior executive and three colleagues from Rio

---

[26] S. Romero, "Brazil Bars Oil Workers from Leaving after Spill," *New York Times*, March 18, 2012, accessed November 24, 2015, http://www.nytimes.com/2012/03/19/business/energy-environment/brazil-bars-17-at-chevron-and-transocean-from-leaving-after-spill.html.

Tinto's mining operation in China in 2009, who found themselves front and center in a high-profile tussle between China and Australia. The employees were held on suspicion of espionage and stealing state secrets after Rio Tinto canceled a $19 billion investment transaction with Chinese state-owned firm Chinalco.[27] The four subsequently pleaded guilty to bribery allegations. Although relations between the countries and companies continued, the incident remained constantly in the background.

Expatriate Risk

This is really perhaps the most important point, which is that business transactions and relationships do not exist in a vacuum and they often help define investment landscapes. The conduct of employees and governments can ultimately have a significant impact on local operations or overall corporate performance, and reputational risk can take a real toll on profitability. All these factors jointly determine how, or indeed whether, an organization invests successfully overseas, yet their potential importance in the cross-border investment process, and in 'managing' relations with a government following an investment, are too often either not deemed important or not paid sufficient attention to throughout the process.

The fact that there may be no immunity for expatriate employees stands in contrast to their otherwise generally privileged status around the world, as anyone who has been privileged enough to be an expat can attest. Yet, depending on the location, the concept of firm neutrality is similarly at risk, when waving a white flag, wearing a blue helmet, or an emblazoned red cross is no more a symbol of safety than wearing a target on one's back. That said, the risk of retrenchment and disengagement from the world—or specific countries—seems far greater than the likelihood of suffering from unfortunate high-profile events abroad. This is especially true if firms and their international assignees take precautions throughout the entire expatriation process.

It is important to remember that, while many expats may get lulled into a false sense of security—commingling with ambassadors and dignitaries in their host countries—they are not afforded a guaranteed *laissez-passer* in the event of an upheaval. The fact is, there is no turning back on fast-growing emerging markets, developing countries, and even disaster and relief areas. Wanderlust has long tempted mankind to take flight, discover new horizons,

---

[27] D. Cimilluca, S. Oster, A. Or, "Rio Tinto Scuttles its Deal with Chinalco," *The Wall Street Journal*, June 5, 2009.

and make home abroad. However, in this century of emerging markets and frontier economies, organizations and their international assignees should take heed to manage risks along the entire risk continuum.

Antiagile

Applying a holistic approach to risk management following the three-dimensional structure outlined above requires as much focus on organizational threats, as it does attention to organizational obstacles that may stand in the way of agility. The era of man-made risks requires a lithe response. Like rigid buildings that do not sway with an earthquake, organizational rigidity can be perilous in the face of a rapidly evolving and dynamic risk landscape. Agile risk managers and organizational leaders alike need to be mindful of organizational effectiveness, for this is the area that produces the most blind spots to potentially preventable losses—the very source of antiagility.

There are a number of specific challenges that need to be addressed. The first is the potential for carelessness to creep into large organizations. Such carelessness may be so large and be sprawled over many countries that personnel may feel a social and organizational detachment. Detachment of this kind fosters risk myopia and the potential concealment of known threats from those empowered to act. Another organizational tendency that stands in the way of agile risk management is stratification and the top-heavy structure that tends to dominate large enterprises.

In effect, only a few people in top-heavy organizations are in a position to respond to potentially serious risks—so many more stand by idly wondering if they should reveal some potentially critical information. Moreover, the decision-making structures are often too slow and rigid. Risks do not recognize when boards meet any more than they wait idly for quorum to be formed in crisis management committees. Therefore, agile, risk-ready enterprises equip all organizational levels with an appropriate degree of latitude in responding to emerging threats, whether they are acute or attritional.

Moral hazard—a term that is mentioned throughout the book—is one of the more pernicious forces driving economic and reputational losses. This threat has as much ability to harm the internal structure of an organization and create indifference among teams as it does to harm either external stakeholders or, indeed, an entire economy. Therefore, building risk agility and organizational resilience depends very much on how firms understand and combat this silent threat. The first line of defense is an organization's value system. The second is the tone at the top and the general permissiveness of the environment. Simply put, if people can get away with murder, the bodies are likely to pile high. Risk agility is therefore naturally intolerant to behaviors that are anathema to organizational values and the overall risk management strategy (within reason, of course). The greatest way to combat moral hazard is to let the accused stand trial and the guilty fall.

Appropriate levels of risk tolerance also need to be devolved throughout an organization. The owner's mentality seen in entrepreneurs needs to be developed in large enterprises. This would replace the entitlement mentality that often dominates large organizations where an implied struggle between the top brass and lower levels often rules the day. In this struggle, the frontlines are "owed something" and their disenfranchisement is both a source of ignorance to emerging risks and a source of deliberate harm. Think of the impact of strikes, locked-out workers, work slowdowns and other manifestations on an organization or industry. These grievances can—and often do—cause more harm to an organization than many of the "scarier" risks described in the book. Part of what drives these grievances is the feeling of indifference and the adversarial climate that dominates many large organizations.

Having so-called *skin in the game*, like the entrepreneur, is the best way of creating the right balance of risk agility on the one hand and traditional command and control on the other. Skin in the game is created through the right balance of incentives, transparency, and organizational accountability. Building skin in the game goes beyond making money and preserving an organization. It also implies that people would put up a fight for what they perceive to be right, and would be prepared to make sacrifices for a greater good.

Skin in the game is also what makes airlines so safe and why so much of risk management has taken a very pragmatic, 'check-the-box' form, with multiple layers of redundancy and verification. Equally, the military (notwithstanding the rise of impersonal drone warfare) has a similar quality of skin in the game, with each man or woman having a sense of responsibility for the others in their squad. Risk agility harnesses our intrinsic, biological predisposition for self-preservation. Rather than having the best people flee a sinking ship, a sign of little skin in the game, imagine the difference to an organization's survival rate if the best people put in their best work to reverse the decline.

Far too often, following one of the many risks plaguing our times, employees may become a liability rather than an asset. This is driven by a risk culture that lacks agility as much as it is driven by the adversarial nature of large firms. Employees have long memories and, populism aside, will recall layoffs, negative or stagnant wage increases, and other forms of real or perceived unfair treatment when the organization may need them the most. One of the compounding woes of cyber risk, for example, is the inevitable battery of lawsuits that will usually follow any unwanted disclosure of sensitive information. It is often easy to form class actions following a data breach, as a common class of current and past employees, customers and other stakeholders may be affected. These 'legacy' risks will come back to haunt organizations that have been disingenuous with their stakeholders at a time when they can least afford it. Hence, the need for, and desirability of, three-dimensional risk management.

While the era of man-made risk is rife with acute potentially large-scale losses, attritional, slow-burn events are equally costly and ever-present. Ironically, most firms, and the financial solutions they leverage, are better equipped to handle acute, surprise events, than to deal with long-term attritional risks. The three-dimensional approach outlined above can also help improve resilience and decision making around long-range planning and strategic options. As noted in the SkyMall example, failure to anticipate market changes can, for example, kill a firm as easily as cyber risk or a terror attack. Yet our tendency is to heed the more visceral, short-term losses than to turn our attention to potentially intangible long-range impacts. Attitudes and responses toward climate change, for example, fall prey to this 'speed blindness'. In many ways, we tend to be so concerned with short-term matters, such as beating quarterly shareholder expectations or winning a mid-term election, that we ignore or negate long-term perils.

Creating a risk culture that fosters agility toward long-range planning is more like running a marathon at a sprint's pace than like establishing a permanent set of rules and policies governing long-term risk. One of the biggest dangers in mitigating long-range risk is the use of static tools in the face of dynamic challenges. For example, many insurance policies that were written before the rise of cyber risk are rarely, if ever benchmarked. Unwittingly, organizations, their boards and officers believe they are covered by insurance or other risk mitigation tools, when in fact they may remain entirely exposed against certain perils.

Long-range risks require long-range scenario planning where a range of possibilities, however remote and unlikely, need to be contemplated. How many firms have contemplated climate contingencies that require the wholesale relocation of low-lying offices and personnel in the event of a hurricane, storm surge, or rising sea level? How many organizations are prepared to (re) onshore or nearshore their far-flung supply chain operations as lower-cost emerging markets bear the brunt of climate risk? This very example played out in Thailand in 2011 during massive flooding that shut down production at facilities across the country, and caused a global shortage of semiconductors.

Climate risk offers a great example for how to improve long-range risk readiness, but it is not alone in underscoring the need for broader thinking about resilience. Other man-made risks are also greatly impacting the long-range landscape firms and governments must grapple with. For example, the closest proximate cause of Europe's migration crisis in 2015 was the lack of development and opportunities throughout the Middle East, North and Sub-Saharan Africa. Migrants from as far afield as Afghanistan have attempted the land crossing toward Europe.

When European policymakers contemplate an era of open borders—free of encumbrances for the transfer of goods, services and people—the specter of a mass migration crisis and transnational terrorism did not factor into their imagination. This very real long-range risk has taken center stage in corporate boardrooms and national war rooms alike. Thinking with agility in the short term and over long imaginative horizons where all manner of possibilities are contemplated will improve risk readiness and organizational resilience. If ever there was a time to reorient strategic planning in the private and public sectors to consider the previously unimaginable, it is in the second decade of the twenty-first century.

Part II

The Global Risk Labyrinth

6

Terrorism

A Globalization Backlash

In 1993 Harvard political science professor Samuel Huntington first published his thesis that the root cause of the terrorism that had been unleashed
against the West was the "Clash of the Civilizations". Huntington asserted
that the economic rise of East Asia in the 1980s and 1990s, coupled with a
population explosion in Muslim countries over the same period, had changed
the nature of the global political system. According to him, the predominance
of Western ideals over the rest of the world in the modern period had been
challenged, leaving a "clash" between the fundamental tenets of Islam and the
West.

Central to Huntington's thesis was his belief that the growth of Islam was
a manifestation of anti-westernism and a rejection of the West's direct and
indirect dominance over the political, economic, and social systems of the
world, leading to an inevitable conflict between the peoples of the West and
the Islamic nations. In his prescient words: "During the coming decades,
Asian economic growth will have deeply destabilizing effects on the Western-
dominated established international order, with the development of China,
if it continues, producing a massive shift in power among civilizations. In
addition, India could move into rapid economic development and emerge as
a major contender for influence in world affairs. Meanwhile, Muslim population growth will be a destabilizing force for both Muslim societies and their
neighbors. The large numbers of young people with a secondary education

© The Author(s) 2016
D. Wagner, D. Disparte, *Global Risk Agility and Decision Making*,
DOI 10.1057/978-1-349-94860-4_6

will continue to power the Islamic Resurgence and promote Muslim militancy, militarism, and migration. As a result, the early years of the twenty-first century are likely to see an ongoing resurgence of non-Western power and culture, and a clash of the peoples of non-Western civilizations with the West and with each other."[1]

Huntington's analysis provides an important piece of the puzzle when deciphering the source of the rabid hatred of the West (especially the U.S.) among a large number of people in the world. The "clash" is also fed by the sense of disenfranchisement felt not only by some Muslims, but by a huge swathe of the world's population who feel they have not benefited from globalization, but have instead become ever more economically marginalized. The West's battle against terrorism is therefore largely a battle to win the hearts and minds of billions of people at the bottom of the economic pyramid, who believe they have very little to lose and everything to gain by opposing the West.

An important aspect of risk agility is the ability to understand and acknowledge perspectives and opinions that may differ radically from one's own. This does not mean, of course, that one is obligated to accept or agree with those views, but the ability to internalize them is very helpful in arriving at solutions to otherwise intractable problems. From the perspective of a great many people around the world, the globalization process has further enriched the world's most prosperous nations, and in the process has contributed to the enhanced marginalization of at least one-third of the world's population. Many in the developing world believe that the integration of most of the world's countries into the "global" economy has at best been an uneven process in terms of trade and investment flows, and at worst the perpetuation of a system that distributes assets and advantages in a biased fashion, favoring the wealthiest, most developed countries while leaving the poorest and least developed countries further and further behind.

An examination of trade and investment statistics from 2000 to 2014 shows that, in general, developed and developing countries followed a similar pattern of peaks and troughs based on global economic conditions, due in large part to the globalization process. While the economic performance of the majority of developing countries is inconsistent, and even gyrates wildly over time, a variety of developing economies performed better than developed economies during this period (coinciding with the peak of the modern globalization process). It may come as a surprise to some in the developing world to know that the trend line for GDP growth for the G7 nations between 1960 and 2013 is actually downward (see Graph 6.1), so the argument that glo-

---

[1] S. Huntington, *The Clash of the Civilizations and the Remaking of the World Order* (New York: Simon & Schuster 1996), 121.

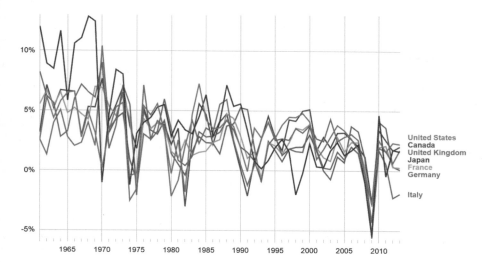

Graph 6.1 GDP growth rates of the G7 (1960–2013) (*World Development Indicators (GDP)*, Google, accessed November 24, 2015, http://www.google.ca/publicdata)

balization has primarily benefitted the most developed nations is actually not accurate, based purely on economic performance. If that were the case, the long-term G7 GDP growth rate would positive instead of negative.

Yet for hundreds of millions of people in the developing world, the wealth generated as a result of globalization that was *supposed* to be created has not trickled down to them. Whether the contributing causes are endemic corruption, hopelessly inefficient state enterprises, inadequate infrastructure, or an insufficiently developed private sector, this perceived inequality is helping to fuel the rise of global terrorism. What many in the developing world fail to recognize is that income inequality is a global phenomenon—not one that is just negatively impacting the developing world—and reducing the gaps in resources, personal and national incomes, infrastructure and technical skills will be one of the great challenges of the future.

A 2016 Oxfam report has noted that the richest one percent of individuals will own more than 50 percent of the world's wealth.[2] When people who have difficulty feeding themselves or their family hear this statistic, it must make them very angry. The sense of hopelessness and powerlessness that follows has to be considered a primary reason for why anti-western extremist groups proliferate today. As the gap between the haves and have nots grows wider in

---

[2] L. Elliott and E. Pilkington, "New Oxfam Report says Half of Global Wealth Held by the 1%," *The Guardian*, January 19, 2015, accessed November 24, 2015, http://www.theguardian.com/business/2015/jan/19/global-wealth-oxfam-inequality-davos-economic-summit-switzerland.

the era of globalization, the number of people likely to find the rhetoric being spewed by the anti-western extremist groups appealing can only grow.

The Importance of Perceptions

Spreading the benefits of globalization, and having them be perceived as being more equitable around the world, is a daunting challenge. Equally challenging is how to encourage a change in the way many Muslims around the world feel about the U.S. and the West, and vice versa. Those opposed to the U.S. in the Muslim world far outnumber those who are sympathetic. In a 2002 Gallup poll of 10,000 residents in nine Muslim countries[3] (*before* the U.S. invasion of Iraq, the Arab Awakening, or the Global War on Terrorism), most respondents called the U.S. "ruthless and arrogant," with most describing themselves as "resentful" of the U.S. Some 67 percent saw the 9/11 attacks as morally justified and 77 percent saw the U.S. military action in Afghanistan as morally unjustified. Most respondents said they thought that the U.S. was aggressive and biased against Islamic values, specifically citing a bias against Palestinians.

Many in the Muslim world follow the teachings of Mohammed Khaldun, a fourteenth-century Muslim historian who had an encyclopedic knowledge of the Islamic world. He was renowned for his historiography, which is widely referenced in the Muslim world today. Khaldun saw solidarity (in the form of consciousness of communal or blood ties) as the source of civilization. He believed that prosperity leads to corruption and, ultimately, the abandonment of religion, which inevitably results in self-destruction. He believed that all civilizations pass through stages, beginning with youth and ending in senility, and that all civilizations last no more than 200 years—once its power has peaked, it begins an irreversible decline.

By this measure, the U.S. is due for a fall, and many followers of Khaldun believe this to be preordained. It is ironic that without the shockwaves that resulted from the multiple terrorist attacks on the West—starting with 9/11—policymakers and ordinary citizens of the West might have remained oblivious to the increasingly serious threat posed by Islamic extremist groups. The attacks themselves may have provided the pillar of solidarity among Western governments that Khaldun's followers may ultimately judge them by.

---

[3] F. Newport, *Gallup Poll of the Islamic World*, Gallup, February 26, 2002, accessed November 24, 2015, http://www.gallup.com/poll/5380/gallup-poll-islamic-world.aspx.

Table 6.1 Characteristics associated with Muslims and Westerners (percent)[a]

	Muslim views of Westerners	Western views of Muslims
Selfish	68	35
Violent	66	50
Greedy	64	20
Immoral	61	23
Arrogant	57	39
Fanatical	53	58
Respectful of women	44	22
Honest	33	51
Tolerant	31	30
Generous	29	41

[a]Median percentage of Muslims across seven Muslim countries who say each of these traits describes people in Western countries. The median percentage of non-Muslims across the U.S., Russia and four Western European countries who say each of these traits describes Muslims.
Muslim-Western Tensions Persist, Pew Research Center, July 21, 2011, accessed November 24, 2015, http://www.pewglobal.org/2011/07/21/muslim-western-tensions-persist/.

Western culture, religion, values, upbringing, and history all contribute to Westerners' views of the world. Depending on one's frame of reference, what may be considered an abhorrent behavior by one person may be considered perfectly acceptable to another. Keep this in mind when considering the subject of terrorism, for it weighs heavily on the nature of the debate, and where a solution resides.

In 2011 the Pew Research Center conducted a comprehensive study[4] examining why Muslim–Western tension persists. The results are surprising. As noted in Table 6.1, Muslims and Westerners share similar concerns about each other, albeit to differing degrees. Most Muslims see Westerners as selfish, violent, greedy, immoral, arrogant, and fanatical. Most Westerners see Muslims as both fanatical and honest. Both sides see the other as nearly equally fanatical and tolerant. While only so much can be gleaned from such an exercise, it appears that, according to most Muslims, it is Westerners who have the most work to do. This has presumably not occurred to most Westerners.

In 2008 Gallup published another survey,[5] which was started just after 9/11 and included more than 50,000 Muslims from 35 nations (Gallup contends this represents the views of 90 percent of the Muslim world of 1.3 billion

---

[4] *Muslim–Western Tensions Persist*, Pew Research Center, July 21, 2011, accessed November 24, 2015, http://www.pewglobal.org/2011/07/21/muslim-western-tensions-persist/.

[5] "Most Muslims 'Desire Democracy'," BBC News, February 27, 2008, accessed November 24, 2015, http://news.bbc.co.uk/2/hi/americas/7267100.stm.

people). According to Gallup, the majority of average Muslims seek democracy consistent with religious values. Those polled also said the most important thing the West could do to improve relations with Muslim societies was to change its negative views towards Muslims, respect Islam, and that *decision makers on both sides need to listen to the voices of moderation* and consider new policies to prevent extremists on both sides from becoming more popular.

It is becoming apparent that the task at hand will require a sea change in how policymakers around the world address issues such as development and poverty alleviation, as well as terrorism and intelligence sharing. Winning the battle against terrorism will require truly revolutionary thinking in terms of how to strengthen weak political systems, how to change perceptions and realities about income inequality, and how to enhance genuine opportunities for the average person in any country to get ahead in life, while creating an environment that will invite people in the Muslim and developing worlds to reconsider their views about each other.

Some Surprising Terrorism Facts

Many of the statistics that follow are truly surprising, and should shatter a lot of preconceptions many of us have about what terrorism is, and what it is not. What is unsurprising is that there has indeed been a dramatic rise in the number of terrorist acts and deaths since the start of the twenty-first century. In 2014 the total number of deaths from terrorism increased by 80 percent when compared with the prior year—the largest yearly increase in the last 15 years. Since the beginning of the twenty-first century, there has been over a ninefold increase in the number of deaths from terrorism.[6] However, in 2013, 66 percent of all deaths resulting from terrorist attacks were claimed by just four terrorist organizations—Al Qaeda, Boko Haram, the IS, and the Taliban—with more deaths being the result of attacks from Boko Haram than any other such organization. While much of the increase has been due to the ongoing Syrian conflict, it is worth pointing out that more than *82 percent of those killed in terrorist attacks died in just five countries*: Afghanistan, Iraq, Nigeria, Pakistan, and Syria.[7] The majority of deaths from terrorism do not occur in the West. Excluding 9/11, between 2000 and 2015 only 0.5 percent

---

[6] Institute for Economics and Peace, *Global Terrorism Index 2015,* 2015, accessed November 25, 2015, http://economicsandpeace.org/wp-content/uploads/2015/11/Global-Terrorism-Index-2015.pdf.

[7] Institute for Economics and Peace, *Global Terrorism Index 2014,* 2014, accessed November 25, 2015, http://economicsandpeace.org/wp-content/uploads/2015/06/Global-Terrorism-Index-Report-2014.pdf. Additional statistics in this section were derived from the Index.

of all deaths have occurred in Western countries, and only 5 percent of all terrorist deaths have occurred in OECD countries since 2000.[8]

Some other facts that are worth noting[9]:

- Approximately 50 percent of terrorist attacks actually claim no lives, and the remainder claim less than ten lives
- Terrorism incidents that kill more than 100 people represent only .001 percent of total incidents
- Homicide claims more lives globally than terrorism
- The primary target of terrorism has consistently been *private* property and citizens
- Religion as a driving ideology for terrorism has risen dramatically since 2000
- The number of political and separatist movements have changed little since 2000
- 70+ percent of terrorist attacks occur within a country during periods of major conflict
- The number of countries that experienced more than 50 deaths in a year reached an all-time high in 2013
- Thirteen times as many people are killed globally by homicides as die in terrorist attacks

Some statistics specifically related to the U.S.[10]:

- Between 1970 and 2012 more than 2,600 terrorist attacks took place, resulting in more than 3,500 fatalities
- Approximately 86 percent of all deaths from terrorist attacks during this period were the result of 9/11
- While attacks occurred in all 50 states, 50 percent occurred in California, New York, and Puerto Rico
- More than half of the attacks took place in the 1970s
- Nearly 80 percent of all terrorist attacks involved no casualties
- Businesses were the most frequent type of target, with nearly a third directed against banks or businesses involved in commerce. 22 percent were directed toward retail entities

---

[8] Institute of Peace, *Global Terrorism Index 2015*.

[9] Statistics derived from the *Global Terrorism Index* for 2014 and 2015.

[10] National Consortium for the Study of Terrorism and Reponses to Terrorism (START), *Global Terrorism Database*, 2015, accessed November 24, 2015, http://www.start.umd.edu/gtd/.

- Between 2000 and 2012, there were fewer than 20 attacks per year, on average, with the majority of those identified being environmental or animal rights extremists groups

Most of us have been acutely aware of the threats posed by global terrorism for many years, but even after everything we have seen, heard, and experienced for ourselves, we still have ill-conceived notions about the exact nature of terrorism. Some of these are simply wrong. For example, there is no strong statistical link between poverty and terrorism, and many individuals who join terrorist groups in wealthy countries are well educated and come from middle-class families.

A study from 2005[11] found, in an examination of 75 terrorists behind the four most significant attacks against Westerners at that time, that 53 percent had either attended college or had a college degree (as a point of reference, only 52 percent of Americans have attended college). The authors of the study also found that, contrary to the belief that most terrorists have attended Islamic Schools (madrassas), less than one percent of Pakistani students attend madrassas. (By contrast, more American children are home-schooled than Pakistani children who attend madrassas. Of course, that figure may be higher in other countries). Many of the 9/11 hijackers came from middle class families. Also, even with substantial resources, most large terrorist organizations have had very limited success in launching large-scale attacks.

The more you know about terrorism, the more you will realize that you are personally unlikely to become a victim of terrorism even if you live in one of the countries cited above—unless you happen to become a target by virtue of your nationality or profession. That should not serve to minimize the global threat, or how important it is to be prepared for the eventuality an act of terrorism occurs in your country, but it should prompt everyone to think about other preconceived notions they may have, about anything, and how becoming educated about the issue may change their approach to problem solving.[12]

---

[11] P. Bergen and S. Pandey, "The Madrassa Myth," *New York Times*, June 14, 2005, accessed November 24, 2015, http://www.nytimes.com/2005/06/14/opinion/the-madrassa-myth.html.

[12] Much of the material presented in this section was derived from the *Global Terrorism Index 2014*, 2014, Institute for Economics and Peace, accessed November 25, 2015, http://economicsandpeace.org/wp-content/uploads/2015/06/Global-Terrorism-Index-Report-2014.pdf.

The Economic Cost of Terrorism

While it is true that terrorism is a 'global' phenomenon, as noted, its most direct impact is felt in only a few countries. While a terrible tragedy for these countries, their people, and others that suffer from the direct impact of terrorism, the impact felt by most of the world is limited to the ongoing metamorphosis in the manner in which governments address security protocols, how average citizens around the world are impacted by those protocols, and the economic costs associated with terrorism.

The economic cost of terrorism is estimated to have reached its highest-ever level in 2014, at $53 billion—a 61 percent increase from the previous year and a tenfold increase since 2000.[13] When considering these economic costs, bear in mind that the long-term *indirect* costs of terrorism can easily be up to twenty times larger than the *direct* costs. While the 'direct' cost of the 9/11 attacks was estimated at $55 billion, the secondary effects of increased security totaled nearly $600 billion, decreased economic activity was nearly $125 billion, and the cost of the Iraq War that followed has been estimated to cost in excess of $3 trillion.[14] The Australian government estimated in 2013 that extra trade costs associated with terrorism could be as high as $180 billion per year. These indirect costs are truly significant.

While trade and investment are negatively affected by acts of terrorism, most such events do not seem to have a significant impact on Foreign Direct Investment (FDI). Some studies have shown that the 9/11 attacks had relatively little impact on U.S. FDI,[15] while FDI in Spain actually increased by $6 billion in 2005 following the 2004 Madrid train bombings, and an additional $11 billion in 2006.[16] The effects can be dramatically negative, however. In Nigeria, it is estimated that FDI dropped by $6 billion following acts of terrorism by Boko Haram in 2010.[17] So it does not make sense to generalize about the impact terrorism has on national economies.

Perception of risk clearly plays a significant role in how investors react to acts of terrorism. Terrorism in countries perceived to have higher risk can

---

[13] Institute of Peace, *Global Terrorism Index 2015.*

[14] J. Stiglitz and L.J. Bilmes, *The Three Trillion Dollar War: The True Cost of the Iraq Conflict* (New York: W.W. Norton & Company, 2008), accessed November 25, 2015, http://threetrilliondollarwar.org.

[15] W. Enders et al., *The Impact of Transnational Terrorism on U.S. FDI*, Create Homeland and Security Center, Paper 55, 2006, accessed November 25, 2015, http://research.create.usc.edu/cgi/viewcontent.cgi?article=1035&context=published_papers.

[16] World Bank data, 2015, accessed November 2015, www.worldbank.org.

[17] A.A. Adbayo, "Implications of Boko Haram Terrorism on National Development in Nigeria," *Mediterranean Journal of Social Sciences*, Volume 5, Number 16 (2014), accessed November 25, 2015, http://www.mcser.org/journal/index.php/mjss/article/view/3330/3284.

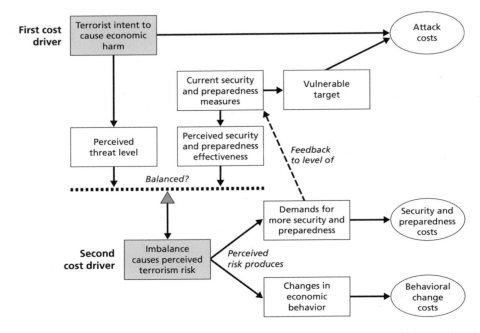

Illustration 6.1 Framework for examining economically targeted terrorism (*Source*: Jackson, B. A. *Economically Targeted Terrorism*, Rand Center for Terrorism Risk Management Policy, 2007, accessed November 25, 2015, http://www.rand. org/content/dam/rand/pubs/technical_reports/2007/RAND_TR476.pdf)

prompt a severely negative result with traders, investors, and lenders. This is an important reason why acts of terrorism often have a more negative impact on developing countries than developed countries. The average amount of FDI in the countries most affected by terrorism is less than half that of Organization of Economic Cooperation and Development (OECD) countries. Trade as a percentage of GDP is 51 percent in these countries, whereas it is 87 percent on average for the OECD.[18]

The Rand Corporation produced a useful study[19] that classifies the economic costs associated with terrorism into three categories: the actual cost of an attack; the costs associated with implementing security and preparedness measures; and the costs resulting from changes in behavior in anticipation of future attacks. Illustration 6.1 shows the interrelationship of all three categories in the decision-making process for protecting against economically targeted

---

[18] World Bank data.

[19] B. A. Jackson, *Economically Targeted Terrorism*, Rand Center for Terrorism Risk Management Policy, 2007, accessed November 25, 2015, http://www.rand.org/content/dam/rand/pubs/technical_reports/2007/RAND_TR476.pdf.

terrorism. Attack costs are driven by the group's intent to cause economic damage, and whether or not vulnerabilities in security systems can be exploited. If an attack is perceived as a result of this, security/preparedness costs typically rise, and the net result may be behavioral changes by the organization and its employees. Any additional investment in security adds to the potential costs of an attack, coinciding with a rising perception of risk of an attack.

Since individual, business, and government interests and approaches to addressing the problem of terrorism are often not in synch, all parties need to have incentives and guidelines in place so that they may act in a consistent fashion. Increasing safeguards and enhancing general awareness is a precursor to meaningfully addressing the persistent nature of the threat, but this can have unintended consequences. For example, for many years following the 9/11 attacks, the U.S. was on 'orange' alert, meaning that an attack could happen at any time and that the threat was ever-present. However, after the passage of some time, average citizens began to pay little attention to the threat, particularly if no attacks occurred.

Reducing the possibility that an attack can occur, will be successful, or will cause significant damage by enhancing security protocols goes hand in hand with changing perceptions that an attack may occur at all. There is a risk that constantly increasing security spending may both enhance the cost of an attack and reverse any gains made in changing risk perceptions if an attack occurs and is successful.

Spending the money necessary to reduce vulnerability to attack is perhaps the greatest challenge currently facing developed country governments and businesses. This is because most developed countries did not construct infrastructure and buildings with the type of security safeguards necessary to thwart terrorist attacks. When power transmission substations, railroad tracks, or office buildings were constructed, security was not a foremost consideration. Most developed countries remain highly vulnerable to attack on "soft" targets. One of the ironies in the terrorism landscape is that this same type of concern is not the case in many developing countries, where security was always a primary concern.

Cost-Effective Terrorism

For less than $1,000, Al Qaeda succeeded in 2004 not only in its objective of punishing the Spanish people and government for their support of the war in Iraq, but also in achieving a Spanish pullout from the American-led coalition forces in Iraq. The added bonus was the ability to ultimately change the

Spanish government, as the Socialists were swept to power in the wake of the bombings. By any measure, that is good value for money. The fact that so much can be achieved by spending so little is certainly an incentive for other terrorist groups to want to try to do the same thing.

Spending so little to achieve so much was nothing new for Al Qaeda. It is estimated, for example, that for less than half a million dollars they achieved all that was 9/11, which resulted in the loss of some 3,000 lives and more than $50 billion in property and related damages. Similarly, for $74,000 Jemaah Islamiah (an Al Qaeda affiliate) killed more than 200 and temporarily ruined the tourist industry in Bali in 2002. Other examples of cost-effective terrorism include:

- The IS spending less than $10,000 to finance the Paris attacks in 2015, resulting in the deaths of more than 130 people;
- The twin truck bombings of U.S. embassies in Kenya and Tanzania, which killed more than 200 people in 1998 for $10,000; and
- The bombing of the USS *Cole* in Yemen, which killed 17 people in 2000 for less than $10,000.[20]

One of the lingering impacts of these attacks was that Al Qaeda and the IS permanently raised the price of travel-related security, at great cost in terms of time and effort expended by millions around the world who travel by train or bus, and also the firms that operate those modes of transportation. The cost and resources required to implement airport-style security for trains and buses is simply too prohibitive and unrealistic. The George W. Bush Administration announced an effort to make an attempt at enhancing security on trains in the U.S., however, doing this after the Madrid bombings is actually a central part of the reason why terrorist groups continue to be successful—just as businesses are more often than not reactive rather than proactive when it comes to risk management, governments tend to be *re*active rather than *pro*active on the subject of terrorism.

We have known for some time now that terrorist groups prefer to attack "soft" targets because they usually lack proper security and also because there are many more soft targets than hard targets. Hotels (Mombasa), restaurants/night clubs (Bali), museums (Tunis), places of worship (Istanbul), trains (Madrid), and buses (Israel) are among the targets of choice. This being the case, there should be greater effort made to implement at least minimal secu-

---

[20] http://www.nbcnews.com/storyline/paris-terror-attacks/terror-shoestring-paris-attacks-likely-cost-10-000-or-less-n465711.

rity for soft targets that have any reasonable appeal to terrorists. If this can be done in developing countries with meager financial resources, it should be achievable in the developed world.

In Manila, for example, the entrances to the metro rail system are checked in the same way as entrances to department stores, office buildings, and shopping centers. Security personnel check everyone's bag or purse as individuals enter. Is this a guarantee that a bomb will not be smuggled onto a train? Of course not, but apart from providing some peace of mind, it is a sufficient deterrent to prevent would-be bombers from attacking with impunity. Had such a system been implemented in Madrid, most, if not all, of the bombs would likely have been detected. Surely, Manila would have experienced many more bomb attacks on its metro rail system if it did not have this rudimentary system in place.

Will countries in the developed world that currently do not have security systems and personnel in place in shopping malls, movie theaters, office buildings, and other public places ultimately need to do as the Philippines has done? The answer is more than likely "yes." And businesses should be prepared to share the cost of implementing these measures. The threat from terrorist groups is certainly not going to diminish in the near or medium term. We have a decades-long problem on our hands. This is just the beginning.

The Impact of Terrorism on Foreign Direct Investment

One would think that acts of terrorism would have a negative impact on FDI flows to affected countries. Common sense dictates that the loss of foreign investor confidence following acts of terrorism would prompt large outflows of capital in affected countries, and that once a country is branded a terrorist target, it would attract reduced levels of FDI. As previously mentioned, sometimes this is in fact the case; however, foreign investor sentiment is not always dictated by common sense. The lure of profit and desire to establish trade partnerships is often a stronger motivational force than perceived political risk is a disincentive to invest.

Although the growth of global terrorism is indeed on the minds of some corporate decision makers when contemplating whether or not to make overseas investments, it did not prevent many of them from deciding to invest in the post-9/11 developing world. According to the United Nations Conference on Trade and Development (UNCTAD), FDI flows to the developing world *surged* 200 percent between 2000 and 2004, up from 18 percent to 36 percent of global FDI. During the same period, FDI flows to *developed* countries *plunged* 27 percent, from 81 percent to 59 percent of global FDI.

In every category of *developed* countries cited, the inward FDI trend was *down* significantly, while in every *developing* country category, the inward FDI trend was *sharply higher.* Although the vast majority of terrorist attacks take place in developing countries, the post-9/11 FDI trend was clear. That trend has continued. UNCTAD's updated statistics show that *developed* countries' share of FDI continued to *drop*, reaching 52 percent of the total in 2011 and just 39 percent of the global total in 2013. Over the same period, *developing* countries' percentage of the global total experienced the opposite trend—it *rose* to 43 percent in 2011 and to 54 percent in 2013.[21]

Does this mean that perceived terrorism risk has a more general negative affect on FDI decision making? That undoubtedly depends, of course, on where one intends to invest. Clearly, a company considering investing in Iraq would have far greater concerns about terrorism than one investing in Canada. Interviews and surveys of executives in multinational corporations in the 1960s and 1970s[22] found political events to be one of the most important factors influencing foreign investment decisions. This was no doubt due in large part to the Cold War and the perception that "regime change" could have stark implications for foreign operations.

It appears that times have changed, however. Consulting firm A.T. Kearney produces an annual publication, the *FDI Confidence Index*, in which it polls top decision makers in the world's largest 1,000 companies and asks their opinions on a range of FDI-related issues. The conclusions are similarly surprising. In 2003 corporate leaders' top pick for global event most likely to influence their investment decisions was the recovery of the U.S. economy. Terrorism and security concerns were tied with the Middle East conflict for number 7 on the list of 11 concerns. In 2013, the top three concerns according to Kearney were political volatility in developing countries, slowing consumer demand in developed countries, and regulatory barriers in developing countries.

Empirical studies examining the link between perceived political risk, terrorism, and FDI flows have yielded contradictory results. Some have found linkages, whereas others have not. Those that have found linkages tended to be older studies. Some of the newer studies challenge the previous results.

---

[21] United Nations Conference on Trade and Development, *World Investment Report 2014*, 2014, page xiv.
[22] H. Schollhammer, *Locational Strategies of Multinational Firms* (Los Angeles: Pepperdine University, 1974).

More recent empirical studies have tended to put more emphasis on macro-economic variables as explanatory factors in FDI flows, while others stress the importance of political variables. In practice and in theory, it appears difficult to make a clear-cut distinction between political and economic variables as definitive sources of influence, and it is reasonable to conclude that FDI decisions in developing countries are determined by both political and economic factors.[23]

A Harvard study[24] states that higher levels of terrorism risk are associated with lower levels of net FDI. In an integrated world economy, where investors are able to diversify their investments, terrorism may induce large movements of capital across countries. Another academic study[25] took this a step further and examined the impact of terrorist attacks on capital markets. The authors researched the U.S. capital markets' response to 14 terrorist/military attacks from 1915 to 2001 and concluded that present-day U.S. capital markets make faster recoveries from such events than they did a century ago. This is largely attributed to a stable banking/financial sector that provides adequate liquidity in times of crisis and thereby promotes market stability.

In their largest ever decline, the U.S. markets dropped 21 percent over an 11-day period when Germany invaded France in 1940 and they took 795 days to recover to their pre-event level. After 9/11, the markets dropped just 8 percent over an equivalent 11-day period and thereafter they took just 40 days to recover. Other financial markets were not as resilient. For example, over the 11-day period following 9/11, Norway's stock market dropped 25 percent and took 107 days to recover. One possible reason for the favorable U.S. performance is that the U.S. Federal Reserve took steps to provide liquidity throughout the banking and financial sector. This serves to emphasize that post-event investor perceptions can to a limited degree be managed by effective government response.

In a study done at Pennsylvania State University[26] (PSU), the effect of economic globalization on transnational terrorist incidents was examined statisti-

---

[23] G. Leopold and K. Wafo, *Political Risk and Foreign Direct Investment*, Faculty of Economics and Statistics, University of Konstanz, 1998, accessed November 25, 2015, http://kops.uni-konstanz.de/bitstream/handle/123456789/12070/161_1.pdf?sequence=1.

[24] A. Abadie and J. Gardeazabal, *Terrorism and the World Economy*, Harvard University/NBER and the University of the Basque Country, October 2005, accessed November 25, 2015, http://www.hks.harvard.edu/fs/aabadie/twep.pdf.

[25] A. Chen and T. Siems, *The Effects of Terrorism on Global Capital Markets*, Cox School of Business and the Federal Reserve Bank of Dallas, August 2003, accessed November 25, 2015, http://dev.wcfia.harvard.edu/sites/default/files/ChenSiems2004.pdf.

[26] Q. Li and D. Schaub, "Economic Globalization and Transnational Terrorism," *The Journal of Conflict Resolution* 48, no. 2 (April 2004), 230–258.

cally using a sample of 112 countries during the period 1975 to 1997. The strong results show that FDI, trade, and portfolio investment have *no* direct positive effect on the number of transnational terrorist incidents among countries, and that the economic development of a given country and its trading partners reduce the number of terrorist incidents in a given country. *To the extent that FDI and trade promote economic development, they have an indirect negative effect on transnational terrorism.* Perhaps the decision makers polled in the A.T. Kearney studies knew intuitively what the PSU study proved statistically—that economic development is a deterrent to terrorism.

A related study done at PSU[27] probed whether democratic forms of government reduce the number of terrorist attacks. In this case, 119 countries were examined between 1975 and 1997. Contrary to some earlier academic studies on this subject, which promote the idea that terrorist groups are more often found in countries with democratic forms of government than authoritarian forms of government, the authors found that some aspects of democracy (such as higher electoral participation, which produces a high degree of satisfaction among a general population) tend to reduce the number of transnational terrorist incidents, while other aspects of democracy (such as a system of strong checks and balances and the ability to restrict press freedoms) often serve to increase the number of such incidents.

The argument made in both PSU studies make sense and are backed up by statistics, yet they do not address the fact that many countries with vastly different histories and forms of government have experienced long-term terrorism[28] on their soil (for example, Algeria, Colombia, India, Israel, Nepal, Pakistan, the Philippines, Spain, Turkey, and the United Kingdom). The same is true of countries with newer terrorism problems (such as Thailand and the U.S.). More often than not, it appears that countries with significant terrorist acts tend to have *democratic* forms of government. Terrorism does *not* appear

---

[27] Q. Li, "Does Democracy Promote or Reduce Transnational Terrorist Incidents?," *The Journal of Conflict Resolution* 49, no. 2 (April 2005), 278–297.

[28] An "Academic Consensus" definition of "terrorism" is: "Terrorism is an anxiety-inspiring method of repeated violent action, employed by a (semi-) clandestine individual, group or state actors, for idiosyncratic, criminal or political reasons, whereby the direct targets of violence are not the main targets. The immediate human victims of violence are generally chosen randomly (targets of opportunity) or selectively (representative or symbolic targets) from a target population, and serve as message generators. Threat- and violence-based communication processes between terrorist (organization), (imperiled) victims, and main targets are used to manipulate the main target (audience(s)), turning it into a target of terror, a target of demands, or a target of attention, depending on whether intimidation, coercion, or propaganda is primarily sought" A. Schmid et al., "Political Terrorism: A New Guide to Actors, Authors, Concepts, Data Bases, Theories, and Literature", (Amsterdam: North Holland, Transaction Books, 1988), 28.

to occur with great frequency in countries with authoritarian or communist forms of government.

The same PSU author produced another interesting empirical recent study examining the impact of political violence on FDI.[29] The study posits that terrorist incidents do not produce any statistically significant effect on the likelihood that a country will be chosen as an investment destination, or on the amount of FDI it receives. Further, it states that unanticipated acts of terrorism do not generate any changes in investor behavior, in terms of either the choice of investment location choice or the amount of investment.

However, a study on the impact of terrorism and FDI in Spain and Greece[30] arrived at a completely different conclusion—that acts of terrorism did have a significant and persistent negative impact on net FDI. They concluded that one year's-worth of terrorism discouraged net FDI by 13.5 percent annually in Spain and 11.9 percent annually in Greece. On this basis, it was concluded that *smaller countries that face a persistent threat of terrorism may incur economic costs in the form of reduced investment and economic growth.*

Related to this, the same authors of the previously cited Harvard study produced a case study on the economic costs of the Basque conflict[31] and concluded that there is evidence of negative economic impact associated with terrorism in the Basque region of Spain. On average, over a two-decade period the conflict resulted in a 10 percent gap between the per capita GDP there and that found in a comparable region which was not experiencing terrorism. Moreover, changes in per capita GDP were shown to be associated with the level of terrorist activity. The authors also demonstrate that once a ceasefire came into effect (in 1998–1999), Basque stocks outperformed non-Basque stocks. When the ceasefire ended, non-Basque stocks outperformed Basque stocks.

An interesting corollary is the research done by the Asian Development Bank (ADB) when it created a terrorism insurance facility for investors in Pakistan.[32] The ADB learned that in nearly every instance, acts of terrorism in Pakistan were directed at government and/or military targets, and that commercial loss (if any) was nearly always the result of collateral damage. A sur-

---

[29] Q. Li, "Political Violence and Foreign Direct Investment," *Research in Global Strategic Management*, volume 12, 2006, 231–55.

[30] W. Enders and T. Sandler, "Terrorism and Foreign Direct Investment in Spain and Greece," *Kyklos* 49, no. 3 (1996), 331–52.

[31] A. Abadie and J. Gardeazabal, J. *The Economic Costs of Conflict: A Case Study of the Basque Country*, Harvard University/NBER and the University of the Basque Country, July 2002, accessed November 25, 2015, http://hks.harvard.edu/fs/aabadie/ecc.pdf.

[32] Based on Daniel Wagner's own on-the-ground research.

vey of local insurance companies in Pakistan revealed that the incidence of commercial loss due to acts of terrorism was almost zero. This was in sharp contrast to the image of Pakistan that prevails in the global media, where it is portrayed as a poor place to invest because of perceived terrorism risk. Yet 9/11 produced more than $50 billion in commercial losses in the U.S., and it remains one of the top FDI destinations. This demonstrates just how flawed common perceptions of risk can be.

Perceptions of terrorism risk have a great deal of influence on some investment decisions, and very little on others. Among the factors that influence decision makers are the economic health of the investment destination, the difficulty associated with doing business in a given country, the existence of rule of law and good corporate governance, the existence of corporate and government connections, and, of course, the cost of production. Investors may also distinguish between "perceptions" of the existence of a terrorism threat in a given FDI destination and "acts" of terrorism, or between "domestic" acts of terrorism and "international" acts of terrorism.

However, one factor often not considered when contemplating making a cross-border investment is consumer behavior, and its linkage to the political process. Perceptions are also important here. Predicting consumer behavior correctly can be as important in determining the success of an investment as predicting whether terrorism will have an impact on operational capability. For example, one would think that the rise in hostility toward the U.S. by a variety of Europeans in response to the Iraq War would have resulted in fewer European sales of goods by American companies. Interestingly, however, one of the first detailed empirical studies on consumer behavior post-2003[33] noted that although up to 20 percent of European consumers consciously avoided purchasing American-made products, sales by American companies in 2000–2001 and 2003–2004 grew at least as quickly as those of their European rivals in Europe. In the case of Coca-Cola, McDonald's, and Nike, European sales grew 85 percent, 40 percent, and 53 percent, respectively, for the period. Apparently, Europeans make a distinction between the actions of the U.S. government and the products of American companies.

Short-term corporate costs directly or indirectly linked to acts of terrorism can be substantial, but the potential long-term costs of terrorist threats to national economies can be devastating. A study by Australia's Department of Foreign Affairs and Trade[34] found that developing countries stand to lose the

---

[33] P.J. Katzenstein et al., *Anti-Americanism in World Politics* (Ithaca, NY: Cornell University Press, 2006).
[34] Australia Department of Foreign Affairs and Trade, *Combating Terrorism in the Transport Sector— Economic Costs and Benefits*, 2004.

most because of their dependence on FDI and export-led growth, with the developing economies of East and Southeast Asia having been deemed to be the most vulnerable. The study estimated that economic growth in the region could decline by 3 percent after five years of ongoing terror threats, and by 6 percent in the case of persistent threats lasting more than ten years. The attacks of 9/11 are estimated to have significantly reduced global investment levels.

The research indicates that there is no single answer as to the impact terrorism has on FDI, as it is dependent on numerous variables. The empirical evidence shows that terrorism does indeed impact levels of FDI to varying degrees, and persuasive arguments have been made on both sides. Similarly, some theorists maintain that democratic political systems are a breeding ground for terrorism, while others claim just the opposite. And some earlier studies concluded that corporate executives consider political and terrorism risk to be among the most important factors influencing the decision-making process, while later studies minimize their importance.

It does not make much sense to generalize about what motivates foreign investment decisions. Existing theories and arguments fail to explain the rationale behind what motivates many foreign investment decisions. One is left to speculate about such motivations, although the A.T. Kearney surveys lead one to conclude that economic motivations are stronger than political deterrents in influencing foreign investment decisions.

Yet to be addressed is the question of whether certain sectors of an economy are more sensitive to the negative effects of terrorist attacks than others, why some countries experience protracted terrorism over time, and its impact on FDI decision making. A lengthy history of terrorism has not prevented foreign oil companies from making, and continuing to make, long-term investments in such countries as Algeria and Colombia, with long histories of terrorist or 'freedom fighter' movements. Angola continued to receive huge foreign investments in its energy industry at the height of its civil conflict.

Of course, investment in all these countries would presumably have been much higher in the absence of recurring terrorism or civil conflict. The U.S. continues to be one of the world's top foreign investment destinations, even though it remains most global terrorist organizations' number one target. Although the level of FDI was down significantly in the U.S. post-9/11, it is hard to say for certain whether this was due primarily to a changed perception of the U.S. as a "safe haven" destination, or whether the prevalence of low interest rates prompted capital investors to seek more lucrative alternatives.

Some companies are concerned primarily with profit maximization while other companies are more concerned with risk management and loss minimization. The impact of government-to-government relations on the FDI equation can be an important factor motivating FDI flows, as can the desire to establish and maintain international trade links. Experienced foreign investors may discount terrorism risk automatically because they will have had good historical experience or strong corporate and government relationships locally. Other foreign investors may never pursue cross-border investment opportunities because of the absence of prior experience or meaningful corporate and governmental relationships.

The question of what would happen in the event of a truly catastrophic terrorist event must also be considered. Would new construction-related investment flow in, as is the case when natural disasters occur? Would the explosion of a dirty bomb, for instance, make a city so dangerous that the replacement of damaged buildings would prove impossible? It is questions like these that serve to reemphasize the limited value of generalizing about terrorism's impact on FDI. Theorists can speculate all they want about what "may" happen if such an event were to occur, but theories and complicated forecasting models have been proven wrong many times in the past.

Depending on the investment destination, terrorism either already is, or has the potential to become, a primary consideration in the formulation of investment decisions. Much will depend on the motivations, experience, and resources of a given foreign investor. As the Pakistan example noted above demonstrates, it is vitally important not to rely solely on widely held perceptions about the nature of terrorism risk in a particular country. A wise foreign investor will separate fact from fiction to arrive at an investment decision based on the reality on the ground that is consistent with its investment objectives.[35]

Why It Is so Difficult to Stop the Funding of the IS

The governments of Kuwait, Qatar, and Saudi Arabia have for many years funded anti-Shia political and military movements in the Middle East without any substantial resistance from the international community. After the IS morphed into the monster that is has become, and the group was widely condemned by governments around the world, these nations either stopped or substantially reduced their official funding of the organization. Yet wealthy

---

[35] Adapted from D. Wagner, *The Impact of Terrorism on Foreign Direct Investment*, International Risk Management Institute, February 2006, accessed November 25, 2015.

individuals in these countries picked up where the governments left off. The IS continues to be funded from sources within these countries unabated.

The original intention of the governments of Kuwait, Qatar, and Saudi Arabia in deciding to provide funding for the IS was simple: covertly fund an organization that promoted their collective anti-Shia regional agenda while maintaining plausible deniability with regard to their involvement. While the leaders of these governments could not have predicted what IS would turn into, they have failed to implement and enforce regulations that would prevent their citizens from funding the IS.

A contributory reason for this failure is disillusionment with the U.S. government's response to the Arab Awakening, and U.S. policy toward Iraq and Syria. Regional governments feel disengaged from America and believe that the only way they may influence the course of events is by crafting their own solution. They believe that the threat of failure on that basis is comparatively smaller than taking no action and leaving the fate of the region to what they view as an ineffective American strategy.

Some regional governments now understand that the monster they helped to create is coming to attack them in their homeland. Indeed, this is a growing fear in Saudi Arabia, where in 2015 the IS claimed responsibility for two mosque bombings in the Eastern Province in as many weeks. The IS has specifically targeted the governments of Jordan, Lebanon, and Bahrain, among others, and its ability to strike, seemingly at will, has been impressive.

As reprehensible as IS is to many of the region's governments and citizens, it represents stability in some of the territory it controls, having established its own form of government, collecting taxes and providing government services. While many residents of the cities it controls may abhor the IS, many of them admit that the IS is doing a better job of providing basic services than either the Iraqi or Syrian governments. Given the ongoing political mosaic in the region, it can certainly be argued that some stability and provision of basic services is better than none.

Given all this, why has the U.S. government not leaned harder on the region's governments to cut off the IS's financing? America's ability to influence either regional or domestic policy is probably more limited today than at any time since the end of World War II. The U.S. government wishes to maximize the leverage that it still possesses—in terms of oil purchases and military sales—and does not wish to rock the boat more than necessary, thereby risking further disenfranchisement among the regional powers. In addition, the U.S. wishes to maintain whatever goodwill it still has in the region so that its foreign policy and military objectives can be implemented, to the extent any regional governments wish to do so.

Yet the hands of America—and indeed the whole of the West—are really tied with respect to either applying pressure or stemming the flow of regional funds to the IS. Not only is its ability to influence foreign policy in the region more limited than before, but its ability to impact domestic policy is even more limited. Hard power failed to work in Afghanistan and Iraq, and with respect to stemming the tide of the IS through air strikes, and the West is forced to apply 'soft' pressure in an environment where doing so is not conducive to being effective. Regional governments appear to be even less inclined to view the world through an American or Western lens.

The moment for the U.S. to be able to effectively influence the flow of funds from the Gulf to the IS has essentially passed. The IS funds itself, through a combination of oil sales, tax collection, kidnapping, extortion, cross-border smuggling, and other means. It is precisely because the IS has so many potential sources of revenue that the U.S. and the international community must grapple with the fundamental question of whether it is realistic to completely cut-off external sources of funding for the IS, and whether doing so (if it were even possible) would make a substantial difference in degrading the organization's operational capabilities. Based on the IS's operational model as a multi-headed hydra, and the West's performance in combatting it, the answer to these issues had been "no".[36]

The October 2015 bombing of the Russian Metrojet flight in the Sinai, the November 2015 IS attacks in Beirut and Paris, and the December 2015 IS attack in San Bernadino prompted the West, and indeed the wider world, to change the manner in which they thought about the group, and the tactics it employs to fight it. Until then, each country opposed to the IS was, more or less, pursuing its own independent policy. Following these events, many of the world's major powers finally began to share more intelligence and coordinate a joint approach to eradicating the group. However, the success of these attacks only emboldened followers of the IS, and prompted more young and impressionable individuals from around the world to join the group, with nations struggling to identify who traveled to Iraq and Syria to fight for the group, and who returned home as a potential lone wolf terrorists or part of domestic terror cells.

The Arab Awakening, the risk of extremist Islamic jihadist groups, the propensity of *ancien régimes* in the Middle East and North Africa to cling to power, the birth of the IS, and its ability to attack major powers all coalesced into a cauldron of long-term instability, creating what may very well turn out

---

[36] Adapted from: D. Wagner and A. Stout, "Why There is No Stopping the Funding of the Islamic State," *The Huffington Post*, June 5, 2015, accessed November 25, 2015, http://www.huffingtonpost.com/daniel-wagner/why-there-is-no-stopping-_b_7518012.html.

to be a thirty-year war. Given its dramatic rise to prominence in the space of little more than a year, the viciousness and ferocity of its terror methodology, and its ongoing appeal to disenfranchised and bored youth throughout the world, the IS may be expected to remain a long-term presence in the global landscape.

What the IS Is Teaching the West About Social Media

The IS was initially written off by numerous governments as a disparate group of lunatics with little direction or focus, who relied on a range of illegal activities for their financial and military livelihood. That characterization stands at odds with what the IS has morphed into—an increasingly sophisticated organization that, far from being on the verge of extinction, has grown in terms of numbers, influence, and territory. A significant reason for its expansion is its mastery of messaging through the use of social media.

What started as an adept approach using social media has turned into a phenomenon, as the IS demonstrates its prowess in attracting new recruits through the production of slick videos with a variety of themes. In the process, the IS has become skilled at integrating dichotomies to broaden its appeal—portraying itself, for example, as both "ruthless killers" but also being "just like everyone else". Its cutting-edge approach to creating well-choreographed state-of-the-art videos with sophisticated messaging has made the West's response look sophomoric and lame by comparison.

The IS first targeted potential recruits in the Muslim world by emphasizing its anti-establishment and conservative credentials, which had instant appeal to young men who were poor, disenfranchised, and angry. The steady flow of gruesome decapitation videos drew fighters from across the Muslim world. The IS used brutality and fear as a form of seduction, appealing to a sense of individualism while glamorizing death and martyrdom. For young men living on the margins of society, with no job and little or nothing to lose, joining the IS gave their lives meaning and enabled them to fight their perceived oppressors.

The organization has gradually shifted both the focus of its targeted recruits and its messaging in order to draw in a larger pool of devotees. Later videos changed the focus away from the 'unfamiliar', individualism, and anonymity toward themes that are more instantly alluring to a broader range of people, with a more personal flair. For example, one round of videos showed IS members enjoying themselves at a swimming pool, gathered around a fancy car, eating delicious looking food, and portraying themselves as hedonists.

This is odd, as it stands in such dramatic contrast to the IS's original portrayal of itself as highly conservative and anti-materialistic, but its objective is to attract new recruits. Those recruits are likely to be grossly disappointed when they learn that joining the IS does not in fact translate into sipping champagne by the pool. A number of stories have emerged of recruits finding the IS distasteful and wanting to escape, only to be held prisoner, tortured, or killed. This 'IS lite' approach may end up backfiring.

The mystery surrounding who the IS actually is also is being used by the organization to enhance its perceived appeal. Joining a mysterious organization and not knowing the members' identities is undoubtedly also part of the attraction for some of the recruits—in the quest to be different and do something unusual with unknown people in an unknown place. Part of IS's ability to project an aura of power is to be associated with so many unknown unknowns. The dual appeal of individualism while being part of a larger group gives the IS the ability to manipulate peer pressure, especially among younger recruits, as an incentive to become part of 'the cause'.

As more people from a more diverse range of countries join the ranks of the IS, the group's 'cheerleaders' are increasing exponentially. The western media is, ironically, greatly assisting the cause by dutifully broadcasting IS's videos to its own brethren—ironically, including such outlets as Fox News—generating some unlikely new devotees. This is part of what accounts for so many of the young westerners seeking to join the organization. If the western media did not play along, many of these western recruits may not even know what the IS is. Doing so is like giving an unintended seal of approval for those who do not know better. When will western news organizations stop this practice?

The greatest threat to western countries now comes from home-grown terrorists, either those returning from the IS front lines, or sympathizers who never left their homes. Some of the IS's videos and social media are specifically targeted at these individuals, urging them to wage jihad at home. While governments can monitor their citizen's online presence, they cannot read their minds or control their actions. The 2015 attacks in Paris and Tunis offer two good examples of why this risk is growing, and why all governments should be concerned. As previously noted, it may not be too long before westerners begin to fear going to the malls.

Western governments have taken a page out of the IS playbook and started to produce videos with similar appeal—by directly countering the variety of messages being produced by IS, in a similarly slick and sophisticated manner. As long as the IS has the upper hand in the social media arena, it will not matter how much military might is devoted to their defeat, because the recruits will continue to line up to join them.[37]

Implications for Business

How did Al Qaeda originally become so successful? It was not through the use of high-tech gadgetry or the possession of greater resources than its foes. Rather, Al Qaeda masterfully combined low-tech means of destruction (TNT, ammonium nitrate, and fuel oil) with low-tech communication methods (handwritten notes, couriers, and cash-driven transactions) so as to avoid detection by the West's high-tech gadgetry. More importantly, Al Qaeda and other terrorist organizations' success is due in large part to their ability to tap into the concerns of hundreds of millions of people who either are, or perceive themselves to be, disenfranchised, poor, hungry, left behind, and with little to lose by joining a cause.

Tens of thousands of terrorists now populate most of the world's countries, many of whom have the ability to operate independently with their own objectives, resources, capabilities, and special modes of operation. They have become adept at marshalling personnel and resources from a variety of countries to achieve their attacks, and, most importantly, they are staying a step ahead of law enforcement and intelligence agencies. There are simply too many of them and they have overwhelmed government tracking capabilities.

If governments and businesses are to win the battle against terrorism, they must adopt a completely proactive stance—anticipating likely targets, substantially beefing up security, and establishing countermeasures to prevent future attacks from occurring. While Western governments have, by and large, done a credible job of enhancing their ability to identify terrorists and prevent them from entering countries through airports, they have not, as previously mentioned, done as good a job at creating and implementing effective countermeasures in the host of soft targets. It really is surprising that more attacks have not happened in the West and the U.S. since 9/11. Undoubtedly, part of that is due to the intelligence agencies being very good at what they do, but part of it also is that the terrorists have fumbled more than once (the Shoe Bomber and the Underwear Bomber being prime examples).

---

[37] Adapted from D. Wagner, "What the Islamic State is Teaching the West about Social Media," *The Huffington Post*, March 23, 2015, accessed November 25, 2015, http://www.huffingtonpost.com/daniel-wagner/what-the-islamic-state-is-teaching-the-west-about-social-media_b_6918384.html.

What is to be done? We have little choice now but to tighten the vise on terrorists more firmly. Border controls and identity checks must be stepped up and intelligence sharing across borders must become easier and more comprehensive. Following the Paris attacks in 2015, the EU has started to tighten up its 'Open Borders' policy—something neither European policymakers nor Europeans had anticipated when the EU was created.

Unfortunately, countries that are unaccustomed to implementing strict security precautions on their citizenry will need to begin to do so. Since it appears likely that the days of being able to casually walk into a shopping center without undergoing a security check are numbered, it would be sensible for business leaders to engage immediately in a discussion with governments about how this is going to be achieved. It is not just governments that have an obligation to stay one step ahead of the terrorists, it is business as well. The sooner this is done, the sooner the general public will be conditioned to accept the inevitable—we face a future when security will permeate many aspects of our daily routines.

Terrorism has affected—and will continue to affect—how business operates, and how much it costs to operate. It has become necessary for businesses to begin to bear the cost of installing security systems and personnel to safeguard their customers and employees on a widespread basis. Governments cannot possibly bear this cost alone. Governments and businesses must reach an understanding about how the problem of providing meaningful security is going to be addressed in the future.

A coordinated, purposeful, definitive approach to creating a more secure environment must be achieved if the West is to stay one step ahead of the terrorists. It is surely in the mutual interest of business and government that this should occur sooner rather than later. Businesses may as well embrace the reality that providing funding to achieve meaningful security to protect their physical operations, employees, and customers is not only necessary, but also a smart investment. All one needs do is consider the enormous costs implied in enduring political violence, terrorism, or business interruption losses—and the heightened insurance premiums that result from them—to realize that preventive action is prudent, sensible and cost-effective in the long term.[38]

---

[38] Adapted from D. Wagner, *The Implications of Recurring Terrorism for Business*, International Risk Management Institute, May 2004, accessed November 25, 2015, http://www.irmi.com/articles/expert-commentary/the-implications-of-recurring-terrorism-for-business.

7

Economic and Resource Nationalism

Economic Nationalism's Rise

Expropriation of foreign-owned assets—ruled passé in the 1990s—became *en vogue* in the first decade of the twenty-first century, as the race to control national energy supplies and gain market share prompted an increasing number of governments to nationalize or renationalize strategic assets. The rise in economic nationalism in such countries as Argentina and Russia was good evidence of this, and has resulted in negative consequences for many international businesses, which succumb to host nations forcibly taking extraordinary stakes in their business.

These and other governments have demonstrated an increasing propensity to renege on contracts, which international law gives them the right to do, as long as compensation is fair, prompt, and adequate. As the Spanish company Repsol learned in 2014, when its ownership stake in Argentine oil company Yacimientos Petroliferos Fiscales (YPF) was expropriated, a decade of profitable operation is no guarantee of future success, and it can take years to reach an accommodation on compensation. In the meantime companies are left to wonder when or if they will ever be paid, while bearing the cost of expensive negotiations and laboring under the burden of a cessation of revenue generating activities.

If one or two governments were taking such actions, they would be labeled 'rogue' and simply avoided in the future by investors. But the rise in economic nationalism is occurring in tandem with the institutionalization of 'democratic' elections in many developing countries, lending legitimacy to

© The Author(s) 2016
D. Wagner, D. Disparte, *Global Risk Agility and Decision Making*,
DOI 10.1057/978-1-349-94860-4_7

government actions. Bolivia and Venezuela—both 'democracies'—are perfect examples. Presidents Chavez and Morales enforced multiple expropriatory actions of foreign-owned businesses in the natural resource sector in the name of 'the people' when they first came office. The spread of democracy in the new millennium coincided with the rise of extreme political movements on the Left and Right, with particularly negative consequences for international businesses in strategic sectors of a given economy. Companies operating in mining, power, petrochemicals, telecoms and high technology are particularly prized by governments inclined toward expropriatory activity, for their perceived strategic value.

This is occurring at the same time that the 'devolution' process is taking hold in an increasing number of developing countries. As central governments grant greater political and economic rights to provincial governments, the rules under which contracts were originally negotiated may change, leaving many businesses with little option but to forcibly renegotiate or unilaterally accept the changes imposed on them. This is being seen in the coal industry in Indonesia, for example, with greater frequency. The quest for political equality on a local level has thus come to transcend borders, with national and international implications.

In some cases, regional autonomy has increased to such a degree that would-be investors must contemplate the possibility that commencing operation in a single country today may ultimately become two countries at some point in the future. Two good examples of this are Nigeria, which is threatened by ethnic and religious strife between its northern and southern regions, and Iraq, which has been in the process of breaking apart since 2003. The forces unleashed by the Arab Awakening will continue to threaten the sanctity of existing borders throughout the Middle East and beyond.

For foreign-owned businesses operating in strategic sectors of developing economies, the implications are clear. Economic nationalism, fluctuating commodity prices, global competition, extremist political movements, and the propensity to expropriate foreign-owned assets all add up to the potential for an even more challenging international investment climate in the future. Yet market saturation in the developed world makes the desire to invest abroad—even in risk-prone countries and sectors—inevitable.

Yet the risks associated with doing so exist for companies headquartered in developed as well as developing countries. With the rise of emerging market titans such as Petrobras, Cemex, Gazprom, and PetroChina, international companies of all sizes and degrees of sophistication must become smarter about investing overseas, especially given the rise in South/South and East/West investment. In some respects, emerging market multinationals are better equipped to deal with the multitude of challenges associated with operating in developing countries, given their experience in maneuvering through frequently gyrating economic movements, shifting political sands, and legendary corruption. Their challenge is to understand that operating abroad will, by definition, involve new and different norms and standards, with different rules for achieving success.

To maintain their footing, managers must place renewed emphasis on strategic planning and forward-looking risk management—at *all* phases of the trade and investment process. Being 'reactive' is no longer sufficient. Corporate managers must consider how to reevaluate existing activities and analyze new opportunities in light of the rapidly changing global investment climate. The best way to address these risks is to establish risk management procedures that ask the right questions and establish effective methods for managing risk—before it becomes an issue. The ability to conduct realistic and effective scenario planning and stress testing are essential for any international business.

The propensity of natural resource-rich nations to lash out at foreign investors is likely to grow. The world is only going to become a riskier and more complex place in which to do business in the medium and long term, and there is plenty of scope for trouble. What follows are some examples of where economic and resource nationalism have played pivotal roles in how countries relate to each other, and to the companies that invest there. The key message is that economic and resource nationalism exist everywhere and can have a profound impact on a political, economic, financial, operational, legal, regulatory and judicial climate.[1]

Bolivia's Indigenous President

Bolivian President Morales's decision to seize the assets of foreign-owned oil and gas operations, and to nationalize large privately held property holdings in his first year as president (2006), serves as a useful reminder of the political risks associated with owning and operating businesses abroad. Morales's action should have been no surprise, given his indigenous roots, the platform on which he was elected, and that numerous countries around the world—and particularly in Latin America—have a history of nationalizing

---

[1] Adapted from D. Disparte and D. Wagner, *Economic Nationalism's Impact on International Business*, International Risk Management Institute, September 2012, accessed November 25, 2015, http://www.irmi.com/articles/expert-commentary/economic-nationalisms-impact-on-international-business.

natural resources, extractive industries, and large land holdings. Bolivia is no exception.

Bolivia first expropriated foreign-owned assets in 1937 (taking over the Bolivian resources of Standard Oil of New Jersey—ExxonMobil's corporate ancestor) and has seized control of oil and gas assets on numerous occasions since that time. International law permits such actions, as long as investors are given fair, adequate and prompt compensation. Contrary to international law, Morales had ruled out compensation to foreign investors, claiming companies such as British Gas, Petrobras of Brazil, Repsol of Spain and Total of France had recovered their initial investment and made a handsome profit for many years while operating in the country.

Morales subsequently nationalized land holdings of wealthy farmers who held large unutilized tracts of land that were in some cases acquired by force or illegal means. He reclaimed illegally obtained parcels and redistributed them to Bolivia's poor. Since the 1990s, hundreds of Brazilian farmers have invested a total of more than $1 billion to purchase land in Santa Cruz, an eastern region of Bolivia that borders Brazil and has some of the most fertile land in South America. At the time, those farmers grew more than a third of Bolivia's soybeans, which accounted for almost 7 percent of the national economy.

Morales's actions were particularly noteworthy because they came at a time when many in global business thought nationalizations were a legacy of the Cold War. But it also came at a time when economic nationalism was on the rise, and states had demonstrated a growing propensity to become increasingly engaged in business. This has now become a global phenomenon and evidence of the value placed on securing long-term energy and natural resources when commodity prices can fluctuate wildly and energy resources may be more or less plentiful. Hence, can it be any surprise then that the interests of business are taking a backseat to national interests?

The legacies of history are often at the heart of an issue, and history teaches us a lot about the present and the future. In the case of Bolivia, since the nineteenth century it has fought with its neighbors over land and what lies beneath it. Approximately half of the land Bolivia once held, which at one time extended to the Pacific coastline, is gone:

- In 1884 a Chilean attack on Bolivia's Litoral Province cost Bolivia its coastline (the war was ultimately fought over the ability to export dried bird dung, prized at the time for making fertilizer and saltpeter);
- In 1903 Brazil persuaded the large state of Acre to secede. Brazil thus gained a highly productive rubber-growing state, at Bolivia's expense;

- A three-year war with Paraguay, which ended in 1935, was motivated by Bolivia's desire to secure a disputed border in which it hoped to find oil (it wound up losing a region the size of Utah to Paraguay in the process); and
- Argentina and Peru secured additional portions of Bolivia through diplomatic demarcations later in the twentieth century.

It is the cumulative history of land loss, the resulting national humiliation, and Bolivia's desire to secure and control its natural resources that was ultimately behind Morales's actions. Resentment over the war with Chile had stopped plans for a gas pipeline to the Chilean coast, even though financial backing and export markets were already in place.

In the 1980s Bolivia was one of the first Latin American countries to adopt IMF structural adjustment programs, which tied loans to privatization, debt reduction, and a relaxation of labor standards. State-owned companies were sold, government spending and regulation was reduced, and foreign capital was courted with the promise of a new beginning. But twenty years later, the average Bolivian was worse off than before, exports had declined, incomes were stagnant, and half the population was living on less than $2 per day. Bolivia's experience was similar to that of other Latin American countries (such as Brazil, Chile, Uruguay and Venezuela), whose cumulative disillusionment had led to the revival of leftist/populist governments.

It would be easy to dismiss Morales's action as simply the result of his friendship with Hugo Chavez and Fidel Castro. Chavez had, after all, made similar noises with respect to foreign-owned energy businesses and routinely rattled his saber against the U.S., as Fidel Castro has done for decades. But Morales was responding to the long-simmering desire among Bolivians for increased control over their natural resources. This desire was as strong in Bolivia as it was in Venezuela and many other countries that had opened their doors to foreign investment, even though they may owed much of the economic strength at the time in their natural resource industries to these investors.

While nationalization is a short-term 'fix' for national economic aspirations, its longer-term consequences generally prove to be damaging, since foreign businesses often hesitate to invest in countries that demonstrate a propensity to nationalize foreign-owned assets. Brazil, Chile, and Uruguay have achieved what Morales had failed to achieve, having managed to combine fiscal prudence with open trade and investment policies, preserving their options with respect to foreign traders and investors while addressing broader national social needs. By his actions Morales was not only closing the door to foreign investors, but was also violating international law (by not providing fair, adequate, and prompt compensation for his expropriations) in the process.

Another reason why Morales's actions were important is that they came after nearly the whole of Latin America had embraced democracy. Latin American democracy produced not only champions of political legacy busting—as in the case of Mexico's Vicente Fox, who succeeded in finally breaking the PRI party's 71-year uninterrupted rule—but also individuals such as Morales and Chavez, who used a combination of indigenous entitlement, nationalistic sentiment and demagoguery to fire up their political constituencies. Democracy does not always produce the results other countries would like, but to the extent it has enabled the will of the people to be expressed, it has served an important purpose in the political history of all these countries.

The larger lesson to be derived from Morales's actions is that despite the era of globalization, in which all countries are presumed to play by the same rules, and national economies are inextricably linked with the global economy, *human behavior and aspirations remain unchanged*. Citizens of every country want to be able to say they have control over their own destiny and natural resources. The rise of democracy in Latin America has accentuated this desire, for it raises pressure on leaders to ensure that national economic aspirations are achieved. Businesses that embrace globalization by investing abroad would be wise to remember this. A democratic election is no guarantee of friendly foreign investment policies, and may in fact work against it.[2]

Argentina's History Lesson

Created in 1922, Argentine oil producer YPF became the world's first entirely state-run oil company and it came to symbolize the politicization of oil well before today's Middle Eastern oil producers became omnipotent. The nationalization of oil has been a consistent theme in Argentina since the state first established a monopoly on oil exploration in 1928. President Cristina Kirchner's partial renationalization of YPF in 2012 continued a long tradition of gyration between pro- and anti-foreign investment sentiment in the country's oil sector.

Kirchner had consolidated her power by being reelected in a landslide vote the previous year, giving her a mandate to continue the 'Argentina first'

---

[2] Adapted from D. Wagner, *Bolivia's Larger Message*, International Risk Management Institute, June 2006, accessed November 25, 2015, https://edit.irmi.com/articles/expert-commentary/bolivia's-larger-message.

policies she had pursued during her first term. YPF's expropriation was the largest of its kind in the natural resource sector since the Russian government expropriated Yukos in 2003, and a reminder—as if any were needed—that acts of expropriation can occur without notice, even to the largest of multinational corporations.

The Spanish oil major Repsol had purchased a majority stake in YPF in 1999. Surely, Repsol would have been aware of Argentina's history of expropriating foreign investment in YPF over the course of the past century. Given how entrenched oil politics have become in Argentine political culture—and given how deftly both Kirchner and her husband (her predecessor as president) played the nationalism card—only a naïve investor would agree to invest in YPF believing that expropriation was not possible.

Since 2002, there has been a considerable increase in expropriations in the worldwide natural resource sector.[3] At the time of YPF's renationalization in 2012, according to the International Center for the Settlement of Investment Disputes, of the 143 pending cases of investment disputes awaiting an outcome, an astonishing 49 percent (70 cases) were in Central and South America. And the country with the most number of cases lodged against it was Argentina—it had even more cases than Venezuela. The 'bad boys' of Latin America—Bolivia, Ecuador, Nicaragua and Venezuela—had all either said they will no longer resolve disputes through international forums of arbitration, or had already pulled out of them.

Argentina had enjoyed an impressive period of growth since the resolution of the Argentine Crisis a decade prior. The country had more than tripled its GDP since 2002,[4] which no doubt gave Kirchner the confidence to take such a bold action. But history has shown that those countries that go down the path of expropriation end up paying the price in terms of reduced FDI. This has certainly been the experience of Bolivia, Ecuador, Nicaragua and Venezuela. FDI in Argentina declined from a high of $1.4 billion in the first quarter of 2012 to a low of under $300 million in the second quarter of 2014, and had only started what may be referred to as a 'steady' recovery by the second quarter of 2015.[5]

---

[3] C. Hajzler, "Expropriation of Foreign Direct Investments: Sectoral Patterns from 1993 to 2006," Department of Economics, University of Otago, Economic Discussion Papers, No. 1011 (September 2010), accessed November 25, 2015, http://www.researchgate.net/publication/225719365_Expropriation_of_Foreign_Direct_Investments_Sectoral_Patterns_From_1993_to_2006.

[4] GDP Data, Google, 2015, accessed November 25, 2015, http://www.google.ca/publicdata/explore?ds=d5bncppjof8f9_&met_y=ny_gdp_mktp_cd&idim=country:ARG&dl=en&hl=en&q=argentina+gdp.

[5] Argentina Foreign Direct Investment 2003–2015, Tradingeconomics.com, 2015, accessed November 25, 2015, http://www.tradingeconomics.com/argentina/foreign-direct-investment.

In 2014 Repsol ended up settling with the government of Argentina for $5 billion in reparations—just half of the figure it had sought. The lesson for Repsol—and indeed for all international investors—is that history cannot be erased or ignored, for it is an important teacher. Given history, it was just a question of time until the Argentine government would turn against Repsol. In the end, it didn't matter how supportive Spain had been of Argentina through its difficult times, or how good the company's relations had been with the government. When the time is right, history will recur and reassert itself in the present-day political and economic dynamics of a country. Given Argentina's example, foreign investors would be wise to remember this.[6]

A New Era for Papua New Guinea?

In 2009 a consortium led by Exxon Mobil approved a massive and logistically challenging liquefied natural gas (LNG) project in Papua New Guinea (PNG). The project (PNG LNG) represented a significant vote of confidence in the suitability of PNG's business environment for FDI, and a variety of governments backed the project with export credits. Given its size, scope, and foreign government support, the project became a litmus test for future investment in the country.

PNG LNG's $19 billion initial phase expenditure was, by a considerable margin, the largest single foreign investment in the country. The project is set to produce an estimated $35 billion in revenue over its economic life, may double the country's gross domestic product and triple its export earnings. On completion in 2014, PNG LNG's production capacity was expected to reach 6.9 million tons of liquid fuel per year, generating more than $100 billion in sales over a period of twenty years. By comparison, the Ok Tedi mine, the country's largest in terms of export earnings, earned approximately $1.5 billion in 2007, representing 32 percent of total exports and 22.9 percent of GDP. Exxon concluded sales agreements with China, Japan, and Taiwan, and believes the PNG LNG will help meet increased demand, estimated to triple globally by 2030.

PNG's gas fields are considered the best underdeveloped reserves outside Qatar—dwarfing the potential of its oil fields—with earning potential exceeding that of PNG's already substantial gold and copper earnings. Given

---

[6] Adapted from D. Wagner, "Argentina's Expropriation and the Lessons of History," *The Huffington Post*, April 17, 2012, accessed November 25, 2015, http://www.huffingtonpost.com/daniel-wagner/argentinas-expropriation-_b_1431288.html.

the lack of indigenous extraction and processing facilities, the sector possesses tremendous potential for additional foreign investment. Many prospective natural resource investors in the country are wondering whether a project of this scale and visibility will help PNG transform itself into a more developed country with a more responsible government, or whether it will simply encourage more of the same type of behavior investors have come to expect of PNG over the past several decades. Unfortunately, the latter seems more likely.

It Takes Two to (Con)Tango

While LNG has the potential to create a windfall for the country, it is also a test. PNG's economy has long been heavily dependent on the proceeds from resource extraction. The government will likely realize a total $5.6–7.5 billion revenue stream from PNG LNG—more than from any existing single revenue source. However, the country has a long history of government interference with FDI, and of squandering the gains thereof. The government has not always competently implemented or managed natural resource projects in the past. Such mismanagement has proven to lead to significant downside risks for investors.

The primary risk facing investors is not the outright expropriation of foreign investments, but rather 'blowback' from local stakeholders regarding the manner in which resources are developed and how proceeds are allocated, causing significant operational disruption. By their nature, large-scale extractive projects impact the environment and result in hearty revenue streams whose distribution tends to be politicized. Environmental degradation and a perception that proceeds have been distributed inequitably has led to violence and investor loss numerous times in the past.

The most infamous episode of political violence by indigenous groups against FDI is Bougainville. The Panguna copper mine, operated by a subsidiary of Rio Tinto, opened on the island in 1969. Bougainville leaders claimed mining activity was causing environmental degradation, while the local population received neither compensation nor a share of the mine's revenue. The local populace perceived the government's policy as promoting colonial exploitation. In 1989, guerillas began a campaign of sabotage against the Panguna mine and civil war erupted. More than 20,000 people were subsequently killed. The mine never reopened. Although Bougainville eventually became autonomous in a 2001 settlement, it has become the best known and most commonly referred to example of what *not* to do as a natural resource sponsor and responsible host government.

Other projects faced similar problems. The Ok Tedi copper mine and Porgera gold mine—two of the largest in the country—faced disruption from both local stakeholders and international groups resulting from land rights disputes or environmental issues. Ok Tedi is widely believed to have created an environmental disaster by releasing mine tailings into the Fly River system. BHP Billiton, the mine's operator, transferred its 52 percent equity share in the mine to an offshore trust for PNG's people in 2002, and paid a $26.8 million settlement to local groups. Barrick Gold's operation of the Porgera mine has attracted international attention: the Norwegian government disinvested $230 million from the company due to "irreversible environmental damage" related to mining operations. In January 2010, Amnesty International criticized Barrick for complicity in the government's illegal destruction of homes and forced eviction of villagers near the mine.

In response, a number of foreign investors have adopted internationally recognized best practices to minimize adverse environmental impacts and foster a spirit of cooperation with indigenous communities. Chevron and Oil Search's management of the Kutubu oil field—the largest in the country—provides a good example: a World Wildlife Fund observer described the operation as the most rigorously controlled national park in Papua New Guinea. Newcrest Mining's Lihir Gold mine (originally developed and owned by Rio Tinto) was created with heightened sensitivity to local customs and a desire to create the perception of legitimacy and fair play among the indigenous population. Other firms have tried to address the concerns of all stakeholders from the planning stage forward. By building relationships with local leaders outside government, such project sponsors have gained an understanding of local concerns, built trust, and provided a conduit through which concerns can be negotiated and addressed before they can erupt into violence.

Instead of relying on the government to ensure local economic development, many companies now commonly include the construction of schools, roads, and hospitals in project budgets. One such example is Harmony and Newcrest Mining's Hidden Valley mine. After consultation with landowners, the group facilitated the creation of a dedicated business development vehicle owned by local landowner communities (NKW Holdings). NKW supplies a range of goods and services to the Hidden Valley mine, thereby providing immediate local economic benefits. In addition, Harmony is assisting NKW to establish long-term sustainable businesses beyond the life of the mine, including agriculture, tourism, and food processing. Not only are such efforts laudable from a moral standpoint, they also benefit the bottom line by minimizing conflict and disruption.

Systemic Corruption Creates Risk

Following a boom in investment in the 1990s from the Porgera and Kubutu operations, the government borrowed heavily and spent inefficiently. The resulting economic calamity resulted in severed ties with international donors until 1999, and erased all of the country's growth in per capita income since its independence in 1975. At a regional level, the Southern and Western Highlands—the two richest provinces and the location of the Kutubu fields— witnessed gross corruption and fiscal mismanagement. While revenue from the project flowed, previously sound service provision collapsed, and violence increased as politicians treated the revenues as personal slush funds.

Corruption is strongly correlated with unsound governance. In 2014, Transparency International's Corruption Perceptions Index, a measure of corruption, ranked PNG 145th of 175 surveyed countries, lower than Nigeria, and on a par with the rankings of Bangladesh, Kenya, and Laos. Despite a modern constitution, range of laws, and institutions designed to attract FDI, much of the government continues to perform poorly due to years of cronyism, underinvestment in basic services, and graft.

The national government transfers revenue to landowners in a variety of ways, including direct payments, the allocation of funds earmarked for infrastructure to provincial governments, and the partial local landownership of a given project. Each is vulnerable to the diversion of funds at the national, provincial, and local levels, and such diversion is common. When landowners feel exploited, there is a real potential for project-disrupting violence, which foreign investors in PNG will say is often their top concern.

Lessons Unlearned

The government managed to improve its performance at the national level in the first decade of the twenty-first century, investing much of its windfall revenue from the 2005–08 commodity price boom in trust funds with the support and encouragement of multilateral development banks. However, that did not prevent the government from raiding the windfall funds for social and infrastructure spending in 2009 and 2010. More recent spending programs—such as the District Services Improvement Program—were criticized for the same problems that plagued regional spending in the 1990s: a lack of transparency and oversight. At the beginning of the PNG LNG project, oversight of the government's investment in the project was concentrated in the hands of the Public Enterprises Minister, who was the prime minister's son—a move the opposition described as immoral and irresponsible. At provincial and local levels, government corruption and mismanagement continues to run rampant.

As scrupulous and thoughtful as ExxonMobil has publically claimed to be with respect to engaging the indigenous population and adhering to internationally accepted environmental guidelines, PNG LNG encountered many of the familiar historical problems. Landowners as a group sought a larger equity share in the project. Others claim to have been unfairly frozen out of benefits or not to have seen those that were promised. Unhappy landowners were blamed for disrupting the construction of an international airport at Komo to service the project site. Among other incidents, tribal disputes related to PNG LNG contracts or funds resulted in multiple fatalities in the Hides district and Central province, areas where extraction and processing had been planned.

PNG LNG is evidence that improved performance by the PNG government vis-à-vis implementing and maintaining meaningful FDI will require changes to existing laws and stronger oversight mechanisms. The current system of revenue sharing from FDI needs reform. Former Prime Minister Julius Chan tried for many years to get PNG's legislature to approve substantial changes to the 1992 Mining Act, which largely governs the current revenue distribution structure. Chan proposed to transfer revenue from FDI directly to an investment vehicle owned by landowners, who would then release a portion to the national government for discretionary spending, essentially reversing the current arrangement. As of 2015, the changes had yet to be made.

Politics in PNG has been dominated by the so-called "big men", who have run a succession of nepotistic and corrupt governments since its independence in 1975. The odds of PNG LNG ushering in a new era of government accountability and sustainable economic development are therefore slim in the near or even the medium-term. The onus is on the government to transform itself; outsiders are neither capable of doing so, nor responsible for doing so. Most likely, PNG LNG will have a neutral or somewhat negative effect on the political environment, if its revenue is used to entrench corrupt politicians, as can be expected. If that is the case, then the same 'natural resource' curse that has bedeviled previously poor/newly rich developing countries will continue to do so with PNG.

Given PNG's natural bounty of gas and mineral resources, it remains a lucrative investment opportunity despite the downside risk from the political environment. Gas and minerals will continue to find their way to market. The Exxon Mobil consortium and future investors will apply a range of best practices to minimize risk, working around the government as much as through it. It is regrettable that after all these years the government of PNG may continue to squander the considerable wealth being generated from the country's natural resource operations. Fortunately, most foreign natural resource sponsors have received and understood the message delivered repeatedly to them by PNG's people have the ability to deliver what the PNG government cannot and will not—economic development—and have taken matters into their own hands, providing infrastructure and housing needs, and education and medical benefits to mine workers and their families.[7]

Pakistan's Message to Foreign Investors

In 2008 Pakistan de facto expropriated of one of the country's largest-ever foreign investments—the Reqo Diq mine in Baluchistan—which sent a shudder through the foreign investment community. The government claims that the deal, which was originally agreed with the provincial government and BHP in 1993—was unfair, and not in the country's long-term interests. In so doing, Pakistan sent foreign investors a mixed message with respect to the sanctity of contracts and the unpredictability of government actions. At the same time, there is an important lesson to be learned for foreign investors everywhere: *cutting a deal that is perceived as fair by all parties will enhance the chance that a project will proceed, and be successful in the long term.*

A Bad Deal

In 2010 the government of Pakistan threatened to cancel the $3 billion Reko Diq copper and gold project led by Canadian investor Barrick Gold and the Chilean mining company Antofagasta in the country's resource-rich Baluchistan province, citing the need to protect is strategic national interests. With an economic life of 25–30 years, the project was expected to earn $240 to $260 billion[8] from the extraction of raw copper and gold, making it one of the largest mines of its kind in the world. It was therefore not a surprise that the government was interested in maximizing its potential long-term monetary benefits from the project.

---

[7] Adapted from S. Goldsmith and D. Wagner, *A New Era for PNG*, Project Finance International, May 19, 2010.

[8] S. S. Hassan, "Analysis: Reko Diq's Billion Dollar Mystery," *Dawn*, January 23, 2015, accessed November 25, 2015, http://www.dawn.com/news/1158808.

Given that the mine is located near the Afghan border, in a province that has been at odds with the central government over revenue sharing from natural resource projects, and the base of a nationalist insurgency for decades, the original investors were given some incentives to proceed with exploration and feasibility studies—and were in return given a lucrative deal. The government came to believe that the deal was too lucrative, and that some of the terms agreed upon were unfavorable to Baluchistan and the central government. Baluchistan was required to provide 25 percent of the project funding in order to get a 25 percent return—a tall order for a province with limited financial resources. Baluchistan decided it wanted 80 percent of the proceeds.

The project was controversial from the beginning. Proponents claimed it would generate much-needed jobs and, if successful, would act as a magnet for future investment in the province and the mining sector. They argued that proceeds from the project should help fund infrastructure development in the area. Opponents cited concerns about by-product pollution from mining, security issues, whether revenues would actually lead to infrastructural development, and whether corruption among local government officials would actually be exacerbated as a result of the project.

These were all legitimate issues, but the bottom line is that Baluchistan stood to gain from Reqo Diq, as did the central government. Without the seed capital to fund exploration and produce a feasibility study from private investors, the project would never have gotten off the ground. Having accomplished that, Pakistan jeopardized the 15 years it spent to get to the pre-investment stage by playing the nationalism card, while FDI fell from more than $5 billion per year in 2007 to negative numbers since 2010,[9] and it currency, the rupee, fell 75 percent between 2005 and 2015.[10]

Extractive Enterprises Are Particularly Vulnerable

Extractive industries are generally more vulnerable to adverse action on the part of host governments for several reasons:

- They tend to be large and high-profile;
- They employ thousands of people and have a significant impact on local communities;

---

[9] *Foreign Direct Investment—Net (BoP—US dollar) in Pakistan,* Tradingeconomics.com, 2015, accessed November 25, 2015, http://www.tradingeconomics.com/pakistan/foreign-direct-investment-net-bop-us-dollar-wb-data.html.

[10] Bank of Canada, *Foreign Exchange Rate: US Dollar and Pakistani Rupee*, 10-year Currency Converter, 2015, accessed November 25, 2015, http://www.bankofcanada.ca/rates/exchange/10-year-converter/.

- They are strategically important to host countries;
- They are subject to a wide range of legal regimes and laws, which change frequently; and
- They involve the production of waste products, which makes them more highly scrutinized for environmental compliance than other forms of investment.

Mining projects, in particular, have been the object of numerous instances of expropriatory-type actions on the part of host governments since the 1970s. Interference by local or national authorities and the revoking of mining or export licenses are the most common ways in which governments indirectly expropriate mining investments. Although rare, outright acts of expropriation on the part of host governments do still happen, as noted above.

In addition, mining projects are prone to sustainable development-related operational complications, largely as a result of the impact they have on local environments and populations. Many mining companies have made great progress in taking the initiative to avoid conflict with NGOs and indigenous peoples by taking care to be inclusive in the planning, construction, and operation of mines. Many recent investors in the sector (as noted in the PNG example) have taken care to establish strong relationships with tribal leaders and negotiate agreements that give all participants a sense of fair play.

Lessons Learned

That Pakistan made such a bold move with Reqo Diq, when it was facing one of the worst political and economic crises in its history, implies that governments can take action contrary to investors' interest at any time. Pakistan has demonstrated that it will not hesitate to look after its own interests, for its own benefit. International investment law states that governments have a right to seize foreign investments in the national interest, providing that investors are given fair, equitable, and timely reimbursement. Even if a government does not have the financial means to provide such reimbursement, it can still take unilateral action, leaving investors to seek legal recourse through arbitration or other means. This is indeed one of the risks of engaging in cross-border investment.

One of the biggest mistakes international investors make is to focus on their own future net income generation, rather than on what makes sense from all parties' perspective. Experienced investors in natural resource projects know that when they agree to tariff-, tax-, or revenue-sharing arrangements that give local participants and host governments a real sense of fairness, their projects generally proceed with minimal conflict. This minimizes operational complications and usually ends up contributing handsomely to the bottom line. Yet, in the negotiation process, too many companies make the mistake of being too focused on their own well-being, and not enough on what benefits local communities and the long-term health of provincial and host governments.

The message in the Reqo Diq case is that a cash-strapped government with limited sources of revenue will naturally seek to protect its long-term interests by either seizing high-profile foreign exchange-generating projects, or renegotiating them. Barrick and Antofagasta would have been wise to consider the possibility of a change in Baluchistan's posture over the long term and negotiate a deal that gave the province a better sense of genuine benefit and fair play. Many other mining companies have learned the same lesson the hard way.[11]

The Impact of Energy Resources on Bilateral Relations

China and Myanmar have had a mixed post-war relationship, ranging from warm to hostile, with Burma being the first non-communist country to recognize China in 1949, before expelling much of its Chinese population following anti-Chinese riots in the 1960s. Since the repression of pro-democracy riots in 1988, Myanmar has generally sought a closer relationship with China, as the government wished to strengthen itself in the process. Today, the relationship is characterized by strong bilateral trade and investment, but also a similar sensitivity to that observed in other Southeast Asian nations to China's growing strength and influence in the region.

Myanmar's Strategic Energy Play

The acquisition of energy has become a dominant influence in China's foreign policy orientation more generally, and has been a driving force in its relationship with Myanmar in recent years. While Myanmar is resource-rich, it does not have particularly noteworthy hydrocarbon reserves, but, because of its geostrategic position, has punched above its weight in perceived significance

---

[11] Adapted from D. Wagner, *Pakistan's Message to Foreign Investors*, International Risk Management Institute, February 12, 2010, accessed November 25, 2015, https://www.irmi.com/articles/expert-commentary/expropriation-pakistan's-message-to-foreign-investors.

to China. China views Myanmar as important to its own strategic objective of having access to more ports in the Andaman Sea and the Indian Ocean.

The Shwe Gas Pipeline (otherwise known as the Myanmar–China Oil and Gas Pipeline), which commenced operations in 2013, certainly symbolizes China's willingness to invest in Myanmar in order to achieve its energy acquisition objectives. Although built by a consortium that includes India and South Korea, the pipeline is an important component of China's ability to deliver oil and gas to southwest China.

China took a big step further with the inauguration in 2015 of a 478-mile crude oil pipeline the runs the length of Myanmar and transports oil from the Middle East and Africa to southwest China. China's CNPC (China National Petroleum Corporation) owns 50.9 percent of the venture, which included the construction of a new deep-water port and oil storage facilities on Myanmar's Maday Island. The completion of the port was also a long-held strategic ambition of China in terms of its ability to project its military power in the region. It is fair to say, therefore, that bilateral economic relations has never been better between the two countries, and that Myanmar's economic significance for China has never been higher.

You Can't Always Get What You Want

It didn't get that way without a few bumps in the road, however. The government of Myanmar halted the construction of the Myitsone Dam in the country's north, a large hydroelectric project that was under construction along the Irawaddy River and was to be completed in 2017. It was to generate up to 6,000 megawatts of power for Yunnan Province in China. The project had been highly controversial, largely because of the environmental damage it would cause and the thousands of people who would have needed to be displaced to build the dam.

A broad range of interests and organizations opposed construction of the dam, which was being built by the China Power Investment Corporation—one of China's largest power producers—and Sinohydro—one of the world's largest hydropower contractors. In 2011, Myanmar's then president, Thein Sein, succumbed to pressure and canceled the project, to the great consternation of the Chinese government, and to the great delight of environmentalists.

Cancellation of the dam raised a whole host of issues which China had never previously had to address in public. It is extremely rare for a developing country government with a long history of friendly relations with China, and seeking its investment, to publicly challenge the Chinese government in such a manner. Even more interestingly, it was one of the first instances when major Chinese government-owned companies had been forced to deal with issues related to contract cancellation and de facto expropriation of Chinese assets related to its FDI.

As a result, the Chinese government learned that it ultimately has no more power than any other government when contracts are canceled in another country—even Myanmar. While China sought to portray the project in a more favorable light, local environmental activists remained adamantly opposed to the project, and allegations of fraud, corruption, and lack of transparency remained rife. A rather thought-provoking dichotomy—the dam and the pipeline—since it may certainly be presumed that many of the objections Myanmar had to the dam may surely also be said of the pipeline, but are not.

The Myanmar people raised objections to the manner in which a variety of Chinese-sponsored and -funded projects have evolved. Although there are more than two dozen mega-dams either in place or planned in Myanmar, 90+ percent of the electricity generated by them ends up being exported either to China or Thailand, meaning that the benefits for Myanmar's citizens are 'limited'. Power shortages are still common in Myanmar as a result. How will this simmering enmity manifest itself in terms of domestic politics?

The Lady and the Dragon

Myanmar's growing economic interdependence with China would appear to be consistent with its backtracking on 'democratization', which had been evolving since 2010, when the period of military rule was succeeded by a military-backed civilian government. Then opposition leader Aung Sung Suu Kyi (otherwise known as "The Lady") distanced herself from the ongoing love-fest with China, which saved her from inevitable domestic criticism given the anti-Chinese sentiment. The Lady had in the past experienced a backlash when she appeared to be too close to Chinese interests. How she plays her China card may ultimately prove to be important in how her evolving political power unfolds.

In 2012 a violent crackdown was initiated by police on peaceful protestors at the Letpadaung copper mine in northwestern Myanmar. The mine was at the time a joint-venture between the military-owned Myanmar Economic Holdings and Wanabo Mining, an affiliate of China North Industries Corporation. The project was marred by accusations of land expropriation and environmental damage, and was temporarily shut down.

Suu Kyi was appointed to head a parliamentary commission tasked by President Thein Sein to investigate the crackdown and whether the project should continue to operate. In March 2013 the commission issued a report recommending that the government permit the mine to resume operations. The backlash was swift, in part because the entire incident was seen as a test of Myanmar's will to stand up to China, but also because of allegations it had failed to meet requirements for transparency vis-à-vis its compliance with environment and health standards. Anger among opponents of the mine continued to simmer, fueling the notion that Myanmar has served as a pawn for Beijing, taking a back seat to its economic interests. Establishing stronger ties would probably have been beneficial for the NLD's longer-term future, but could have put Suu Kyi at risk of appearing too close to Beijing. The last thing she would want is to end up being compared to the political establishment she fought so hard to unseat.

Suu Kyi had a dilemma—whether to embrace Beijing or continue to distance herself from it—and had to grapple with the political implications of choosing between the two courses of actions. Prior to the 2015 elections the international community had called for amendments to be made to Myanmar's 2008 Constitution—in particular, to section 59(f), which bars anyone whose spouse or children are foreign nationals from holding the position of president. Since Suu Kyi had two sons from her marriage to the late British academic Michael Aris, she was ineligible under the existing Constitution to become president. Many of her constituents lost faith in "The Lady" as a result of her failure to more forcefully speak out against human rights abuses occurring in Myanmar against the Muslim Rohingya minority, which had been persecuted for decades.

Suu Kyi has learned that the seemingly limitless praise she has received in the past—either from her constituents or from the international community—can no longer be taken for granted. The price of becoming an icon of international human rights is that she has little room to maneuver outside of the personae she has established. The Lady learned that with political power comes the need for compromise in order to get to the finish line. Her embrace of Beijing was therefore inevitable, with economic and political implications for both nations, and the region.

Beating the West at Its Own Game

That said, China has gotten used to getting its way, one way or another. If it were not for the West's preoccupation with achieving a higher moral standard and adherence to international standards of acceptable behavior, China would not have been as successful as it has been in securing its own FDI in the developing and emerging world. China is in the process of beating the West at its own game—identifying what is sees as the West's 'weakness' on the grand chessboard and filling in the gaps left behind.

If the West played the game the same way, China's investment ambitions would have been restricted and would presumably have been more difficult to achieve. But the West is not going to change its stripes any more than China will be changing its own. In some respects, China is outmaneuvering the West in the "great game" invented by the West. As the Myanmar example has shown, China is learning quickly about the potential benefits of establishing more equitable and genuinely mutually beneficial bilateral economic relationships, as well as being more sensitive to environmental issues and the concerns of host country inhabitants. Perhaps that will be one enduring legacy of Myanmar's growing importance to the economy of China.[12]

China and the Rule of Law

As a rising global power, and as the largest and most important economy and military power in Asia, China has had the luxury of being able to do more or less whatever it wants in terms of challenging its neighbors over disputed land, and oil and gas claims—knowing that, in all likelihood, it would not be challenged. That dynamic is now changing, with Japan having vigorously contested China's claim over the Senkaku Islands and the Philippines having taken its claim over the Spratly Islands to international court.

The Philippines pursued "compulsory process" under Article 287 of the United Nations Convention on the Law of the Sea (UNCLOS). The "Notification and Statement of Claim"[13] initiated arbitral proceedings under UNCLOS over the merits of China's claim to much of the South China Sea (which the Filipinos know as the West Philippine Sea). The suit was immediately recognized as the first "legal case" against China over a number of territorial and maritime disputes with its neighbors, many of them members of the Association of Southeast Asian Nations (ASEAN). In initiating arbitration, the Philippines government stressed that the Philippines had exhausted virtually all political and diplomatic avenues for a peaceful negotiated settlement since 1995, requiring the commencement of the arbitral suit.

---

[12] Adapted from D. Wagner, "The Impact of Energy on China/Myanmar Relations," *The Huffington Post*, April 14, 2015, accessed November 25, 2015, http://www.huffingtonpost.com/daniel-wagner/the-impact-of-energy-of-c_b_7061318.html.

[13] Philippines Department of Foreign Affairs, *SFA Statement on the UNCLOS Arbitral Proceedings Against China*, January 22, 2015, accessed November 25, 2015, http://www.dfa.gov.ph/newsroom/unclos.

The threshold question really was whether China could be bound by UNCLOS courts and tribunals, including its arbitral panels. China ratified UNCLOS in 1996, but in 2006 the Chinese government filed a statement with UNCLOS saying that it "did not accept any of the procedures provided for in Section 2 of Part XV of the Convention with respect to all the categories of disputes referred to in paragraph 1 (a), (b), and (c) of Article 298 of the Convention." These provisions of the Convention refer to "Compulsory Procedures Entailing Binding Decisions" issued by at least four venues: the International Tribunal on the Law of the Sea, the International Court of Justice, an "arbitral tribunal" which may refer to the Permanent Court of Arbitration (PCA), and a "special arbitral tribunal".

While there are venues available for the resolutions of disputes under the UNCLOS regime, China did not wish to be bound by its compulsory processes—including the International Court of Justice (ICJ) and PCA. In essence, it wanted to be able to pick and choose which statutes of the treaties it had voluntarily signed it wished to adhere to, and be free to ignore those that it found "inconvenient". Can a state remain a party to a treaty or convention without being bound by its rules? Can contracting states adhere to an international legal regime and simultaneously opt out of any binding force required or to be required by that regime?

China's position effectively served as a "reservation" against any binding outcome of UNCLOS's grievance procedure in the future. It is worth pointing out that international law does accord states the freedom to disclaim whole corpuses of treaty rules through irreducible principles of self-determination, state independence, and state sovereignty. In short, China can decide to opt out of treaty rules which it considers to be inconsistent with national or domestic policy, and it did so in the manner required by the treaty.

The Philippines' attempt to haul China before an international tribunal was a problem because it invoked the very compulsory jurisdiction which China has disavowed since 2006. For the Philippines, an important point is that cases can be fielded in other forums. If military activity were to flare up, the same case can be brought to the UN Security Council—the principal repository of enforcement powers under the UN system. A state can be found to be in violation of a substantive legal norm even without a coercive or compulsory judgment in a given venue.

While China disavows UNCLOS against the Philippines, it expressly invoked UNCLOS provisions in its claims against Japan—so it wanted to have its cake and eat it too. In 2009, China submitted a claim over the Senkaku Islands (which, like Scarborough Shoal and the Spratlys, are believed to be fuel-rich) and turned to UNCLOS rules in defining and delineating its continental shelf beyond the 200-nautical mile exclusive economic zone, again within the meaning of UNCLOS. There is some international legal doctrine supporting the view that a state's acts in one place can be used as an admission and adversely bind that state in another set of circumstances.

The larger point is that China did not personify the Rule of Law in this case, or in others related to maritime borders, and that it wanted to be able to 'cherry pick' which provisions of international treaties it will willingly comply with, and which it will not. That is behavior unbecoming of a rising global power and will make states which are signatories to treaties with China wonder if its signature is worth the paper it is printed on. This cannot be in China's long-term interest, but has not prevented China from taking an approach which it would surely object to if another state tried to do the same thing.[14]

Do Sanctions Actually Work?

Economic sanctions have long been used as a foreign policy tool—sometimes being perceived as the tool of choice for nations where diplomacy has failed to yield desired results. Yet as widely used as they are, and despite the fact that some sanctions may remain in place for years, they generally fail to achieve their objectives. One of the most definitive studies[15] on the effectiveness of sanctions—covering the period from 1915 to 1990—has shown that comprehensive sanctions are partially effective at best 34 percent of the time, and that the more comprehensive the level of sanctions, the lower their degree of success. In spite of this, sanctions remain one of the few internationally accepted means (short of military conflict) of attempting to change the behavior of national leaders.

Sanctions have actually had a mixed record in terms of success. While they worked well in South Africa, for example, they failed to stop Iran from pursuing its nuclear weapons program. In the case of Russia, sanctions certainly did hurt the economy, but not nearly as much as the collapse in global oil prices, and they did not ultimately bring about any change in Russia's behavior

---

[14] Adapted from E. Tupaz and D. Wagner, "China, the Philippines, and the Rule of Law," *The Huffington Post*, January 23, 2013, accessed November 25, 2015, http://www.huffingtonpost.com/daniel-wagner/china-philippines-rule-law_b_2533736.html.

[15] G.C. Hufbauer et al. *Economic Sanctions Reconsidered*, 3rd edition, Peterson Institute for International Relations, 2007, 23.

vis-à-vis Ukraine. The ruble basically rose in tandem with the rise in global oil prices, in spite of sanctions.

While sanctions are often intended to impact a country's leadership or elite, they sometimes have the opposite effect. For instance, in Russia, Mr Putin's popularity had rarely been as high as it was following the imposition of sanctions, and in Iran, the country's elite largely benefitted from the implementation of sanctions as a result of the country's ability to engage in widespread illicit trade conducted under the international radar. Sanctions rarely succeed in banning products from a country, merely supporting black market trade and driving the price of those products significantly higher. Currency speculators are often the greatest beneficiaries of sanctions. By contrast, it is almost always the case that a country's poorest and most vulnerable suffer the most when sanctions are imposed.

In Haiti in the 1990s, more than 10,000 Haitians tried to leave the country to get to the U.S. as a result of sanctions, and up to a quarter of a million children are estimated to have died when sanctions were imposed on Iraq during the same period. It is the unintended consequences associated with sanctions that are perhaps the greatest argument against them, yet they may receive the least amount of attention in the press.

Among the other unintended consequences of the imposition of sanctions are the knock-on effect sanctions can have on trade and investment between the impacted nations and their neighbors and business partners. The value of the EU's trade with Russia is 13 times that of the U.S., so while the U.S. isn't impacted greatly by the sanctions it has led against Russia, some European nations have experienced considerable difficulties. While the U.S. Congress must specifically remove sanctions against Russia in order for them to no longer remain in force, European nations must actually reimpose them, and there is ongoing pressure on Germany and other nations that have been hit hard by the sanctions not to renew them. So, what happens if Europe decides not to renew them? The U.S. put itself in a difficult position, as the conservative-dominated U.S. Congress is unlikely to remove the sanctions.

Will sanctions against Russia ever have the desired impact? Probably not. It is unusual for sanctions to be imposed on a country with the large foreign exchange reserves, which implies that Russia has a lot more wriggle room than most countries would have under similar circumstances. That said, financial sanctions often have a lot more impact on the target country than trade sanctions. In Russia's case, the price of oil is likelier to be the most important determinant on whether sanctions will have a profound impact on the country's economy.

If the U.S. wanted to damage Russia, it would exclude it from the Society for Worldwide Interbank Financial Telecommunication (SWIFT) banking clearance system. Yet doing so would risk that Russia could initiate an alternative system to SWIFT with China and other countries, which could end up doing the U.S. more harm than good. Perhaps that is why the U.S. has rarely done so in the past. It did so against Iran, but one of the results was that Iran became one of the most self-sufficient countries in the world.

In today's world of shifting alliances and alternative currency arrangements, sanctions face ever-greater challenges in being effective. Drawing on the issue of Russia again, Mr. Putin could change the terms of its gas trade with Europe to insist that future sales be made in rubles. And while the dollar remains the reserve currency of choice, much of the world's trade is also conducted using bilateral currency swaps. The dollar may end up becoming one of several currencies used as a 'basket' of trade currencies going forward, particularly if the yuan were to become fully convertible.

If regime change is an objective of sanctions, how can the West know that whatever comes next will be preferable to what it has replaced? Using the example of Ian Smith, who was ultimately removed from power in Rhodesia in 1979 in large part as a result of the impact of sanctions, his long-term replacement ended up being the arguably more detestable Robert Mugabe, who has been in power since 1987. Who might replace Vladimir Putin, Ali Khameini or Bashar al-Assad? In each case it is unlikely to be anyone more moderate. The West should be careful what it wishes for.[16]

How Sanctions Can Backfire

Western sanctions on the Russian energy sector in retaliation for the annexation of Crimea and support for rebels in Ukraine did not have the intended devastating impact on Russian energy companies. Sanctions certainly did restrict the flow of Western investments, equipment, and technology necessary for the development of oil and natural gas fields in Russia, and the overall impact on Russia's energy sector was indeed significant. The Russian government estimated the annual loss for Russia due to sanctions to be approximately $40 billion, in addition to the $100 billion in revenue lost as a result of the fall in global oil prices in 2014 (which would have happened anyway). Unlike the sanctions against Iran, the main intention of sanctions against

---

[16] Adapted from D. Wagner, "Do Sanctions Work?," *The Huffington Post*, May 4, 2015, accessed November 25, 2015, http://www.huffingtonpost.com/daniel-wagner/do-sanctions-work_b_7191464.html.

Russia was not to stop its oil and gas from getting to the world market, but to significantly diminish prospects for the sector's development in the long run.

Reserve depletion levels are the more pressing issue for Russia. According to studies[17] of Russia's oil industry, the depletion level of recoverable reserves exceeds an average of 50 percent of the developed oil fields, while the degree of depletion is critical in the Urals and Volga region (in excess of 60 percent) and stands at nearly 80 percent in the North Caucasus. To improve long-term production capabilities, Russia needs to develop new fields (including deep-water, Arctic offshore or shale projects), but the country lacks both the technology and the investment necessary to do so. Western sanctions targeted exactly these types of ventures, prohibiting the provision, exportation or re-exportation of goods, services, or technology in support of exploration or production. In that regard, the sanctions hit Russia hard indeed.

Russia was able to seek other sources of support for its energy sector, primarily from China and other Asian countries. Following ten years of negotiation, in 2014 Russia and China signed a major pipeline gas deal that will transfer Russian gas to China through the new pipeline "Power of Siberia", starting in 2019. The thirty-year agreement enabling the creation of the Power of Siberia will likely withstand any short- or medium-term economic or political pressures that may appear bilaterally. There is a question, however, about whether the fundamental assumptions and projections behind the agreement will endure for the coming three decades.

For example, will China's demand for oil continue to grow at the pace that it has over the previous three decades and continue to provide it with an incentive to purchase oil from Russia? Given China and Russia's history, there is no guarantee that their bilateral relations will remain conflict-free, particularly as there remain some unresolved border disputes between the two nations. It is important to bear in mind that both powers seek to enhance their ability to achieve their economic and military objectives in overlapping parts of the world, such as Central Asia, the Middle East, Southeast Asia, and Japan. They are, therefore, not natural political allies.

Russian–Asian cooperation is also evident from gas pipeline consultations with South Korea, Japanese support for a pipeline from the Sakhalin fields to Japan, and interest on the part of the Indian government for the building of a pipeline from Russia. Russia's oil and gas continues to be imperative for Europe and will grow in importance to Asia, as Europe remains dependent on

---

[17] T.I. Ruzhinskaya, "Russian Oil and Gas Industry's Investment Potential and Problems," IPSIonline.com, December 2011, accessed November 25, 2015, http://www.ispionline.it/it/documents/Analysis_90_2011.pdf.

Russian gas. The development of shale gas fields in Europe has almost completely stopped due to high production costs and environmental concerns. When U.S. shale begins to be exported, it is likely to be sold mainly to Asia, where prices are higher and the infrastructure to import the product already exists. Other plans to supply Europe with natural gas—including pipelines from the Caspian region—remain at the inception stage, despite nearly two decades of European interest and support from the U.S.

Russia has no choice but to diversify its energy exports, and its ability to do so beyond its current initiatives appears to be good, which will give Russia the necessary flexibility to hedge against future risks. Russian political and business leaders have stated recently that despite the sanctions and other issues with Europe (such as the EU's third energy package), Russian energy companies will continue to work there, which means that it will be harder for Western sanctions to stand the test of time[18]. As noted above, European nations must renew sanctions annually, whereas the U.S. must specifically stop them.

Western nations' decision not to impose sanctions blocking the exportation of Russia's oil and gas to the world market was deliberate, as the world needs Russia's resources. If Russia had been restricted in this way, the result would have been significantly higher global oil and gas prices, so they were crafted so as not to have global repercussions. If the West were to try impose broader sanctions, such as to restrict Russia's banking sector, it could easily blow back, as noted, in the form of an alternative currency union between Russia, China and its trading partners that excludes the use of the U.S. dollar.

The entire situation has essentially evolved into a game of "chicken," where Russia's energy sector and economy are pressured while the West risks creating an even more tenuous global energy security. It is one thing to impose sanctions on the 29th-largest economy in the world (Iran), and another to do the same on the world's 10th-largest economy by GDP (Russia), which has extensive trading relationships around the world. In truth, the negative impact on the Russia economy and oil industry has much more to do with the decline in the price of oil than it does on the imposition of sanctions. The more important issue is the evolving landscape among the oil-producing nations regarding oil production levels.

As the West was imposing sanctions on Russia, it appears not to have considered the potential consequences for Europe, and the changing dynamic on global trade in oil and gas. Having only partially succeeded in its original objective, and in turn having resulted in closer relations between Russia, China, and

---

[18] European Parliament, *Economic impact on the EU of sanctions over Ukraine conflict*, Briefing, October 2015, http://www.europarl.europa.euhttp://www.europarl.europa.eu

other Asian nations, lawmakers on both sides of the pond may have regretted having imposed the sanctions in the first place, particularly as it has served to emphasize how very different relative costs and benefits of sanctions have had on Europe versus the U.S., based on their levels of trade with Russia. Europe has been significantly negatively impacted, while the U.S. has not.

As a result of the sanctions, arms control between Russia and the U.S. is virtually dead, as is important potential cooperation between Russia and the U.S. on a host of other issues, ranging from Syria and Iran to the global fight against terror. A generally heightened state of tension has emerged, prompting military chest-thumping on both sides, alleged cyber-attacks, and the death knell of any thought of another 'reset' in bilateral relations. When faced with an external threat, Russian nationalism goes into hyperdrive, and the average Russian circles the wagons in response. Mr. Putin's rankings have remained above 80 percent in national polls for years. Which western leader can say that?

Did Western policymakers consider all this before they pushed for the imposition of sanctions? Apparently not, for if they had they presumably would not have pursued them. They should know that Mr. Putin does not blink and that he is a worthy adversary. The truth remains that the U.S. needs Russia for its own foreign policy objectives a lot more than the reverse. That will continue to be true for the foreseeable future. Lawmakers should consider cost/benefit analyses before imposing sanctions, and, after they have been in place, before considering next steps. Too often, sanctions are imposed too quickly, without taking the time to really think about their implications.[19]

Conclusion

The debate over the existence of economic and resource nationalism, the need for it, and its relative worth, is certainly not new. An essay on the future of economic nationalism published in 1932 (when the world was in the midst of the Great Depression) could easily have been written today:

> So long as credit flows freely, economic developments and policies that are ulti-
> mately inconsistent and impracticable can long continue without apparent
> disaster. The arrest of credit brings us face to face with realities ... A country
> cannot be at once a successful creditor from the past and a successful exporter,

---

[19] Adapted from N. Pakhomov and D. Wagner, "How Western Energy Sanctions Against Russia Have Backfired," Russia Direct, June 23, 2015, accessed November 25, 2015, http://www.russia-direct.org/opinion/how-western-energy-sanctions-russia-have-backfired.

and at the same time a reluctant importer and a reluctant new lender. But when this lesson has been learned, the conclusion remains uncertain.[20]

In short, we continue to repeat old mistakes, take actions that may have contradictory consequences, and have difficulty integrating the lessons of history into present-day behavior. As has been shown by the examples noted above, it is our natural inclination to want to retrench when times get difficult economically, to want to protect what we perceive to be ours (even if international law says otherwise), and to want to lash out when we believe an action counter to our own interests appears to have been taken. In every cited example, history played a critical role in triggering a chain of events that led to economic or resource nationalism, but so did the prospect of monetary gain (or loss), boundary considerations, and something as simple and ancient as national pride.

Interestingly, the recent rise in economic nationalism has occurred in tandem with the institutionalization of 'democratic' elections in many developing countries, which has lent legitimacy to government actions that may not have existed otherwise. The spread of democracy over the past several decades is now also coinciding with the rise of extreme political movements on both the Left and Right, and at the same time that the 'devolution' process is taking hold in an increasing number of developing countries. As central governments grant greater political and economic rights to provincial governments, the rules under which contracts were originally negotiated at a national level naturally change, leaving many businesses with little option but to forcibly renegotiate or unilaterally accept the changes imposed on them. The quest for political equality on a local level has thus come to transcend borders, with both national and international implications.

All of this implies negative consequences for international businesses, particularly in the strategic or sensitive sectors of a given economy. What has changed over time is the degree to which some governments have become emboldened by 'democratic' mandates or their willingness to test the limits of international acceptability, which is becoming more prevalent in an era with limited national budgets, military resources, and political resolve. Economic and resource nationalism is here to stay. History will always collide with the present day. The challenge is to understand why, how, and when.

---

[20] A. Satler, "The Future of Economic Nationalism," *Foreign Affairs* (October 1932), accessed November 25, 2015, https://www.foreignaffairs.com/articles/1932-10-01/future-economic-nationalism.

8

Climate Change

"There is a strong, credible body of evidence, based on multiple lines of research, documenting that climate is changing and that these changes are in large part caused by human activities. While much remains to be learned, the core phenomenon, scientific questions, and hypotheses have been examined thoroughly and have stood firm in the face of serious scientific debate and careful evaluation of alternative explanation.[1] Some scientific conclusions or theories have been so thoroughly examined and tested, and supported by so many independent observations and results, that their likelihood of subsequently being found to be wrong is vanishingly small.[2]" This is the judgement of the U.S. National Academies of Sciences, whose members include more than 300 Nobel laureates.

The debate about whether climate change is man-made, the result of El Niño, or a combination of man-made and natural phenomena has captivated scientists, policymakers, NGOs, and individuals across the globe for decades. With every year that passes, more record-setting hurricanes are generated, glaciers continue to melt, sea levels continue to rise, and more people are impacted by them. There is no doubt that climate change is real. The scientific evidence proving it exists in ample supply. Yet with every early snowfall or exceptionally cold winter, some constituencies in the business and government arenas deny it exists, or claim that it is a creation of tree-hugging

---

[1] "Advancing the Science of Climate Change," *America's Climate Choices: Panel on Advancing the Science of Climate Change, National Research Council* (Washington, DC: The National Academies Press, 2010), 1.
[2] National Research Council, *Science of Climate Change*, 21, 22.

© The Author(s) 2016
D. Wagner, D. Disparte, *Global Risk Agility and Decision Making,*
DOI 10.1057/978-1-349-94860-4_8

environmentalists. Their reasons for doing so often boil down to the thirst for money, which has a short-term benefit, but, given the mountain of evidence to the contrary, doing so is little more than gambling with our collective long-term interests.

The battle lines that have been drawn between climate change believers and deniers could not be more distinct. However, outside of the halls of power and beyond diplomatic chatter there is a subtle, if imperceptible shift heralding a potentially more impactful climate awakening. Those favoring grandiose solutions to the very real challenges posed by climate change will find no victory with the advent of true market-based solutions to climate risk. For one, these shifts in the market are brought forth without presidential decree and signing ceremonies. Rather, they are brought about the same way each economic revolution occurs—with the inexorable drive among entrepreneurs, engineers, planners and risk managers tackling the immediacy of problems, with practical solutions for which the market is willing to pay.

The former moniker for climate change—*global warming*—which was popularized in the 1980s, failed to capture the world's imagination and trigger significant action. As a result, opponents of climate science used the reminder of winter and imperceptibly small temperature changes as evidence that the world was not warming, but in fact remaining in an eons-long natural oscillation. The absence of evidence is not the evidence of absence, however, and the increasingly frequent climate impacts in major cities around the world is changing both attitudes and perceptions. Although more urgent action is needed to combat climate change, there is reason for guarded optimism.

When one person tries to blow up an airplane with his shoes, the whole world takes their shoes off at airports. No matter how infinitesimal the risk of a shoe-borne airline calamity, we will subject ourselves to inconvenience on a grand scale when the economic costs are linear and fear becomes real.[3] One such climate change shoe dropped when New York's emergency managers put sandbags around the floor of the New York Stock Exchange during Hurricane Sandy and had to contemplate the very real costs of building flood walls to protect New York City from once unimaginable risks.[4] Sandy, and Irene before it (a near miss of Sandy's proportions), are now part of our climate data sample and are not merely consigned to the labs or weather models

---

[3] T. Connor, "Shoe Bomber has 'Tactical Regrets' Over Failed American Airlines Plot," NBC News, February 3, 2015, accessed November 25, 2015, http://www.nbcnews.com/news/us-news/shoe-bomber-has-tactical-regrets-over-failed-american-airlines-plot-n296396.

[4] C. Marshall, "Massive Seawall May be Needed to Keep New York City Dry," *Scientific American* (May 5, 2014), accessed November 25, 2015, http://www.scientificamerican.com/article/massive-seawall-may-be-needed-to-keep-new-york-city-dry/.

of scientists, but on the streets of Manhattan and up and down the East Coast of the U.S. As is often the case, it takes a calamity in a center of power and influence (New York rather than the Maldive Islands) to focus attention on an issue that some people around the world live with on a daily basis.

Across the planet we are reminded just how fragile we are in the face of accelerating climate change, from super typhoon Haiyan (which wreaked havoc on the Philippines in 2013) to the monster hurricane Patricia (which hit Mexico in 2015), the two most powerful cyclonic storms in recorded history.[5] Climate deniers would say (with a point, albeit a small one) that recorded history of these storms only dates to the 1850s for the U.S., and that, therefore, it is likely that certain weather events will be *out of sample*.[6] Denial or not, it does not bode well for humanity's future that we mostly have our heads in the sand and go about our merry lives without change. Europe has been awakened by increasingly severe storms, floods, and other weather events. Australia routinely battles drought, and California has been tormented by persistent droughts and wildfires that have triggered water rationing on a scale the state and, indeed, the U.S., has not seen since the Dust Bowl dried up much of the country during the 1930s.[7]

As is the case with so many aspects of man-made risk, climate change, which is certainly exacerbated by man and our inexorable carbon footprint that is leaving a deep impression on the planet's complex ecosystem, is most felt by the most vulnerable. Climate change is driving the very real specter of water wars and the rise in piracy in East Africa, which is, in no small measure, triggered by the scarcity of resources on land. Indeed, the perennial conflicts in the Middle East are not only fueled by centuries-old animus, but by control of natural resources, from fresh water in the Golan Heights to access to open water in the Mediterranean.[8] Indeed, the *casus belli* for Russia's annexation of Crimea from Ukraine in 2014 was so that Russia could regain control of its erstwhile Soviet-era access to the warm water port in Sevastopol.[9] The

[5] D. Rice, "Patricia Tops List of the World's Strongest Storms," , October 24, 2015, accessed November 25, 2015, http://www.usatoday.com/story/weather/2015/10/23/hurricane-patricia-strongest-hurricane/74461754/.

[6] National Hurricane Center, *The Most Intense Hurricanes in the United States 1851–2004*, 2015, accessed November 25, 2015, http://www.nhc.noaa.gov/pastint.shtml.

[7] V. Davis-Hanson, "California is Becoming a Dust Bowl," *Newsweek*, July 1, 2015, accessed November 25, 2015, http://www.newsweek.com/california-becoming-dust-bowl-349255.

[8] G. Barbati, "World Water Wars: In the West Bank, Water is just Another Conflict Issue for Israelis and Palestinians," *International Business Times*, July 1, 2013, accessed November 25, 2015, http://www.ibtimes.com/world-water-wars-west-bank-water-just-another-conflict-issue-israelis-palestinians-1340783.

[9] P. N. Schwartz, *Crimea's Strategic Value to Russia*, Center for Strategic and International Studies, March 18, 2014, accessed November 25, 2015, http://csis.org/blog/crimeas-strategic-value-russia.

veritable 'land grab' taking place under rapidly shrinking polar ice caps is a sign of how climate change is no longer a distant prognosis, but rather a current malady desperately in need of treatment. Shipping firms, as well as nations, are contemplating the impact on global trade using northern open ocean routes.[10] We will derive little comfort from receiving goods faster if our cities are under water, however.

Some mega trends are working in humanity's favor. Living in the age of *homo urbanus*, with more of humanity living in dense urban environments, produces strains on the environment and intensifies resource demands, but it also exposes many more people—in whose adaptation climate resilience lies—to the line of site of extreme weather events. Just a few years ago it would have been unthinkable to see London's ubiquitous 'Boris Bikes' on the streets of Washington, D.C. or New York, but for economic and business model reasons (not for the sake of being 'green') this mode of urban transport is now flourishing. Uber, Zipcar, and 'hives' of Smart cars and Cars2Go are enabling urban dwellers to forgo car ownership altogether. Hybrids and all-electric innovations are not only common place, they are also increasingly seen as 'cool' and part of the evolving mainstream. While diplomats and politicians grapple with not losing climate's upcoming 'big moment' at the next climate change conference, a 'climate standards war' is being waged in the market of ideas, giving the world hope that at last we are marshaling market-driven ingenuity to make a business out of solving climate change—or at least containing its impact.

Once upon a time the words corporate social responsibility (CSR, discussed at length in Chapter 10) were shunned in corporate boardrooms, whose members naively regarded dealing with myopic shareholder interests as the only business interest worthy of discussion. Today, CSR (or, more simply, managing against a triple bottom line of economic, social, and environmental factors) is not only commonplace, it is the source of competitive advantage for many firms.[11] The garment industry serves as a prime example of how quickly an interconnected and flat world can punish corporate social irresponsibility—even when it is carried out by sub-suppliers in a global chain.

As previously noted, the collapse of the Rana Plaza building in Bangladesh not only impacted share prices in the garment industry worldwide, it also triggered 'vicarious liability' because the garment tags of global brands were

---

[10] B. Plummer, "Climate Change Will Open Up Surprising New Arctic Shipping Routes," *The Washington Post*, March 5, 2013, accessed November 25, 2015, https://www.washingtonpost.com/news/wonk/wp/2013/03/05/climate-change-will-open-up-surprising-new-arctic-shipping-routes/.

[11] D. Disparte, and T. Gentry, "The Rise of Corporate Activism: From Shareholder Value to Social Value", *CSR Journal*, June 30, 2015, accessed November 25, 2015, http://csrjournal.org/the-rise-of-corporate-activism-from-shareholder-value-to-social-value/.

found in the rubble.[12] Even though these firms did not own the building or the subcontractors, they could not wash their hands of the lost lives and reputation risk. Similarly, Apple showed leadership by being among the first global technology companies to publish worker and other standards for its global supply chain. This not only defanged some of the pressures Apple had been facing from lax or nonexistent standards at Foxconn (its inextricably intertwined Chinese supplier), it also (in a rare move) used transparency as a source of competitive advantage.[13] Transparency itself will increasingly become a source of competitive advantage in the future.

In short, just as it took about 30 years to move CSR from a 'nice to have' to a 'need to have' in business, being 'green' is on a similar pathway. Entire industries have cropped up, devoted to energy efficiency. Economic break-throughs are afoot in virtually every aspect of the so-called 'green' economy. And while many of these micro-battles in the climate standards war may have gone unnoticed in the 2015 Paris Climate Change Summit,[14] and other global summits to come, a whole generation of business leaders and entrepreneurs stand to create a lot of real economic value by making it a business priority for the world to stay in business. The world's ability to tackle climate change is the tragedy of the commons of our times, and perhaps the greatest risk we have ever faced—as well as being our greatest risk management failure.[15]

Climate change was once managed as an 'externality' by risk managers and quants—one so big that they could not conceivably factor it into their pricing models or risk-hedging strategies. There has also been a very dangerous remoteness to climate impacts, once seen as primarily the scourge of unfortunate and ill-prepared developing countries. Ethiopia's dire famine, triggered by persistent droughts, claimed the lives of nearly a million people across the Horn of Africa.[16] As recently as 2011, severe draughts revisited the region, killing at least 260,000 people, most of whom were in war-torn Somalia.[17] Until recently, these grim images served to form so many peoples' misconcep-

---

[12] J. Drennan, "Laundering the Global Garment Industry's Dirty Business," *Foreign Policy*, April 24, 2015, accessed November 25, 2015, http://foreignpolicy.com/2015/04/24/laundering-global-garment-industrys-dirty-business-rana-plaza-bangladesh-factories/.

[13] *Supplier Responsibility*, Apple.com, 2015, accessed November 25, 2015, http://www.apple.com/supplier-responsibility/.

[14] Known officially as the 2015 United Nations Conference on Climate Change.

[15] D. Nuccitelli, "Is Climate Change Humanity's Greatest-ever Risk Management Failure?," *The Guardian*, August 22, 2013, accessed November 25, 2015, http://www.theguardian.com/environment/climate-consensus-97-per-cent/2013/aug/23/climate-change-greatest-risk-management-failure.

[16] M. Wooldridge, "World Still Learning from Ethiopian Famine," BBC News, November 29, 2014, accessed November 25, 2015, http://www.bbc.com/news/world-africa-30211448.

[17] L. Smith-Park and N. Elbagir, "Somalia Famine Killed 260,000 People, Report Says," CNN, May 2, 2013, accessed November 25, 2015, http://www.cnn.com/2013/05/02/world/africa/somalia-famine/.

tions about the African continent, so as to relegate its massive growth potential to more astute investors and countries. Today, climate change and the urgent need to find solutions to it, is no longer being treated as an externality, but rather something for which organizations and societies must be prepared.

Global leadership, consensus, and, above all, agility are needed to begin arresting the rate of climate change and improving global resilience to it. For sensible solutions to emerge from governments around the world, the economic costs of climate risks can no longer be 'transferred' away by government subsidies and national catastrophe insurance programs that amplify moral hazard—which is risk taking without risk bearing. Why would anyone build a house in Tornado Alley (a wide swath of the U.S. Midwest with a propensity for frequent tornadoes) with anything other than 'ricks and sticks' if, following each instance of severe damage from tornadoes, the true cost of rebuilding is not carried by the communities that are affected?

A broad commitment among the world's governments and businesses to quickly raise and diversify spending on research and development on energy technologies is urgently needed. While this would clearly be costly, it is important to remember three things. First, spending to reduce grave risks is reasonable. Second, some current climate policies cost a lot more than would a greatly expanded research effort. Third, battling climate change can actually raise money, the best example being well-designed carbon prices, which can boost green power, encourage energy saving, and suppress the burning of fossil fuels more efficiently than subsidies for renewables.[18]

While the Paris Climate Change Summit did perhaps represent a tipping point, there was a sense that, as was the case with each previous climate change summit, an opportunity had been missed, and should be tempered by the gradual climate awakening that is occurring in the market. "Thinking caps should replace hair shirts and pragmatism should replace green theology. The climate is changing because of extraordinary inventions like the steam turbine and the internal combustion engine. The best way to cope is to keep inventing."[19] The real opportunity cost is for business leaders *not* to seize this moment.

De-risking Climate Change

The first step toward financially de-risking the impacts of climate change on advanced economies is to change the economics. Shared government-sponsored risk pools, such as flood insurance, mis-price these increasingly

---

[18] "Clear Thinking Needed," *The Economist*, November 28, 2015, 11.
[19] "Clear Thinking Needed," *The Economist*, 11.

commonplace catastrophes. The result is that there is little innovation in making human habitation more resilient (where financial wherewithal is not a limitation) while at the same time moving further inland from flood prone areas. The banks of the Mississippi River will brim and eventually spill over, as will the Yangtze and the various rivers along the Ganges Delta. So much of humanity live near major waterways, oceans and shorelines that they enhance our collective vulnerability. In advanced economies, while government subsidized catastrophe insurance programs help to ease the financial consequences relatively quickly, they also stem the tide of innovation and climate adaptation that could have started when the first flood programs were created in the U.S. in 1968.[20] Similarly, allocating tax dollars to rebuilding flood-damaged areas takes financial resources away from taxpayers who do not live in such areas, but must in essence subsidize those who do.

During a time when the U.S. economy was largely driven by agriculture and interstate trade through waterways, and there was no extensive transportation alternative by rail or road, the need to be near major bodies of water and flood zones was an economic imperative. Today, however, with an economy that is highly diversified, coupled with multiple modes of industrial transport and global trade, the economic need to be a riparian society is no longer essential. Similarly, subsidies in wind storm insurance coverage, which would contemplate hurricanes, severe storms and tornadoes, have debased advancements in more wind-resistant or wind-proof housing. The toll from these programs, which, year after year and calamity after calamity, have reconstructed homes back to their prior state, mostly of "brick and stick" construction (the principle of indemnity in insurance holds that claimants cannot profit from a loss, thus getting the same asset they lost). This naturally begs the question, how long until the next insurance claim is filed by the same claimant in the same home, and raises the issue of how sensible it is to continue to rebuild and devote increasingly limited financial resources in disaster prone areas.

In 2013 the real toll of this perpetual weather induced creative-destructive cycle was felt by Moore, a small town in Oklahoma. In May that year a Category 5 tornado, the highest class on the Enhanced-Fujita scale, ripped through the center of town and plowed right through two poorly constructed elementary schools. The elementary schools did not have adequately reinforced walls and were hardly a durable construction class for Tornado Alley[21] (no building not constructed of concrete would have withstood such

---

[20] National Association of Insurance Commissioners, *National Flood Insurance Program*, 2015, accessed November 25, 2015, http://www.naic.org/cipr_topics/topic_nfip.htm.
[21] Associated Press, "Moore Schools Destroyed in Tornado Poorly Built, Civil Engineer Says," *Tulsa World*, February 22, 2014, accessed November 25, 2015, http://www.tulsaworld.com/news/local/moore-

a hit). Tragically, and criminally, the buildings had no tornado safe rooms.[22] At a minimum, public buildings in Tornado Alley should be mandated per the construction code to have protected basements or safe rooms sufficient to shelter the building's occupants (this would imply a complex if impossible retrofitting process for existing structures). That there has been little financial incentive to adapt building standards to predictable natural risks is partly driven by the moral hazard created by state-sponsored programs, (to make insureds whole again). In Moore, 24 people perished and hundreds were injured with little more protection than the walls of the flimsy buildings they cowered in to shelter from nature's wrath.[23] Sadly, millions of children around the world face a similar fate from weather-related events, but so many of their governments have little or no money or resources to protect them.

Whether this type of tragedy is caused by climate change or by nature's usual rhythms matters less than the fact that the direst of consequences—the loss of nine young lives and their future potential—were in that case entirely preventable. If the U.S. does not take leadership in confronting the impacts of climate risk at home, global admonishments about reducing emissions and protecting natural resources will continue to fall on deaf ears. Just as Tornado Alley goes through a predictable annual drama of destruction and reconstruction, the Florida Panhandle and Gulf States undergo a similar repetitive annual drama. Here too, miscalculating not only the effects but the reconstruction costs following a storm, hurricane or flood, continue to plague the region.

Just as climate risks take the heaviest toll on those who can least afford it, the entire city of New Orleans was devastated by the effects of hurricane Katrina—which was a Category 3 storm in the Saffir–Simpson scale when it made landfall in the Gulf.[24] As described earlier, the entire city, state, regional and federal response apparatus froze in the face of what was an entirely predictable calamity. These government agencies did not freeze for lack of resources or knowledge of what to do and how to respond; they froze for the

---

schools-destroyed-in-tornado-poorly-built-civil-engineer-says/article_a00b291d-3b1c-5fe5-a4dd-9d575c5a1323.html.

[22] D. Muir and A. Castellano, "Oklahoma Tornado: 2 Devastated Elementary Schools had no Safe Rooms," ABC News, May 22, 2013, accessed November 25, 2015, http://abcnews.go.com/US/oklahoma-tornado-devastated-elementary-schools-safe-rooms/story?id=19230427.

[23] B. Brumfield, "Moore, Oklahoma, Looks Back on Tornado That Killed 24 One Year Ago," CNN, May 20, 2014, accessed November 25, 2015, http://www.cnn.com/2014/05/20/us/oklahoma-moore-tornado-anniversary/.

[24] K. A. Zimmerman, "Hurricane Katrina: Facts, Damage and Aftermath," Live Science, August 27, 2015, accessed November 25, 2015, http://www.livescience.com/22522-hurricane-katrina-facts.html.

lack of risk agility and the capability to act with the type of speed this era requires.

Part of the failure to act is that the system, much like public bailouts of banks, is 'internalizing' the consequences of external events. Clearly, there needs to be a model in which aid and disaster relief work with speed—thus the reason to protect organizations like FEMA and its global counterparts. However, reducing moral hazard requires that the price of risk and its consequences are borne by those who create or take the risk in the first place. In this case, climate change is being produced and accelerated by advanced markets. Now advanced markets are beginning to feel the consequences.

The Readiness Dilemma

Building climate change resilience in developing and emerging countries is, of course, a much more difficult task. It combines propping up often-decrepit national institutions and financial systems, spurring infrastructure investment, and creating first-response capabilities. Above all, the development of nascent local insurance capabilities is critical, and this must be combined with greater interconnections with global insurance and reinsurance capacity. Many protectionist government policies preclude greater integration with global insurance markets, relegating many countries to offset local and regional risks with often shallow, opaque and unregulated insurance pools.

Following a natural disaster, developing and emerging countries most often depend on global aid, development agencies, and their disaster relief response. As climate change becomes increasingly severe, developing regions of the world are facing the most consequential impacts. Few developing regions are spared—and fewer still are able to go it alone. Much like an individual U.S. state's resources will be strained following a crisis, organizations like FEMA help coordinate a whole-of-government response. As is the case with so many of the topics addressed in this book, effective international and regional responses are needed.

In Pakistan in 2010, record-breaking floods triggered by unseasonably heavy rains caused widespread damage, claimed thousands of lives and affected 18 million people (more than one-tenth of Pakistan's population).[25] The hardest-hit regions were in the Swat Valley and the disputed Kashmir region, a tradi-

---

[25] *Pakistan Floods*, Thomson Reuters Foundation information website on Pakistan flooding, April 8, 2013, accessed November 25, 2015, http://www.trust.org/spotlight/Pakistan-floods-2010.

tionally restive region and safe haven for separatist and terrorists.[26] As a result, the Pakistan government has a deep distrust of international aid efforts, fearing that international intelligence agency operatives may be embedded among relief workers, and that the groups it opposes may take advantage of relief operations to pursue their own agendas. In Pakistan, as well as the Indian-controlled side of Kashmir, disaster relief efforts were led by the military, the strongest institutions in each country.

Similar to how the world only gained the upper hand on the Ebola outbreak that started in West Africa and eventually affected 11 countries when its response was militarized,[27] militarizing aid and disaster relief quite literally has a double-edged sword. On the one hand, military intervention is most often associated with armed conflict or a takeover, and people in developing countries have long memories of occupation. On the other hand, many of the worst disasters (man-made or naturally occurring) can only be tackled with the superior logistics and personnel capabilities controlled by the military. Duplicitous actions blurring humanitarian neutrality—such as Colombian paramilitary units posing as Red Cross workers to free Ingrid Betancourt from the *Fuerzas Armadas Revolucionarias de Colombia*—severely hamper global relief efforts, especially when there is no other choice but to bring in the cavalry.[28]

From the desertification of already arid lands caused by persistent droughts to torrential downpours and a record-breaking El Niño weather phenomenon, the worst impacts of climate change are being felt south of the equator. Latin American countries, such as Panama, which did the right thing by the renewable energy 'manual' of the 1980s with its reliance on hydropower, is now facing energy shortages and a head-on collision course with climate change. This confluence of events comes at a time when the country's growth and energy needs are outpacing its capacity to produce sufficiently diversified energy. The adoption of greener energy alternatives the world over is partly suppressed by persistently cheap oil prices and a standard 'war' that is being waged by the alternative energy market. There is no "one size fits all" approach to powering economies.

The unbalanced impacts of climate change are heavily weighted against emerging and developing countries. Not only are most developing countries'

[26] I. Khan, "Flood Brings Chaos Back to Pakistan's Swat Valley," *New York Times*, August 19, 2010, accessed November 25, 2015, http://www.nytimes.com/2010/08/20/world/asia/20swat.html?_r=0.

[27] "Ebola: Mapping the Outbreak," BBC News, November 6, 2015, accessed November 25, 2015, http://www.bbc.com/news/world-africa-28755033.

[28] M. Tran, "Colombia Apologizes Over use of Red Cross Symbol in Betancourt Rescue," *The Guardian*, July 16, 2008, accessed November 25, 2015, http://www.theguardian.com/world/2008/jul/16/colombia.

net contribution to climate change lower, but their ability to respond and rebuild stricken communities is, as we have noted, greatly impaired. While climate change invariably deals with large-scale science and phenomena, micro-climate impacts are increasingly commonplace. China's smog crisis (caused by severe industrial pollution and coal-fired power plants) is linked to major respiratory ailments and a dramatic rise in lung cancer.[29] Similarly, the temporary ban of vehicle traffic in Paris due to lingering noxious smog that affected the city in 2015, serve as reminders of just how fragile our ecosystem is.[30] Adverse micro-level changes like these provide case studies on both the consequences and societal responses to climate change. Affected cities adopt 'a life goes on' attitude and people and commerce proceed at a slower pace, while donning a surgical mask, popularized as a low-cost adaptation to air pollution and communicable respiratory diseases in Asia.[31] In European cities, where people are already used to pedaling to work, bicycle use has skyrocketed and alternate forms of mass transit are available.

Facing hotter temperatures during summer months, colder winters, drier and longer droughts, and soggier monsoons, the world is being profoundly tested by what scientists would call the 'climate change little league'. The dire consequences of rising sea levels and a world beyond a rise of 2° Celsius promise to make it an increasingly unpleasant place—a veritable hot, flat and crowded hellscape.[32] According to a consensus that has clearly formed among the scientific community, a 'Gladwellian' tipping point is nigh.

Winter storms are now named in the U.S. to raise public awareness and improve safety against increasingly severe winter storms.[33] Terms like 'snowmaggedon' have become a part of the weather-watching vernacular, and cities like Boston find no mirth in being snowed in on an epochal scale. Through the winter of 2014/2015, the northeastern U.S. faced a monster

---

[29] C. Larson, "Rates of Lung Cancer Rising Steeply in Smoggy Beijing," Bloomberg.com, accessed November 25, 2015, http://www.bloomberg.com/bw/articles/2014-02-28/rates-of-lung-cancer-rising-steeply-in-smoggy-beijing.

[30] E. Izadi, "Paris Tries to Fight Smog by Banning Half its Cars From the Roads," *The Washington Post*, March 23, 2015, accessed November 25, 2015, https://www.washingtonpost.com/news/worldviews/wp/2015/03/23/paris-tries-to-fight-smog-by-banning-half-its-cars-from-the-roads/.

[31] J. Aleccia, "Flu-fighting Masks May Help, But Don't Bet on It," NBC News, April 30, 2009, accessed November 25, 2015, http://www.nbcnews.com/id/30464365/ns/health-cold_and_flu/t/flu-fighting-masks-may-help-dont-bet-it/#.VlYQg_mrRX8.

[32] R. Leber, "This Is What Our Hellish World Will Look Like After We Hit the Global Warming Tipping Point," *New Republic*, December 21, 2014, accessed November 25, 2015, https://newrepublic.com/article/120578/global-warming-threshold-what-2-degrees-celsius-36-f-looks.

[33] T. Niziol, "The Science Behind Naming Winter Storms at the Weather Channel," The Weather Channel, October 13, 2015, accessed November 25, 2015, http://www.weather.com/news/news/science-behind-naming-winter-storms-weather-channel-20140121.

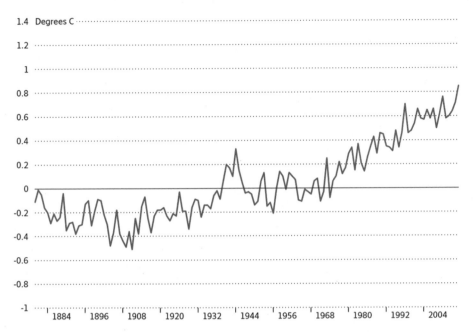

Illustration 8.1   Global land and ocean temperatures (1880–2015, January–July) (*Weather time series*, 2015, National Oceanic and Atmospheric Administration, National Centers for Environmental Information, accessed November 25, 2015, http://www.ncdc.noaa.gov/cag/time-series/global/globe/land_ocean/ytd/7/1880-2015)

snow storm named Juno.[34] Boston was particularly hard-hit throughout the winter, with a record shattering 108.6 inches of snowfall,[35] resulting in massive snow drifts. The economic impact of this calamity, preceded by Super storm Sandy, and Irene before it, served as a stark reminder of the GDP at risk in cities, where climate-risk is one of the chief perils.

As is noted in Illustration 8.1,[36] according to the U.S. National Oceanic and Atmospheric Administration (NOAA), the trend line since the beginning of the twentieth century shows a slow and steady increase in average land and ocean temperatures, rising more than one full degree Celsius during

[34] T. Ghose, "Why Monster Storm 'Juno' Will be So Snowy," *LiveScience*, January 26, 2015, accessed November 25, 2015, http://www.livescience.com/49571-northeastern-snowstorm-juno-causes.html.

[35] D. Abel and N. Emack-Bazelais, "Boston's Winter Vaults to the Top of Snowfall Records," *Boston Globe*, March 15, 2015, accessed November 25, 2015, https://www.bostonglobe.com/metro/2015/03/15/parade-day-snow-but-snowiest-winter-record-unlikely-today/BCxfh7yPtIrxtHVzty5sPM/story.html.

[36] National Oceanic and Atmospheric Administration, National Centers for Environmental Information, *Weather Time Series*, 2015, accessed November 25, 2015, http://www.ncdc.noaa.gov/cag/time-series/global/globe/land_ocean/ytd/7/1880-2015.

this period, with a noteworthy rise since the 1960s. The first seven months of 2015 were the hottest in recorded history, and July 2015 was the hottest month ever recorded.[37] It is hard to argue with facts, and the fact is that average temperatures on land and in the oceans have risen consistently for more than a century.

Scientists differ on the projected rise in average temperatures for the remainder of this century, but there is a growing consensus that the projection of a 2° Celsius rise now appears inadequate, given the already observed impacts on ecosystems, food, livelihoods, and sustainable development. According to scientists from the Inter-governmental Panel on Climate Change, a 2° Celsius increase in temperature would result in greater risk of a rise in sea level, shifting rainfall patterns and extreme weather events, such as floods, droughts, and heatwaves, that would severely impact the polar regions, high mountain areas, and the tropics. This would pose a significant danger to disadvantaged populations in many of the world's largest coastal cities, people whose livelihoods are dependent on natural resources, and those at risk from conflicts over scarce resources.[38]

Three reports produced by the World Bank in its *Turn Down the Heat*[39] series warn that, in the absence of concerted action, temperatures are on pace to rise to 4°C above pre-industrial times by the end of this century. The reports paint a dire picture of what the world could be like should such a temperature rise occur, including the inundation of coastal cities, increasing risks for food production (potentially leading to higher under and malnutrition rates), more extremes in temperature and precipitation, unprecedented heatwaves, substantially exacerbated water scarcity in many regions, an increased intensity of tropical cyclones and irreversible loss of biodiversity.[40]

One of the more insidious effects of climate change is the impact it is having on the spread of communicable, warm-weather diseases, and on human health more generally. The spread of mosquito-borne illness across parts of the mid-Atlantic U.S., such as West Nile Virus, which is inexorably spreading

[37] "Two Degree Celsius Climate Change Target 'Utterly Inadequate', Expert Argues," *Science News*, March 27, 2015, accessed November 25, 2015, http://www.sciencedaily.com/releases/2015/03/150327091016.htm.

[38] "Celsius Climate Change," *Science News*.

[39] *Turn Down the Heat: Climate Extremes, Regional Impacts, and the Case for Resilience*, Worldbank.org, accessed November 25, 2015, http://www.worldbank.org/en/topic/climatechange/publication/turn-down-the-heat-climate-extremes-regional-impacts-resilience.

[40] "New Report Examines Risks of 4 Degree Hotter World by End of Century," Worldbank.org, accessed November 25, 2015, http://www.worldbank.org/en/news/press-release/2012/11/18/new-report-examines-risks-of-degree-hotter-world-by-end-of-century.

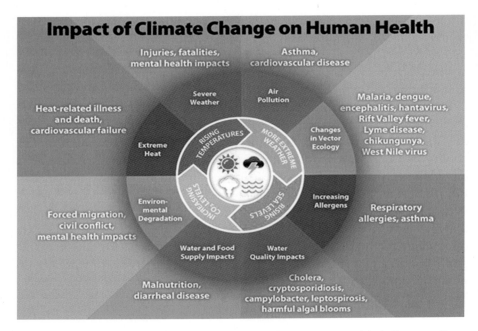

Illustration 8.2   The impact of climate change on human health (*Climate effects on health*, Centers for Disease Control and Prevention, accessed November 25, 2015, http://www.cdc.gov/climateandhealth/effects/)

north, is being attributed to climate change,[41] which will provide little comfort to northern cities when anti-malaria cocktails and mosquito repellents become a part of daily life. While the so-called *maladies of* affluence continue to be the leading killers in developed countries, climate change poses a serious setback to the world's efforts to combat the *maladies of poverty*. From diarrheal diseases to water and food supply challenges exacerbated by forced migration, all development goals are inextricably connected. Other diseases that fester in warm, densely populated environments continue to mutate and spread. While there may be a spurious linkage to climate change, a globally connected world with a new variety of dread diseases, such as H5N1, SARS, Zika, Ebola and vector-borne diseases, pose grave threats to humanity. As noted in Illustration 8.2,[42] the potential impacts of climate change on human health are numerous and serious.

---

[41] Centers for Disease Control and Prevention, *Climate and Health: Diseases Carried by Vectors*, December 11, 2014, accessed November 25, 2015, http://www.cdc.gov/climateandhealth/effects/vectors.htm.

[42] Centers for Disease Control and Prevention, *Climate Effects on Health*, accessed November 25, 2015, http://www.cdc.gov/climateandhealth/effects/.

Undoubtedly, climate change is most visibly affecting the world's food supply chains and agricultural output. From droughts and wildfires in California (one of the largest agricultural suppliers in the world[43]) to $1.25 billion in lost rice production in Thailand, climate change is not only disrupting output, but is also disrupting and potentially eradicating entire crops—and the lifestyles that go with them.[44] Our insatiable appetite in developed countries for a perennial supply of out-of-season fruits and vegetables is pushing agricultural production to new, dangerous limits of industrialization. A year-round bounty of meat, fish, and poultry harvested in massive industrial complexes is having unintended consequences on CO_2 emissions and the world's fish stocks.

Climate-related impacts on food prices and staple crops translate quickly into urgent political and security challenges. Indeed, observers have linked the effects of the rising costs of grain to the spark that ignited the Arab Spring in Tunisia in 2010.[45] One act of self-immolation in protest to the oppressively prohibitive cost of living and the arc of history changed throughout the Middle East, and indeed the world. The security and refugee crisis plaguing Europe and much of the rest of the world are an indirect consequence of this climate-linked spark. Looking ahead, as war-ravaged countries throughout the Middle East face the confluence of low oil prices, desertification, food supply chain pressures, and mass migration caused by sectarian violence, peace and stability face potentially insurmountable odds.

The climate-related risks to food supply chains underscore just how far societal adaptation must go to decelerate the rate of change and eventually establish a new normal, albeit a hotter and more austere normality. Humanity must change its diet, its mobility, modes of transport and, above all, its industrial and residential energy consumption. Policymakers, in turn, must embrace greater diversification and efficiency in energy production and distribution. Governments must implement financial systems that accurately price for the externalities of dirty fuels, following a "polluter pays" principle. Additionally, building up risk hedging that funds crisis and post-crisis reconstruction efforts as if climate change were not a surprise event is key. Perhaps

---

[43] A. Bjerga, "California Drought Transforms Global Food Market," *Bloomberg Business*, August 11, 2014, accessed November 25, 2015, http://www.bloomberg.com/news/articles/2014-08-11/california-drought-transforms-global-food-market.

[44] J. Kawasaki and S. Herath, *Thailand's Rice Farmers Adapt to Climate Change*, Our World, United Nations University, November 19, 2010, accessed November 25, 2015, https://collections.unu.edu/eserv/UNU:1581/journal-issaas-v17n2-02-kawasaki_herath.pdf.

[45] I. Perez, "Climate Change and Rising Food Prices Heightened Arab Spring," *Scientific American*, March 4, 2013, accessed November 25, 2015, http://www.scientificamerican.com/article/climate-change-and-rising-food-prices-heightened-arab-spring/.

the most important step governments can take is to let the thousand flowers of efficiency and climate readiness bloom in the private sector and let the market pick winners and losers.

A Private Sector Response

The internal combustion engine has been in use globally since the 1800s. Over this 200 year period, the basic technology and its basic output—pollution—has been largely unchanged.[46] That the Volkswagen Group, the crestfallen global automotive major claiming the fastest production car in the world, still needed a cheat device to meet regulatory emissions standards speaks to need to leave an old technology to a bygone era.[47] In moving urban dwellers from pestilent horse-drawn carriages, the internal combustion engine and its early mass produced carriage, the Ford Model T, heralded a new era—an industrial renaissance. Mobility over long distance is as essential to economic development as water is to survival. Herein lies the challenge of an antiquated form of mobility: who among the world's advanced countries has the right to dictate the development pathway of emerging and developing countries when they are themselves merely playing a game of 'catch up'?

As in some areas of the economy, people are waiting with baited breath as innovations continue to leapfrog old systems that seem increasingly ill-equipped. Brick and mortar banking, for example, is increasingly being consigned to the history books in countries like Kenya, where the advent of mobile money through the ubiquitous M-Pesa program jump straight to a cash-free and banking branch-free future.[48] Can any 'developed' country say that? Technologies such as Elon Musk's groundbreaking Tesla vehicles serve as a hopeful guidepost for the future of mobility.

Climate change resilience goes far beyond how we get around over long distances and whether we evolve from the internal combustion engine. It also includes our financial readiness and the penetration of insurance and other hedging mechanisms, such as insurance-linked securities. Lloyd's research shows that a 1 percent rise in insurance penetration translates into a 22 per-

---

[46] "The History of the Automobile," About.com, accessed November 25, 2015, http://inventors.about.com/library/weekly/aacarsgasa.htm.

[47] S. Edelstein, *Hold Onto Your Butts: These are the 10 Fastest Cars in the World*, Digital Trends, May 5, 2015, accessed November 25, 2015, http://www.digitaltrends.com/cars/fastest-cars-in-the-world-photo-gallery/.

[48] J. Bright and A. Hruby, *The Next Africa: An Emerging Continent Becomes a Global Powerhouse* (New York: Thomas Dunne Books, 2015).

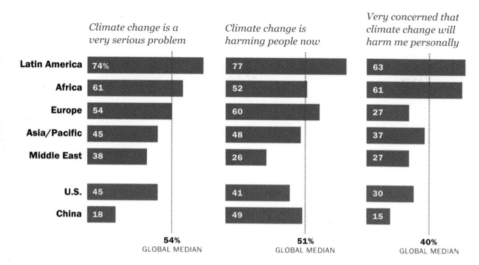

Graph 8.1 Concern about climate change by region (Source: Pew Research Center)

cent decrease in the burden shouldered by taxpayers following a large loss.[49] While some classes of insurance are seemingly more ubiquitous than gravity in advanced economies, underdeveloped or non-existent insurance markets in developing countries greatly hamper post-disaster recovery efforts. Moreover, the lack of insurance penetration across all categories of coverage provides no financial backstop as families that grapple with losing economic breadwinners inevitably slide into hardship or poverty. Increasing insurance penetration, from property programs to life and health, cannot only price climate risk more accurately, it can also help shore up redevelopment and individual household resilience.

Indeed, part of the reason the VW emissions-rigging scandal had such a swift punitive response is that attitudes and enforcement about climate change have evolved the world over. While there remains an unsurprising lag in attitudes among the worst CO_2 emitters, generally the tendency globally is both an acceptance that climate change is real and that the time for action is now. In a compelling read of public sentiment, the Pew Research Center has conducted a global survey on climate change (see Graph 8.1), showing different attitudes worldwide, as well as a dangerous partisan schism in key

[49] C. Edwards, C. Davis, *Lloyd's Global Underinsurance Report*, Lloyd's of London, August, 2012, accessed November 25, 2015, https://www.lloyds.com/~/media/files/news%20and%20insight/360%20risk%20insight/global_underinsurance_report_311012.pdf.

countries, such as the U.S., Germany, and the UK, among others.[50] The more conservative or right-leaning on the political spectrum the less likely one is to believe that climate change is a serious problem that is harming people right now. This schism widens when looking at the partisan difference on climate change among Democrats and Republicans in the U.S., creating a dangerous environment of climate inaction and denial.[51] Political gridlock in the face of such a stark global risk is the very antithesis of risk agility. Therefore, it is largely up to the market and individual households (and voters) to drive change. The more climate impacts are felt in advanced markets, the more likely attitudes will change.

A Return on Misfortune

The global insurance industry is among the most resilient business sectors in the world. Its *raison d'être* is to provide financial coverage against the misfortune visited upon policyholders. The industry is increasingly responding to climate change with innovative financial products that at a minimum provide economic security in the face of more frequent climate-related risks. While financial innovations can mitigate some of the economic consequences of climate change (for a premium), they will, of course, not stop adverse impacts from occurring. Insurance is merely a mechanism for financing unforeseen losses. Risk management is the art and science of stopping (or minimizing) losses in the first place. We need a true climate risk management strategy, for the collision course we are on with the planet's most complex system (the environment) is untenable.

Weather-related crop failures have been a persistent threat since our earliest ancestors gave up their wandering lifestyles in favor of more sedentary farming societies. Today, crop failure insurance has expanded to include other weather-related risks to global food supply chains. Coverage based on specific predetermined rain volumes can help farmers and buyers alike offset some of the economic consequences triggered by weather. Supply chain disruptions continually affect world trade. From the flooding in Thailand that stopped much of the global production of semiconductors in its tracks to disruptions

---

[50] B. Stokes, R. Wike, J. Carle, *Global Concern about Climate Change, Broad Support for Limiting Emissions*, Pew Research Center, November 5, 2015, accessed November 25, 2015, http://www.pewglobal. org/2015/11/05/global-concern-about-climate-change-broad-support-for-limiting-emissions/.
[51] Stokes et al., *Global Concern.*

caused by natural disasters, offsetting economic risks and enabling firms and countries to reposition is a key facet of building resilience.[52]

One of the more promising aspects of financial innovation in building a societal response to climate change borrows from behavioral economics— although many of these innovations are still in their nascent stages. John Hancock, a U.S. subsidiary of the Manulife Financial Corporation, unveiled an innovative whole life insurance policy in partnership with Vitality, a pioneering South African wellness program.[53] In this solution, traditional life insurance customers combat *longevity risk* by wearing a Fitbit device that rewards a healthy lifestyle with a combination of incentives and real savings on premiums, which for a whole life insurance policy can range in excess of $50,000.

This type of behavior-based solution is also occurring in the highly commoditized auto insurance market, where drivers install a device in their car that monitors driving patterns, thereby rewarding good road conduct with reduced premium and other benefits. Further incentives for hybrid or all-electric vehicles add to government-sponsored tax incentives, spurring the adoption of efficient vehicles. Following total loss claims, some homeowner's policies require rebuilding houses with environmentally friendly materials and systems, such as use of natural light and solar panels, among others.[54]

Options for harnessing insurance and financial incentives beyond the traditional indemnification offer a compelling, if underexplored, area for innovation. Similar possibilities also abound to hedge large-scale losses. Harnessing the depth, liquidity, and actuarial capabilities in the private sector will help not only an individual household vis-à-vis climate change, but also highly exposed industries and society writ large. Clearly, the scale of the potential economic consequences are such that it will require true cross-sector collaboration. Bringing government resources to bear, along with global reinsurance and insurance capacity, would create an unbroken chain of financial control that can help communities rebuild following a loss, while helping others gain resilience by incentivizing behavioral change. For developing and emerging

[52] S. Mydans, "Flood Defenses are Overrun in Bangkok," *New York Times*, October 25, 2011, accessed November 25, 2015, http://www.nytimes.com/2011/10/26/world/asia/flood-waters-in-bangkok-shut-domestic-airport.html.

[53] *John Hancock Life Insurance*, Johnhancockinsurance.com, accessed November 25, 2015, https://www.johnhancockinsurance.com/life/John-Hancock-Vitality-Program.aspx.

[54] *Green Insurance*, Insurance Information Institute, November, 2015, accessed November 25, 2015, http://www.iii.org/article/green-insurance.

countries alike, the opportunity lies in creating basic insurance industry capabilities while strengthening ties and access to international markets.

For Rising Tides, Deeper Pools

In the face of climate change—a risk so large that people wrongly believe they cannot have an impact—harnessing all aspects of the capital markets is key. While most climate-related financial innovations are quite sophisticated, insurance-linked securities, such as catastrophe bonds (so-called "cat bonds"), are still underutilized in the world's climate risk management strategy. Given the scale and potential city-wide and broad regional impacts natural disasters can have, catastrophe bonds were developed to bring greater financial liquidity following a large-scale loss, while at the same time giving sophisticated investors an important source of diversification. While catastrophe bonds themselves come with a series of intrinsic risks, they nonetheless offer a compelling approach to improving financial resilience.

While a novel investment vehicle, catastrophe bonds and their constituent funds, like other insurance-linked securities, are prone to a number of risks. Some are inherent in the underlying insurance exposure they are attached to—for example, natural disasters—while other risks may be common to the investment class. The first order of risks materialize if the underlying claim trigger (or covered event) occurs during the maturity of the bond, which is typically three years. This would result in a payout from the special purpose vehicle (SPV) to cover certain portions of the loss that have been transferred to the capital markets. In effect, these specialized financial instruments are the reinsurance of the reinsurance industry.

One of the most attractive features of catastrophe bonds is that they are largely decoupled from the fluctuations of the economy. While it shields investors from potential economic contagion in other asset classes, these bonds expose insurers to highly concentrated regions that are prone to natural disasters, which, as the evidence suggests, are occurring with greater frequency and severity, and are affecting more densely populated/built-up areas.[55] Hurricane Katrina, which decimated the Gulf Coast and inundated New Orleans, Hurricane Sandy, the March 2011 Tohoku earthquake, and the ensuing tsunami and nuclear fallout in Japan, highlight the perils intrinsic in these instruments.

---

[55] World Economic Forum, *Global Risks Report 2015*.

In other words, while seeking diversification from economic risks, investors in catastrophe bonds are largely putting all their proverbial eggs in only a few highly concentrated regions exposed to natural perils, such as East Coast hurricanes, Japanese earthquakes, and European weather events. In short, only those regions that are prone to natural disasters that can be modelled and have a confluence of insurable assets are in the pool—therefore there may be a case of limited gains from diversification. Broadening the geographic scope and the types of natural perils covered in a catastrophe bond will not only help to diversify risks, adding coverage to woefully under-insured regions, but will also help mitigate the effect of adverse selection.

Rating agencies (not necessarily the paragons of accurately rating risk) have noted this so-called 'peak peril' attribute of catastrophe bonds. Consequently, most rating agencies give these bonds low marks in terms of investment grades.[56] In fact, approximately 90 percent of the cat bonds tracked in Swiss Re's Cat Bond Index receive a BB rating or lower.[57] Where other investment instruments have a property of 'slow degradation' and can quickly turn around with the advent of new market information, cat bonds are not dissimilar from a game of Russian roulette: if the event occurs, albeit remote, an investor is likely to lose everything. This feature of being uncorrelated to the 'noise' in the financial markets give cat bonds the quality of being an investment vehicle that is not speculative in nature, but rather associated with the advent of a pure risk.

While climate change is a man-made risk, people will have a hard time *gaming* the acuity of these claim triggers, which can be an important source of stability against Adam Smith's invisible hand. Outside of the 'all or nothing' risk inherent in cat bonds, they face a host of what can be deemed operational risks, which are exacerbated by the fact that there is a limited secondary market for these instruments and many investors do not totally understand this market, as evidenced by its small comparative size, at $24 billion.[58] In addition to potential exposures resulting from issuer insolvency or financial distress, returns to investors are subordinated to other potential obligees. Additionally, reductions in interest rates due to early redemption or extensions in the notes can potentially hamper investor returns.[59]

---

[56] Swiss Re, *Insurance-Linked Securities: Market Update Volume XVI*, July 2011.
[57] Swiss Re, *Cat Bond Indices: Year in Review 2012*, February 2013.
[58] National Association of Insurance Commissioners, *Insurance-linked Securities: Catastrophe Bonds, Sidecars and Life Insurance Securitization*, September 3, 2015, accessed November 25, 2015, http://www.naic.org/cipr_topics/topic_insurance_linked_securities.htm.
[59] Swiss Re, *Insurance-Linked Securities: Market Update,* Volume XIX, July, 2013.

Cat bonds are unequivocally appealing in that they are one of the few investment vehicles that can help diversify away business cycle exposures in a portfolio. While in turn this exposes investors to the often misunderstood peril of natural catastrophes and the opacity of risk modelling agencies, the trade-off is nonetheless attractive. Empirically, cat bonds have performed well in terms of holding their principal and delivering returns to investors that are often superior to other benchmarks.[60] At the same time, there appears to be inherently less volatility in cat bond returns as evidenced when comparing returns to more volatile equity markets. In addition to being uncorrelated to the business cycle, cat bond pricing and returns are relatively straightforward spreads, with returns being a premium spread on treasuries and pricing typically pegged as a spread on LIBOR.

Another attractive feature of cat bonds is that they typically abate counterparty exposures, such as insolvency or financial distress, by being managed through SPVs and collateralized with strong paper. This property makes cat bonds attractive to issuers as well, as SPVs are often off-balance sheet instruments that tend to be bankruptcy remote. Cat bonds aided the broader insurance industry with standardized contracts, a move away from ad hoc reinsurance agreements, providing the vital liquidity of the capital markets following Hurricane Andrew.[61] While many insurers had difficulty coping with this, investors clearly benefited from a degree of harmonization. Finally, with the advent of multi-peril cat bonds, investors have been able to secure an added layer of "natural" diversification, by not being exposed to a single natural catastrophe in a single region, but rather to be invested in a basket of natural perils. While the dominant *force majeure* in cat bonds are U.S. wind-related exposures, the multi-peril approach adds an additional layer of protection.

In terms of disadvantages, cat bonds have largely remained the remit of large institutional investors and insurance and reinsurance firms wishing to tap into the capital markets. This is a by-product of their issuing complexity and costs, combined with their relative novelty. Additionally, cat bonds are heavily dependent on third-party modelling firms, who use historical loss information and various payment trigger approaches (i.e. modelled loss and parametric modelling, among others), thus exposing cat bonds to substantial model risk, as many natural catastrophes, by default, are out of sample.

---

[60] G. Chacko, V. Dessain, A. Sjoman, A. Plotkin, and P. Hecht, *Bank Leu's Prima Cat Bond Fund*, Harvard Business School, December 2004, accessed November 25, 2015, http://www.thecasesolutions.com/bank-leus-prima-cat-bond-fund-10452.

[61] Swiss Re, *The Fundamentals of Insurance-Linked Securities*, June 2011.

This is particularly true when dealing with something as vastly complex as nature and weather systems, which operate on a millennial scale. Another drawback of this asset class is that they are, by design, prone to adverse selection. More clearly, only those regions that are prone to natural disasters and have exposed assets will enter the pool. While this exposure is abated by the various underlying layers of insurance (e.g. underlying insurance limits perform like a deductible, thus they need to be exhausted before the cat bond is triggered), cat bonds offer something of an 'all or nothing' proposition to investors. It should be noted that one major disadvantage of this asset class, to the extent one can rely on rating agencies, is that none receives an investment grade.

Cat bonds are an important financial instrument that are here to stay. As such, they are well placed in large, well-diversified portfolios. In fact, this may be their only home, for the reasons of size and complexity cited earlier. The property of being decoupled from the business cycle is an attractive one. However, fund managers would be well advised to tread carefully into cat bonds, for their 'all or nothing' properties can wreak havoc on an imbalanced portfolio. Crucially, cat bonds not only offer some respite from economic volatility compared to other assets classes, but also tend to demonstrate superior risk-adjusted returns.[62]

A limited secondary market is a business risk that fund managers have to contend with for cat bonds, leaving few exits in instances of investor remorse. Nonetheless, cat bonds offer another avenue in the pursuit of investment yield, with spreads averaging between 300 and 600 basis points over treasuries. There is also evidence, following the Tohoku earthquake in Japan, that even in the face of unforeseen calamity, cat bonds are a surprisingly resilient asset class.[63] This resilience, and, thus, their attractiveness as a source of diversification, is increased with the advent of multi-peril instruments.

Undoubtedly, the risk–return trade-off provided by cat bonds is appealing to investors. While they largely remain an investment vehicle for sophisticated institutional investors, cat bonds offer an attractive value proposition. Multi-peril cat bonds overcome some of the concerns of being exposed to specific regions that are prone to specific natural hazards. Thus, in a given year or over multi-year periods, wherein investors are typically locked into the bond, it is unlikely that a confluence of natural catastrophes will occur. The property of being shielded from the economic cycle lowers the overall volatility of cat

---

[62] Chacko et al., *Prima Cat Bond Fund.*
[63] Swiss Re, *Insurance-Linked Securities: Market Update Volume XVI*, July 2011.

bonds, making them a favorable addition and a potentially stabilizing force in a portfolio.

As investors seek yield, cat bond primary issue yield spreads have stood up reasonably well during the last decade. A period that was both naturally tumultuous—with Katrina, the Tohoku earthquake and Sandy, among others—and financially turbulent, with the Great Deleveraging of 2007–2009 and a historically low interest rate environment. Against this backdrop, and with the evident resilience of cat bond returns, the risk–return trade-off appears to be in balance. This being the case, it is striking how comparatively small the cat bond market is, perhaps signaling that they are only attractive to "practitioners" or those whose investment mandates permit them to dabble in the *below investment grade* "no-go zone".

Nonetheless, there are encouraging signs that these instruments are being cleverly applied to bring liquidity into other areas of the insurance market. For example, life and health insurers are using cat bonds for the 'longevity' calamity that has the potential to upend actuarial models around life expectancy. Likewise, governments are turning to this vehicle as a way of creating a backstop in the event of a natural disaster. Innovative applications in man-made disasters, such as war and terrorism, which already enjoy a vibrant insurance and reinsurance markets, appears to be a logical next frontier.

Financing climate change and natural disasters through a broad-based approach can help improve pre-crisis readiness and post-crisis reconstruction. No matter how innovative pure claims-based solutions are, there is no climate risk management substitute to stopping the accident in the first instance. In order to do this, leveraging financial markets to change behavior through incentives and penalties can have a positive impact on the worst-performing advanced market consumers. Globally, the disequilibrium of impacts from climate change weigh heavily against underdeveloped countries. Calibrating this imbalance through public–private approaches will help stem the tide of forced migration, terrorism, and other global threats in the era of man-made risks.

A 2015 poll[64] found that just 3 percent of Americans viewed climate change as the most important issue facing the U.S. Given the short-term orientation of so many people, perhaps this is unsurprising, but it should be alarming, coming from the world's top producer of carbon dioxide.[65] In the face of overwhelming scientific evidence proving that climate change exists, there is

---

[64] *Fox News Poll: 2016 Matchups*, Fox News, November 20, 2015, accessed November 26, 2015, http://www.foxnews.com/politics/interactive/2015/11/20/fox-news-poll-2016-matchups-syrian-refugees/.

[65] World Bank, *CO_2 Emissions*, 2015, accessed November 26, 2015, http://data.worldbank.org/indicator/EN.ATM.CO2E.PC.

little that can be said to someone who continues to deny this fact. However, much of the world's population clearly believes climate change is here to stay. Whether the result of their own personal experience or their understanding of the science supporting the climate change argument, people around the world are beginning to recognize both the adverse effects of climate change, and the need for urgent action. Reversing the paralysing effects of partisan politics that is stalling serious action on climate change is vitally important. Broad-based resilience on the part of individuals, organizations, and governments will collectively determine our agility in responding to this looming threat.

9

Cyber Risk

Of all the man-made risks transforming our world, cyber risk is perhaps the most problematic to business and political leaders. Cyber risk is a "catchall" for a wide variety of threats to data and systems in the virtual and real worlds. Depending on how cyber risk manifests itself, it can be as destructive to a business as terrorism, with consequences as potentially long-lasting as climate change, if it includes damage to a firm's reputation. This ascendant risk domain continues to mutate and emerge from the shadows of anonymous splinter groups, hacker collectives, and increasingly assertive state-sponsored cyber warfare that is testing military capabilities in an invisible theater of war.

Businesses and countries alike have come to realize that cyber risk is not some temporary fad, but is here to stay, and that no one is invulnerable. Like other risks that lurk in the shadows, the consequences of cyber risk can cause potentially embarrassing exposures to organizations, and is therefore often underreported. As a result, its causes, the identity of the perpetrators, and true scale of its damage remain largely unknown. Entire cottage industries focused on cyber security have emerged in recent years and are flourishing at the same speed that cyber criminals and troublemakers continue to push the boundaries of which computer systems they can successfully breach. Judging by public revelations in recent years, even the safest computer systems in the private and public sectors around the world are vulnerable. This is part of the reason why the Kremlin is alleged to have implemented typewriters for the production of the most sensitive of communiqués.[1]

---

[1] C. Irvine and T. Parfitt, "Kremlin Returns to Typewriters to Avoid Computers Leaks," *The Telegraph*, July 11, 2013, accessed on November 30, 2015, http://www.telegraph.co.uk/news/worldnews/europe/russia/10173645/Kremlin-returns-to-typewriters-to-avoid-computer-leaks.html.

© The Author(s) 2016
D. Wagner, D. Disparte, *Global Risk Agility and Decision Making*,
DOI 10.1057/978-1-349-94860-4_9

High-profile cases of unwanted disclosure can be found all across the globe, the most famous of which being Edward Snowden's release of classified data from the National Security Agency (NSA), a top-secret U.S. intelligence agency where Snowden worked as a contractor. In Snowden's revelations, not only was the scale of the U.S. government's eavesdropping through the PRISM clandestine surveillance program exposed, so was the participation of virtually every major tech company in the U.S.—from Apple and Facebook to Google—were involved in the program. In essence, all the major firms that consumers treat as internet utilities participated in passing metadata on to the government.[2]

How these data were used, and whether civil liberties were breached, is unclear, but the implications are chilling. Attitudes towards these revelations were surprisingly blasé in the U.S. when they were first revealed in 2013. The public seemed to oscillate from acceptance that in the age of global terror-ism there needs to be a public trade-off, to anger at the traitorous manner in which Snowden obtained his information and revealed it to the world. In the post-9/11 world, patriotism trumped anger at the scale and sub rosa nature of the government's prying eyes, which raises fundamental questions about just how far the rules of acceptable intrusions into what was previously considered sacrosanct individual and business information can be stretched.

While the NSA's program was formally abandoned in late 2015, data col-lected up until the end of this controversial program will be retained—span-ning the majority of phone calls made in the U.S. For any suspected cases requiring monitoring, the NSA must obtain approval from a special intelli-gence court, unless there is a pressing security issue, in which case the request can be accelerated.[3] Some may herald the end of this program as a victory for Snowden and civil liberties, however, the general powers to obtain broad information remain in place and the restrictions can in theory be removed by the Congress in the future (which is not hard to imagine). Tacit accep-tance from the public all but condoned this type of modern tradeoff between security on the one hand and the right to privacy on the other. This deep struggle is one of the central themes defining how people, organizations, and nations respond to cyber risk in the twenty-first century. The dilemma of an

---

[2] A. Kleinman, "NSA: Tech Companies Knew about PRISM the Whole Time," *The Huffington Post*, March 20, 2014, accessed on November 30, 2015, http://www.huffingtonpost.com/2014/03/20/nsa-prism-tech-companies_n_4999378.html.

[3] P. Williams, "Massive NSA Phone Data Collection to Cease," NBC News, November 28, 2015, accessed on November 30, 2015, http://www.nbcnews.com/news/us-news/massive-nsa-phone-data-collection-cease-n470521.

omniscient era drowning in metadata is that all things are known and where there is a way in for sensitive data, there is also, invariably, a way out.

The Snowden case revealed the vulnerability of otherwise highly secure computer systems to relatively unsophisticated 'insider threats', in even the best-defended environments. Entire systems, such as Julian Assange's Wikileaks, have been set up to house and disseminate once-private information. While Snowden and Assange have taken on the air and mystique of cyber Robin Hoods, in both cases, the jury is out about whether their revelations have caused more harm to national security and perceived intrusions on civil liberties than their disclosures have helped improve transparency and accountability.[4] Whether or not their motives are akin to so-called 'ethical hackers', the fact that information can be so readily retrieved from the most private and supposedly secure systems is an omen for our collective vulnerability.

Business Models for Ransom

In the early days of mass online consumption, propelled by the rise of Amazon and other online marketplaces, identity theft and credit card fraud were the principal concerns of the internet age. Today, the volume and severity of these risks have largely been reduced to the background noise of a $1 trillion online retail economy.[5] The scale of online consumption so far exceeds the infinitesimally small exposure to noisome credit card fraud that China's Alibaba sold more than $14 billion worth of items on 'Singles Day' in 2015—China's equivalent of Cyber Monday. As if to demonstrate how China has managed to take a Western invention and put it on steroids, Cyber Monday (America's online shopping bonanza) logged a paltry $1.4 billion in sales in 2014 by comparison.[6] Consumer fear of suffering fraud or identity theft while shopping online seems to have diminished, compared to the fear of not being able to find a parking spot, having to battle throngs of shoppers, and needing to wait in long checkout lines in a shopping mall. Convenience appears to have trumped perceived risk, however, with viral videos of shoppers being

[4] J. Assange, "Assange: What Wikileaks Teaches us About How the U.S. Operates," *Newsweek*, August 28, 2015, accessed on November 30, 2015, http://www.newsweek.com/emb-midnight-827-assange-what-wikileaks-teaches-us-about-how-us-operates-366364.

[5] H. Ben-Shabat et al., *2015 Global Retail E-Commerce Index,* AT Kearney, 2015, accessed on November 30, 2015, https://www.atkearney.com/consumer-products-retail/global-retail-development-index/2015.

[6] "China's Alibaba Breaks Single's Day Records as Sales Surge," BBC News, November 11, 2015, accessed on November 30, 2015, http://www.bbc.com/news/business-34773940.

trampled to death by throngs of other shoppers not helping the case for in-person shopping.

Part of the reason these risks have diminished for households is the tacit agreement made between banks and online marketplaces, in which retail consumers have zero liability for their online consumption. With the advent of this policy (which was quickly adopted as the online modus operandi for global banks), online consumption skyrocketed. The rise of social media, with firms like Facebook serving as a de facto census bureau for more than 1.5 billion active users, shows that modern internet users are willing to cede some of their personal privacy in exchange for global connectedness.[7] The assumption of privacy and ownership of a digital second life shows the dangerous naiveté some modern consumers exhibit. It also speaks to why Snowden"s hoped-for cyber awakening from his revelations fell on so many deaf ears. In short, people tacitly assumed that their information was being used, but the presumed value trumped any perceived risk.

The more troubling trend, aside from general tacit acceptance that privacy is now a marketable asset, is the routine occurrence of business models being held for ransom. Just as getting kidnapped by ideologues with no monetary demands usually ends badly, there has been a rise in cyberpolitical ransoms. With increasing frequency, mysterious online groups, from amorphous hacker collectives like Anonymous to state-sponsored cyber terrorists like the North Korea-backed Guardians of the Peace (GoP), are making clear quid pro quo demands. These are happening with increasing frequency, and the few public cases that have appeared for collective consumption illustrate that in the age of cyber risk, even the mighty can fall. What we must assume, then, is that the publicly revealed cases are merely the tip of the iceberg of what is the newest man-made risk.

Cyber Terrorism

As the identity of the sovereign culprit and scale of the damage behind the attack on Sony in 2014 were revealed, the world faced the specter of an uncomfortable new normal emerging in warfare—cyber terrorism.[8] The attack on Sony Entertainment, in response to the planned release of the controversial film *The Interview*, successfully crippled the company's value chain and also

---

[7] *Facebook Online Newsroom*, Facebook corporate website, www.facebook.com, November 23, 2015, accessed at November 30, 2015, http://newsroom.fb.com/company-info/.

[8] D. Disparte, "Welcome to 21st Century Warfare," *The Hill*, January 5, 2015, accessed on November 30, 2015, http://thehill.com/blogs/congress-blog/technology/228510-welcome-to-21st-century-warfare.

that of their downstream partners. Sony's $22 billion market cap could readily absorb the financial consequences of the attack, which some experts estimated would cost between 0.9 percent and 2 percent of the firm's 2014 sales.[9] Sony later said that its financial loss was limited to $35 million, but the longer-term repercussions are substantial.[10] Among the trove of data stolen from its servers were details about the company's strategy in nearly every contractual negotiation it held in recent years prior to the breach, which could have enormous implications for the company. It is also difficult to put a price on the embarrassing revelations in emails that were published in which Sony executives expressed what they thought about some movie stars and writers. The result could be that some actors refuse to work for Sony in the future, or else that they demand higher fees.[11]

A dangerous precedent had been set. As President Obama noted at the time: "If somebody is able to intimidate folks out of releasing a satirical movie, imagine what they start doing when they see a documentary that they don't like, or news reports that they don't like."[12] Imagine the consequences for firms that produce products and services, or for national policies that attract unwanted attention. In the era of man-made risk, firms risk capitulating to an unseen enemy holding a virtual gun to their head, and yet, the threat and perceived risk are just as real as if a real gun were being held to their head.

Prior to Sony's cyber-terror attack, cyber risk was already an insidious drag on the world's economy, with most losses manifesting themselves in identity theft, denial of service (DoS) attacks and silent phishing scams. As of 2014, the average cyber liability loss for a business in the U.S. was $5.9 million. Globally, cyber-crime costs the world economy between $300 and $575 billion.[13] In 2014 the insurance industry collected $2.5 billion in premiums on policies to protect companies from losses resulting from hacks—an increase of more than 20 percent from 2013 (which was less than $1 billion that year).

---

[9] R. Hackett, "How Much Do Data Breaches Cost Big Companies? Shockingly Little," *Fortune*, March 27, 2015, accessed on November 30, 2015, http://fortune.com/2015/03/27/how-much-do-data-breaches-actually-cost-big-companies-shockingly-little/.

[10] R. Lemos, "Sony Pegs Initial Cyber-Attack Losses at $35 Million," February 2, 2015, accessed on November 30, 2015, http://www.eweek.com/security/sony-pegs-initial-cyber-attack-losses-at-35-million.html.

[11] M. Wheatley, "Hidden Costs of Sony's Data Breach Will Add up for Years, Experts Say," February 20, 2015, accessed on November 30, 2015, http://siliconangle.com/blog/2015/02/20/hidden-costs-of-sonys-data-breach-will-add-up-for-years-experts-say/.

[12] J.S. Brady, *Remarks by the President in Year-End Press Conference*, The White House, Office of the Press Secretary, December 19, 2014, accessed on November 30, 2015, https://www.whitehouse.gov/the-press-office/2014/12/19/remarks-president-year-end-press-conference.

[13] T. Kopan, "Cybercrime Costs $575 Billion Yearly," *Politico*, June 9, 2014, accessed on November 30, 2015, http://www.politico.com/story/2014/06/cybercrime-yearly-costs-107601.

The battle against cyber risk is becoming a large business.[14] The industrialization of cyber risk is troubling; For each advance that is made in mitigation, threat vectors always seem two steps ahead of the market. If there is a category of risk to attempt to avoid all together, cyber risk is it.

The attack on Sony not only raises the economic and reputational costs of cyber risk, but demonstrates the ease with which a company's entire business model can be held for ransom. Until President Obama had spoken on the subject, Sony had capitulated by refusing to release the film, which underscored the extent of the informational extortion faced by the the company. The GoP had indicated a lifelong threat to the company should any trace of *The Interview* emerge in the public domain (not to mention the subsequent threat North Korea had issued on the lead actors' lives). Since the film's release in theaters and online, the threat was proven to be bombastic. As outlined earlier in the book, the real lingering harm caused by the breach was not the loss of proprietary information or *The Interview*'s revenues—as it is likely the North Korean drama was actually made into a commercial success *because* of the breach—but, rather, the revelation of internal information not befitting of a value-based culture.

While many have vilified Sony's response to this attack as soft, Sony taking on its aggressor—a state sponsor of cyber terrorism—would have been tantamount to private industry taking on Al Qaeda or the IS. An appropriate governmental response is needed to address such attacks, creating a strong cyber deterrent and exacting a proportionate cost on would-be cyber terrorists, as well as to deter future sponsors of such attacks. Additionally, government programs covering terrorism and war risk should be modified to encompass applications of cyber threats carried out to intimidate or coerce the public or a specific company. In exchange for this type of industry-wide government backstop, the private sector would agree to a rare form of disclosure following a cyber breach. While this would not only help drive much-needed transparency to this otherwise opaque risk, it would help technology security specialists to cover up loopholes. With effort, security specialists can begin to gain the upper hand on how to prevent these risks for occurring.

At the same time, businesses need to advance the standards of cyber risk management by creating greater transparency, more localized encryption, and an early warning system when breaches are detected. Firms fear a public backlash when breaches occur, causing them to share little or no information until the crisis escalates beyond their control. However, improving transparency through a central clearinghouse of cyber breaches would not only improve

---

[14] S. Gandel, "Lloyd's CEO: Cyber Attacks Cost Companies $400 Billion Every Year," *Fortune*, January 23, 2015, accessed on November 30, 2015, http://fortune.com/2015/01/23/cyber-attack-insurance-lloyds/.

system-wide security, it would also help improve the often-mispriced cyber insurance market. Part of the mistrust among consumers, current and past employees is that they are often the last to know about a breach affecting their private information. This goes part of the way to explaining the rise of class action lawsuits for organizations suffering from cyber attacks. In Sony's case, a legal settlement amounting to $10,000 per affected employee was discussed, adding to the firm's overall costs, while creating a dangerous legal precedent.[15]

When combating an asymmetric foe that benefits from opacity, sunlight, transparency, and global coordination are the greatest weapons against simple hacker collectives and state sponsors alike. As new rules of engagement are being drawn up in corporate boardrooms and by policymakers in response to Sony's attack, the nascent cyber insurance market can provide some peace of mind and financial protection in the face of new threats. Although underwriting appetite for large cyber insurance policies is sure to wane as the full scale of these risks continues to evolve, this class of insurance is essential for firms of all sizes operating in the global economy. Large companies like Target, Home Depot and Sony make for headline grabbing targets as they have a sprawling surface area to attack, while they are certainly not the only types of firms vulnerable to cyber risk.

Cyber liability insurance is one of the fastest-growing coverage areas in the insurance arena, suggesting broad-based demand among firms of all sizes seeking to offset the financial and risk mitigation burden of this complex exposure. A survey of business risk conducted in 2015 found that cybercrime, IT failures and espionage were among the top 10 business risks occupying the minds of corporate leaders.[16] Legal liability is rapidly following suit, as the legal system grapples with a growing caseload alleging poor standards of care around privacy and sensitive customer and employee information. Here too, Sony's case stands to set a precedent. Given that the culprit was a foreign nation, perhaps applying the TRIA as a way of shoring up cyber risk underwriting appetite and capping liability will enable further development of this important market.

Time for a Cyber FDIC

In the modern frictionless economy, the principal requisite for growth and order is the free flow of sensitive information utilizing banking transactions and private data on a global scale. With this shift, brought on by the unrelent-

---

[15] "Judge Gives Preliminary Approval of $8 Million Settlement Over Sony Hack," *Time*, November 26, 2015, accessed on November 30, 2015, http://time.com/4127711/sony-hack-security-legal-settlement/.
[16] Allianz SE, *Allianz Risk Barometer 2015*, 2015, accessed on November 30, 2015, http://www.agcs.allianz.com/assets/PDFs/Reports/Allianz-Risk-Barometer-2015_EN.pdf.

ing growth of ecommerce and mobile platforms, comes the nagging reality that cyber risk is here to stay. It is the sort of drag on the market that needs to be priced into the system rather than being treated like something that can be eliminated or perfectly controlled.[17] Market and consumer expectations need to be adjusted accordingly and in the age of hypertransparency, people would be wise to remember that anything can be exposed to sunlight.

Just as banks' zero-liability policies largely defanged identity theft and the risk that consumers would be saddled with large losses due to phishing scams and fraud, so too, aspects of cyber risk need to be neutralized by a broad consensus on how to shift risk and respond appropriately. Perhaps it is therefore time to create a cyber equivalent of the FDIC, to shore up confidence and risk-bearing in the system. This would be particularly relevant in the commercial domain, as smaller firms are saddled with disproportionate exposure for their lack of financial resilience and general systems readiness. Similar to how kidnappers or terrorists move down market (toward soft targets) when their intended targets harden their security, the same pattern holds true with cyber risk.

Coping with this risk requires new standards of practice to emerge around an early warning system, destigmatizing the advent of breaches and capping liabilities when they occur. All of this can be achieved through cross-sector collaboration between the government and the private sector, much like the conditions that gave rise to the FDIC following the Great Depression. In short, cyber risk should not be treated like a discrete risk for individual firms to deal with in isolation; rather, cyber risk needs to be treated like a threat to a country's national security, requiring a strong collective response and measures to shore up the system. In effect, cyber risk should be treated like market risk, especially among certain vulnerable industries and for small to mid-sized enterprises. Organizations will often adhere to known best practices in information security, knowing that they cannot reasonably contain every cyber risk. Capping legal liabilities as a juridical matter is a choice regulators and lawmakers can make, removing some of the more damning consequences of this risk domain.

In order for this model to work, however, attitudes among competing firms must change. Clearly, while a firm like Sony was grappling with its embarrassing breach, rival companies may have taken market share and screening slots at movie theaters that were barred from showing *The Interview*—or were afraid to screen it for fear of a backlash or being embroiled in Sony's drama.

---

[17] D. Disparte, "It is Time for a Cyber FDIC," *The Huffington Post*, April 17, 2015, accessed on November 30, 2015, http://www.huffingtonpost.com/dante-disparte/it-is-time-for-a-cyber-fd_b_7083948.html.

This illustrates something that is understood by leading risk practitioners—that risk management is a source of competitive advantage. While the playing field of a market is littered with the wreckage of affected firms, the agile enterprise (those not similarly affected) are able to chart their own course with the temporary reprieve from looking in the rearview mirror at the movements of their competition. While Home Depot was reeling under a large-scale cyber breach, for example, its bitter rival, Lowes, which was mercifully spared a similar public debacle, made marketplace advances.

The FDIC is regarded as one of the most effective and enduring Federal agencies in the U.S., and its services, like a centralized cyber risk pool, are not free of charge. Banks pay risk premiums based on their balance sheet, solvency, and other factors. These funds are then used to shore up the broad system against bank failures and protect consumer accounts up to $250,000. Following the Great Depression and the wave of bank runs or failures, the FDIC provided a vital safety net and restored confidence in the system. In addition to the powers of restoring trust and capping exposures, a cyber FDIC must also carry strong deterrent powers, in coordination with other government agencies. In short, speaking softly and carrying a big stick is an apt description; cyber thieves should face proportionate applications of power for their misdeeds.

When the TRIA was drawn up following 9/11, cyber risk was in its infancy. As a result, it was not contemplated in the Act, which was designed to shore up mature property and casualty insurance markets and transfer catastrophic losses to the federal government. The fact that Sony's business model was successfully held for ransom by the GoP shows that this risk category has come of age, and that appropriately agile responses are now needed.

The hack of British telecom firm TalkTalk offers guidance on how it is more costly to suffer a breach than to adhere to best practices in preventing them. Initially, TalkTalk reported that up four million customer records (including bank account details) were comprised by a cyber-attack. In reality, the number was closer to 4 percent of the total.[18] This misinformation made an already unfortunate case worse than it needed to be. Adding insult to injury, the hack was apparently carried out by a group of teenagers who were arrested in late 2015, in connection with the case. Estimates show that this breach cost the firm $53 million, in addition to a 20 percent slump in share value following the incident, proving that in the information age, bad

---

[18] D. Thomas, "Teenager Arrested in TalkTalk Hack," *Financial Times*, November 25, 2015, accessed on November 30, 2015, http://www.ft.com/cms/s/0/fdc801ae-936e-11e5-bd82-c1fb87bef7af.html#axzz3t0wJ8n13.

news travels fast.[19] Perhaps the most confounding aspect of this breach from the customer's perspective, is the fact that customers may not be able to 'vote with their feet' and leave TalkTalk, as they are stuck with the usual lock-in associated with telecoms contracts. Here too, a clear ransom demand was made.

Ironically, most firms go it alone when they learn of a breach. This *omerta* is a by-product of fear of a public and employee backlash over the exposure of private data and trade secrets—a silence that makes the overall system weaker, under the guise of risk and crisis management. It is a simple, if painful truth that holding on to bad news does not dissipate its impact, but often makes it worse. Therefore, the final aspect of a cyber FDIC would be to run a central risk reporting clearinghouse that would collect system-wide information on breaches, near misses and comparative data across time. This last function would not only help mitigate cyber risk, it would also help improve private sector risk pricing and appetite as the industry still labors under limited historical data for this relatively new risk domain.

The Internet of Things

With the rise of a wide variety of cyber risk, people often wrongly assume that these threats remain isolated to the virtual world. The truth is that cyber risk is increasingly as much about obtaining data and breaching systems as it is about the risk to tangible assets and physical infrastructure. In Sony's breach, $35 million in hardware and systems were rendered unusable following the GoP attack.[20] Of even greater concern is the vulnerability of critical national infrastructure to cyber risk. All manner of systems, from electricity grids to water control and strategic military systems, are vulnerable and under constant threat. Showing the degree to which cyber-attacks—when deployed correctly—can leap from the virtual world to the real world, Iran's nuclear program was set back many years by the Stuxnet cyber weapon, which was allegedly unleashed by the U.S. in conjunction with Israel.[21]

---

[19] "TalkTalk Hack to Cost up to £35 Million," BBC News, November 11, 2015, accessed on November 30, 2015, http://www.thefrontierpost.com/article/351796/talktalk-hack-to-cost-up-to-35-million/.

[20] T. Hornyak, "Hack to Cost Sony $35 Million in IT Repairs," *CSO*, February 4, 2015, accessed on November 30, 2015, http://www.csoonline.com/article/2879444/data-breach/hack-to-cost-sony-35-million-in-it-repairs.html.

[21] E. Nakashima and J. Warrick, "Stuxnet was Work of U.S. and Israeli Experts, Officials Say," *Washington Post*, June 2, 2012, accessed on November 30, 2015, https://www.washingtonpost.com/world/national-security/stuxnet-was-work-of-us-and-israeli-experts-officials-say/2012/06/01/gJQAInEy6U_story.html.

Deemed to be a relatively sophisticated military-grade cyber weapon, Stuxnet was deployed in 2010. Unlike most cyber-attacks, which aim to retrieve private information or shut down systems through DoS attacks or by overloading servers, Stuxnet was designed to attack physical systems. The virus caused Iranian nuclear centrifuges to spin at an excessive rate, triggering a physical breakdown of thousands of centrifuges.[22] While Stuxnet was heralded as a 'victimless' cyber-attack, it raised the bar on an increasingly dangerous game of cyber brinksmanship being waged between countries. No two countries are locked into this cyber arms race more than China and the U.S., and Stuxnet should be considered an obsolete cyber armament when compared to today's military capabilities.

China is allegedly behind a massive breach of U.S. government systems that exposed sensitive data on more than 22 million people, including a vast percentage of federal employees.[23] Given its scope, what information makes it into the public domain is often the tip of the iceberg. The risk to critical systems, utilities, fail-safes and air traffic control (among others) is a top of mind concern in the cyber security domain. Whether a cyber risk manifests itself through deliberate action, a system error, or a glitch, the outcome is often the same—pandemonium and a rapid break down of order.

While airlines are typically well regarded by risk practitioners, the air traffic control scorecard continues to suffer embarrassingly risky system errors, near-misses, and glitches. One such case opened up the skies over much of the Northeastern U.S. corridor in 2015. A critical air traffic control system at Dulles Airport in Virginia prompted the rerouting and grounding of hundreds of flights. Critical systems are only missed when they are offline—much like autonomic systems in the body, the expectation is 100 percent uptime; it is only when we face shortness of breath that we realize how important our lungs are. To a well-functioning modern economy, connectivity is the air we breathe. From time to time, critical systems will fail and the lights will go out. This can be triggered by foolhardy management or by deliberate malicious action. Whatever the cause, the impact is usually the same.

Showing how complex systems fail in complex and often-inadvertent ways, the largest blackout in North American history occurred in the sweltering summer of 2003. Allegedly triggered by sagging power lines that came into contact with trees in Ohio, a cascade of failures and failed fail-safes sent 50 million people into two days of darkness across eight states and parts of

---

[22] Nakashima and Warrick, "Stuxnet."

[23] J. Sciutto, "OPM Government Data Breach Impact 21.5 Million," CNN Politics, July 10, 2015, accessed on November 30, 2015, http://www.cnn.com/2015/07/09/politics/office-of-personnel-management-data-breach-20-million/.

Canada. The economic toll was estimated at $6 billion, and it is believed that 11 lives were lost due to the blackout.[24] Although this massive blackout was apparently not triggered by malicious events, it underscores the vulnerability of the system, how easily a dangerous cascading effect can occur, and how long it can take to get systems back online following relatively simple failures. Imagine the consequences of a Stuxnet-styled cyber weapon released against North America's creaking electricity infrastructure. The recovery process would be lengthy and the potential software and hardware damage would exact a heavy toll on the U.S. economy.

From the trivial to the severe, cyber risk is here to stay and agility and prudence are required in order to remain a step ahead of it. An increasing number of cases are emerging where otherwise sound products pose serious risks to consumers by virtue of being 'connected'. The alleged hacking of a United Airline's airplane flight controls mid-flight by Chris Roberts, a so-called ethical hacker, takes the possible risk of the 'internet of things' to new heights. The FBI's investigation into whether a backdoor exists through an airplane's entertainment system is ongoing. If so, connected air travel may be considerably more dangerous[25]; All the more so following the U.S. Federal Communications Commission's ruling in favor of mobile phone use through the cruising stage of flight.[26]

GM faces similar risks of vehicle systems allegedly being controlled remotely through onboard systems like OnStar.[27] In this age, even the sanctity of babies sleeping in cribs has been violated by connected baby monitors and video cameras that are web-enabled.[28] Accepting the continued encroachment of a connected world and connected devices and systems comes with a potentially steep price. So, just as parents must weigh the risk–reward trade-off of over-exposing their children to the internet, organizations must weigh their own

---

[24] J.R. Minkel, "The 2003 Northeast BlackoutDOUBLEHYPHENFive Years Later," *Scientific American*, August 13, 2008, accessed on November 30, 2015, http://www.scientificamerican.com/article/2003-blackout-five-years-later/.

[25] "Security Expert Said he Hacked Plan Controls Midflight," *Chicago Tribune*, May 18, 2015, accessed on November 30, 2015, http://www.chicagotribune.com/news/nationworld/ct-flight-hacking-investigation-20150518-story.html.

[26] T. Devaney, "FCC Rule Would Lift In-Flight Call Ban," *The Hill*, January 14, 2014, accessed on November 30, 2015, http://thehill.com/regulation/technology/195358-fcc-rule-would-lift-in-flight-call-ban.

[27] J. Finkle and B. Woodall, "Researcher Says can Hack GM's OnStar App, Open Vehicle, Start Engine," Reuters, July 30, 2015, accessed on November 30, 2015, http://www.reuters.com/article/2015/07/30/gm-hacking-idUSL1N10A3XK20150730#BAV0OC0vC9YCwDty.97.

[28] Y. Grauer, "Security News This Week: Turns out Baby Monitors are Wildly Easy to Hack," *Wired*, September 5, 2015, accessed on November 30, 2015, http://www.wired.com/2015/09/security-news-week-turns-baby-monitors-wildly-easy-hack/.

posture. The best defense in the information age is to have no fear of sunlight. The organizations that gain value in this age care less about data for the sake of information, and more about information for the sake of decision making. The risk to critical systems is ever-present and agile risk and security managers must embark on a zero-failure mission.

Just as secured doors to an airplane's flight deck guarded against the risk of hijacking while creating the risk of a locked-out captain, the advent of near-field communication (NFC) may have a similar outcome. NFC is a widely adopted communication standard between connected devices that enables consumers using Google Wallet or Apple Pay to skip traditional forms of payment for their in-person shopping. As the name suggests, near-field requires that the individual consumer and connected devices are within a certain proximity of each other, in most cases requiring physical contact. When combined with growing biometrics, such as Apple's fingerprint scanning for its late model iPhones, many of the known financial frauds and loopholes are minimized. The question is: what new risks will emerge from this standard? Will financially motivated kidnappings now involve an individual victim who cannot be separated from his or her wallet? Will Apple and Google become the next version of SIFIs?

When Apple Pay was first launched, fawning financial institutions such as MasterCard (among others) boasted that they were among the first to accept this standard. MasterCard went as far as taking out a full-page ad in the *New York Times*, potentially serving themselves up on a platter to the serial disrupters at Apple.[29] A herd mentality is perilous wherever it emerges. That the largest financial institutions in the world were queuing up like so many frenzied shoppers waiting for the latest iPhone shows a lack of circumspection and prudence for what is ultimately a relatively untested financial standard—one that may create new risks, while addressing known security flaws inherent in traditional payment methods. Apple's launch of Apple Pay in China may very well make Apple the world's largest payment provider.[30] Apple clearly falls outside the purview of financial regulators, raising real questions about the speed with which Apple Pay and Google Wallet have disrupted the traditional payment process, and the troubling willingness of the industry's stalwarts to acquiesce. While Apple Pay, Google Wallet and NFC technology generally

---

[29] D. Smith, "MasterCard Takes a Full Page Ad in *The New York Times* to Show Off Apple Pay and The iPhone 6," *Business Insider*, September 12, 2014, accessed on November 30, 2015, http://www.businessinsider.com/mastercard-new-york-times-full-page-ad-shows-apple-pay-and-the-iphone-6-2014-9.

[30] Y. Jie and L. Wei, "Apple Seeks to Launch Apple Pay in China by February," *The Wall Street Journal*, November 24, 2015, accessed on November 30, 2015, http://www.wsj.com/articles/apple-pay-may-launch-in-china-by-February-sources-say-1448340287.

have a strong record, when combined with biometrics, people may quite literally face personal risk from their technological connection.

Flash Crash...and Burn

Since the world abandoned the Gold Standard and adopted the financial standards of the Bretton Woods Agreement in 1944, economic value has been notionally ascribed to the costs of goods and services in the tangible world—thus losing its 'real world' peg. While the economic utility of this move is sound and has facilitated global commerce, when combined with technology, billions of dollars of value can be eradicated in seconds, whether by deliberate man-made risk or inadvertent human or system error. The most glaring case, analyzed earlier in the book, was the collapse of Knight Capital, the result of an untested trading algorithm. Beyond Knight Capital, which was presumably a high-speed accident, flash crashes of critical systems appear to be happening with increasing frequency and substantial consequences in financial markets.

In the spring of 2010, the Dow Jones Industrial Average experienced a 1,000-point drop in a matter of minutes. The triggers of this *flash crash* were high-frequency algorithmic trading firms that were 'supposed' to make their investors safer by responding to new market information in micro-seconds.[31] Instead, high-frequency trading firms undermined market stability during a period of high volatility because 'sell' algorithms were programmed to respond to trading volume, rather than price or time.[32] While traders were able to reconstitute their losses, these types of cases serve as reminder that in the world of notional value and immediate gratification, from time to time the worst-case scenario will occur. When it does, having clear rules of engagement would be helpful. Rather than making the markets safer, high-frequency trading has proven to respond far too quickly to information and misinformation in markets prone to increasing overall volatility.

Even making systems 'upgrades' can lead to large potential losses and restatements of notional value. China's largest securities trader, Citic Securities, had to restate its derivatives business by $166 billion in 2015. While the restatement was eventually corrected, the system 'upgrade' was part of a wider ring

---

[31] B. Korn and B.Y.M. Tham, "Why We Could Easily Have Another Flash Crash," *Forbes*, August 9, 2013, accessed on November 30, 2015, http://www.forbes.com/sites/deborahljacobs/2013/08/09/why-we-could-easily-have-another-flash-crash/.

[32] "What Caused the Flash Crash? One Big, Bad Trade," *The Economist*, October 1, 2010, accessed on November 30, 2015, http://www.economist.com/blogs/newsbook/2010/10/what_caused_flash_crash.

of police and regulatory investigations directed at the top of the organization.[33] When dealing with the incredibly high notional value of complex financial products and markets, adding technological complexity and opacity to the mix can create an environment rife with operational risks and moral hazard. In many cases, the systems are so convoluted that piecing together an adequate history of the terabytes of information that are exchanged in seconds is nearly impossible. Doing so through the global financial system, which still toils without much sensible coordination, is an exercise in futility.

Cyber risk is so amorphous that it can leap from being a virtual threat to causing actual physical damage. By the same token, the physical world can also wreak havoc on the virtual one, including globally distributed data centers. From the simple act of introducing a flash memory drive into a protected system, or using an external mass storage device to download information, the physical world reminds us how fleeting our virtual lives can be. The rare lightning strike that shut down power at one of Google's data centers in Belgium serves as a telling example. While a fractionally small amount of data was permanently lost, for which Google accepted liability, it nevertheless highlighted system vulnerability—even for a company with the largest and most cutting-edge online presence. This case was a stark reminder that in the ephemeral internet age, vast data centers represent the physical backbone of the internet and a critical point of vulnerability, no matter how interconnected and backed up they may be.[34]

Underscoring the risk to physical infrastructure of the world's virtual web, the increasingly common incidence of undersea fiber optic cables being severed by anchors from commercial vessels continues to plague the world, often retarding development goals. The busy east African port of Mombasa (in Kenya), where globalization meets regional demand, was the site of four cable-cutting incidents that disrupted internet and phone services for millions of users across nine African countries—from Zimbabwe to Djibouti.[35] This was not an isolated incident, but rather part of a trend where approximately 100 such cables are severed annually (most inadvertently), showing the perilous exposure of the world's hundreds of subsea fiber optic cables that

---

[33] "China's Biggest Brokerage Citic in $166bn Error," BBC News, November 25, 2015, accessed on November 30, 2015, http://www.bbc.com/news/business-34918510.

[34] M. Brown, "Lightning Strikes Kill Customer Data Stored on Google Compute Engine (GCE) Servers," *International Business Times*, August 19, 2015, accessed on November 30, 2015, http://www.ibtimes.com/lightning-strikes-kill-customer-data-stored-google-compute-engine-gce-servers-2060314.

[35] S. Moore, "Ship Accidents Sever Data Cables off East Africa," *The Wall Street Journal*, February 28, 2012, accessed on November 30, 2015, http://www.wsj.com/articles/SB10001424052970203833004577249434081658686.

make global connectivity possible.[36] In another example, raising the specter of deliberate action, in 2013 Egyptian authorities arrested three people for allegedly sabotaging undersea internet cables, and causing a 60 percent drop in the country's internet speed.[37] Natural disasters, such as earthquakes, tsunamis and increasingly violent electrical storms, all pose ongoing risks to the backbone of the internet. Deliberate harm to this infrastructure caused by activists, terrorists, or military action is a real and present danger in the information age.

In the twenty-first century, knowledge is no longer power, for power lies in the ability to disseminate information and, at times, misinformation. The Arab Spring, for example, has been likened as the *Twitter Revolution*, a moniker the government of Hosni Mubarak latched onto when internet service was cut in Cairo at the height of the conflict.[38] Similar to how protests in Tiananmen Square were sustained by faxes with information from the outside world, this century is already rife with examples of social media being used as a tool to organize and sustain popular movements. The generation of Twitter revolutionaries and subversives owe a debt of gratitude to their fax-machine comrades from 1989.[39] This has been one of the more enduring value drivers for firms like Twitter and Facebook, which continue to flirt with being social utilities on the one hand and massive advertising firms on the other.

What these cases highlight is the need for greater coordination protecting the world's internet 'utilities' while at the same time adopting systems that will improve resilience, access, and recovery. Organizations should not assume the permanence or absolute security of data placed on connected systems. More than 90 percent of all the world's data—2.5 quintillion bytes of data a day and growing at an exponential rate—were produced in just *2013 and 2014*.[40] Agility amid the information inundation sinking the world implies knowing how to connect the dots of critical kernels of information and charting a decisive pathway, rather than being mired by data overload. Knowing how to balance the *filter-to-noise* ratio is a key attribute of achieving a balanced

---

[36] A. Chang, "Why Undersea Internet Cables Are More Vulnerable Than You Think," *Wired*, April 2, 2013, accessed on November 30, 2015, http://www.wired.com/2013/04/how-vulnerable-are-undersea-internet-cables/.

[37] *Wired*, "Undersea Internet Cables."

[38] E. Zuckerman, "The First Twitter Revolution?," *Foreign Policy*, January 15, 2011, accessed on November 30, 2015, http://foreignpolicy.com/2011/01/15/the-first-twitter-revolution-2/.

[39] R. Kluver, "Social Media Wouldn't Have Changed Tiananmen Square," *U.S. News and World Report*, June 4, 2014, accessed on November 30, 2015, http://www.usnews.com/opinion/articles/2014/06/04/revisiting-tiananmen-square-after-25-years-in-the-age-of-social-media.

[40] *Bringing Big Data to the Enterprise*, What is Big Data, IBM, accessed November 27, 2015, https://www-01.ibm.com/software/data/bigdata/what-is-big-data.html.

approach to risk taking and decision making today. While organizational resilience means something very different to IT security professionals, in the context of cyber risk, a true enterprise-wide approach is needed along the value chain. From prevention, to deterrence and financial resilience, few private sector firms have achieved this gold standard, and no area of the economy underscores the needed for greater resilience than healthcare.

The Perfect Storm

Cyber risk proves time and time again that regulatory directives to protect information can be meaningless against rapidly evolving threats. A privacy directive to hackers is the informational equivalent of pollen attracting bees. The fastest way to invite a breach (or an attempted breach) is to advertise adherence to the highest standard of safety and security. Directives such as the Health Insurance Portability and Accountability Act and the Privacy Act often carry financial penalties in the event of a breach and impose massive costs on security and technology, without having measurable improvements in overall safety.

The sheer number of medical breaches (at least, those that make into the public light) is alarming. Virtually all the major players in the U.S. healthcare industry, approximately 81 percent of them, have suffered the consequences of having millions of policyholders' private information stolen by hackers.[41] Such data are among the most sensitive points of information an individual has, which includes personal identification, medical records, and financial information. Healthcare is in many ways the perfect storm of all the cyber risks whipped up into a frenzy in one space.

For example, power outages in a hospital—especially in the absence of any backup power systems—immediately present life and death risks to certain patients, and incalculable malpractice risks to hospitals and doctors. Inadvertent risks to information systems may trigger the loss of critical medical records and prescription information. Health insurers, hospital groups, and patient networks are the persistent targets of cyber risk. Increasing regulatory complexity, privacy impositions, and a continuing volley of large-scale breaches whose costs are ultimately passed on to consumers, puts the industry under great strain to stay ahead of security trends.

---

[41] J. Abel, "At Least 81 Percent of Major Healthcare or Health Insurance Companies had a Data Breach in the Past Two Years," *Consumer Affairs*, September 4, 2015, accessed on November 30, 2015, http://www.consumeraffairs.com/news/at-least-81-of-major-healthcare-or-health-insurance-companies-had-a-data-breach-in-the-past-two-years-090415.html.

With all its fragmentation, opacity, competing interests and systems vulnerability, and set against a persistent series of cyber risks, the U.S. healthcare market is a great space from which to keep a temperature on what government and industry are doing right, and wrong, in cyber space. The industry highlights the often-unintended conflicts that can arise from digitizing information (such as medical records), while at the same time regulating privacy standards with a carrot-and-stick approach. In a time of rampant cyber risk affecting an entire industry, goals and expectations clearly need to be adapted, as well as the implementation of caps on legal and financial liability. The confluence of regulatory complexity, market fragmentation and compliance costs all serve to erode economic competitiveness.

Blurred Lines

In a hyperconnected world, the internet is at the same time a source of both risk and great reward. As easy as it is for tech titans in Silicon Valley to harness technology to create incalculable economic value, it is just as straightforward for criminals, nations, and terrorists to use technology to cause substantial harm. The citizens of nations across the world are in many ways the bystanders and the victims of this drama, and their civil liberties and right to privacy are collateral damage.

On one hand, the creation of stronger cyber security solutions and encryption standards makes financial and legal transactions safer. On the other hand, these hyper-secure tools and methods fall into the hands of tech trouble makers and make everyone more vulnerable. The blurred lines between security and privacy continue getting obscured and the most sacrosanct of Western values are being called into question in the process. Government security and intelligence agencies in the U.S., Europe, and around the world continue weighing the costs and benefits of security and privacy in a digital age—themselves often operating in a vacuum with no public oversight. The terror attacks conducted by the IS in Paris in November 2015 have significantly raised the volume on this debate in Europe's already fraying union.[42]

The Paris attacks illustrate the limitations of the security services and their cyber espionage capabilities, as terrorist and hackers prefer relatively simple, encrypted, and anonymous communications. France's intelligence service, the

---

[42] "Cyber Security: The Terrorist in the Data," *The Economist*, November 28, 2015, accessed on November 30, 2015, http://www.economist.com/news/briefing/21679266-how-balance-security-privacy-after-paris-attacks-terrorist-data.

DGSE, is widely considered to be one of the world's best, yet it failed utterly to detect or prevent any of the Paris attacks. With hundreds of potential terrorist cases being followed at any given time, it is virtually impossible for the authorities to anticipate attacks 100 percent of the time.

As a result, Europeans should prepare themselves for an even greater level of scrutiny from their governments, on all levels. They should expect more closed-circuit cameras to be installed, more police and military presence, and an even greater level of electronic monitoring going forward. That is the only way the security services can hope to gain an upper hand on so-called "lone wolf" terrorists or organized paramilitary cells. Europe is likely to look and feel different in the next decade than it does today. The far right is likely to gain greater political strength in legislative bodies throughout the continent. More resources are likely to be devoted to the military, police, and security services going forward. Tighter restrictions have already been imposed on movement between borders in the EU. The days of open Schengen borders are numbered.

Agile Threat … Agile Response

Just as nations must balance civil liberties, the right of privacy with security, and the need among intelligence services to be omniscient, so too organizations must strike their own balance between information privacy and security. From simple insider threats of mass exposure or data theft to more sophisticated external threats and business models being held for ransom, cyber risk has proven to be an agile and ever-present foe—one that can be triggered into a complex cascade by the deliberate and the inadvertent. In a way, the sprawling surface area of an interconnected global economy is the playground for the dark imaginations of connected malefactors. Even the risk of industrial espionage, where firms attempt to reverse engineer products and processes by hacking sensitive intellectual property, is becoming increasingly common. Perhaps most troubling is the fact that the public seems to have blithely accepted the tradeoff that their privacy and information have become marketable securities in the internet age. This creates an inevitable tension between civil liberties, company boundaries, and privacy directives that often have the unintended consequence of making their subjects unsafe, while adding to regulatory complexity and costs, and not necessarily achieving the degree of safety desired.

As the pursuit for a more perfect balance between such competing objectives rambles on, the stakes for companies that fail to rise to the occasion only grows. In a particularly egregious example of how company liabilities continue morphing in the age of terrorism and cyber risk, Africa's mobile telecom giant, MTN, was fined a record $5.2 billion by Nigerian authorities in 2015. The fine alleges that the sale of ubiquitous and hard to control mobile SIM cards aided Boko Haram's reign of terror.[43] Showing the dilemma of the information age, a tool that at once aids development also hinders security. Mobile density has been heralded as one of Africa's economic advantages, yet, as the MTN case highlights, terrorists also use mobile phones. This dilemma is true the world over. Creating a global information system that equips security services, organizations, and consumers alike with safety, security, and transparency, while protecting privacy, has proven to be one of the most difficult challenges of our times. With each step forward made by cyber security specialists and intelligence services, the threat and complexity of cyber risk moves forward in lockstep.

Is Cyber Risk to Be Feared or Respected?

Clearly, the internet is the thread that holds globalization together, and there is no escaping from the web it has spun, connecting people, markets, and countries. Harnessing cyber risk by having an *eyes wide open* approach to the perils of the information age is not only prudent, it is the source of competitive advantage for many organizations. The co-movement of progress and failure on cyber security raises many important questions for risk practitioners, IT security professionals, and intelligence services. Among them are: How should organizations respond to this rapidly evolving risk domain? What is the appropriate balance of 'collusion' between the private sector and government agencies? How can countries coordinate their response to cyber safe havens and permissive countries that not only sponsor cyber-attacks, but harbor their perpetrators? The list of questions plaguing professionals and boards alike is long and growing. The response to cyber risk, like other risks analyzed in the book, is the same.

An adequate organizational response to cyber risk must match process with process. While a lofty objective, firms should embark on a zero-failure mission when it comes to erecting their counter-valence. Agile risk and security professionals would be wise to remember that insurance is not a substitute for risk

---

[43] S. Tshabalala, "Nigeria is Fining MTN $1,000 Per Illegal SIM Card Even Though Customers Generate Just $5 a Month," *Quartz Africa*, October 26, 2015, accessed on November 30, 2015, http://qz.com/533041/africas-largest-mobile-network-is-being-fined-5-2-billion-for-flouting-nigerias-sim-card-rules/.

management, but merely a mechanism for financing losses. Stopping cyber-related losses in the first place is the key objective of a well-developed cyber risk management strategy. Structurally, so much of the IT security and risk management disciplines labor under the burden of organizational siloes and fragmentation that militate against the required *enterprise-wide* response to fast-moving cyber risks. Instead, most firms silo IT professionals, risk managers and other crisis-related functions, with little or no meaningful interaction with senior organizational leadership, vertically or horizontally in their organizations.

This traditional organizational structure is the very antithesis of the type of speedy, agile structure that is needed to combat cyber risk. Rather than having an early warning system, large, sclerotic enterprises often have silent cyber threats lurking in their systems for many years, entirely unnoticed. When breaches occur or risks are detected, recrimination, political in-fighting and, above all, high costs to repair dated systems follow. While many of the risks falling into this category can be readily absorbed by large firms (even without insurance), the source of irreparable harm is the often-salacious nature of internal communications. Stronger adherence to stated organizational value systems, as outlined earlier, is entirely free of charge, and can help blunt the impact of unwanted disclosure. Firms that are found to be consistent in public and behind closed doors may even benefit from enhanced enterprise value, employee and customer loyalty, and stock market valuations. In the face of cyber risk, agile risk managers have no fear of sunlight.

As in banking and economics, complexity is not only additive in the cyber risk landscape, it is compounding and creating a vast number of interactions and correlations. Risk and security professionals must unite to create a common front against this risk domain. Collaboration across the enterprise is key, as are investments on the front end. In the case of cyber risk, prevention is much less costly than a cure, and organizations will need to collaborate with their industry peers—and, at times, bitter rivals—to stay ahead of these risks. The meteoric rise of cyber insurance, while not enhancing overall financial readiness following a loss, is no substitute for prevention, although some insurers are providing vulnerability assessments to close loopholes as a part of their loss prevention efforts.

While trite, if there is a way into a critical system, there is usually a way out. The sprawling surface area of global enterprises makes for millions of points of vulnerability. From the simple introduction of flash memory drives, to employees downloading malware or using easily deciphered passwords, the task of the best of breed cyber risk manager is not only sleepless—it is often thankless. Much like the questioning of intelligence agencies following a terrorist attack, IT security professionals are the silent sentinels of the modern economy.

In facing down cyber risk it is important not to conflate insurance with deterrence. While the existence of insurance has a catalytic property across many domains, cyber risk is much more punishing than other insurable risks. The risk of contagion across systems can literally handicap a company's business model or an economy's lifeline. Irrecoverable losses of this nature are hard to price and the threat is very real, although there have only been comparatively mild cases so far. Rather than adopting a reactive posture, agile risk managers should go on the offensive against cyber risk. Like all risks, cyber risks will continue to evolve, especially as the virtual and real worlds continue their inexorable collision course merging into one new frontier. The tacit acceptance that privacy is marketable and that the state should be omniscient are but two of the dangerous new trends shaping our times. How organizations build resilience in the face of this new normal will have a profound and lasting impact on how the world thinks in future about everything from information to intellectual property to notional economic value. The challenge is ongoing and never-ending, and will haunt agile risk managers and decision makers for decades to come.

10

Corporate Social Responsibility

Shades of Grey

CSR is not easy to define. It is simultaneously a philosophy, a key component of conducting business, and an integral element of corporate culture. It is so multifaceted as to warrant reference to the following definitions, which do a good job of capturing CSR's complexity and significance:

1. CSR is a management concept whereby companies integrate social and environmental concerns in their business operations and interactions with their stakeholders. CSR is how a company achieves a balance between economic, environmental and social imperatives (the so-called "Triple Bottom Line Approach"), while at the same time addressing the expectations of shareholders and stakeholders.[1]
2. CSR is a pact between a company and society whereby the company agrees to pursue goals in addition to profit maximization, and a pact between the company and its shareholders whereby stakeholders hold the firm accountable for its actions in pursuit of those goals.[2]
3. CSR encompasses not only what companies do with their profits, but how they make them. It goes beyond philanthropy and compliance to address the manner in which companies manage their economic, social

---

[1] United Nations Industrial Development Organization (UNIDO), *What is CSR?*, 2015, accessed November 25, 2015, http://www.unido.org/en/what-we-do/trade/csr/what-is-csr.html.
[2] D. Chandler and W.B. Werther, *Strategic Corporation Social Responsibility* (New York: Sage Publications 2011).

© The Author(s) 2016
D. Wagner, D. Disparte, *Global Risk Agility and Decision Making*,
DOI 10.1057/978-1-349-94860-4_10

and environmental impacts and their stakeholder relationships in all their key spheres of influence: the workplace, the marketplace, the supply chain, the community and the public policy realm.[3]

CSR is an intricate and expanding landscape of interconnected pursuits and obligations that include environmental management, eco-efficiency, responsible sourcing, stakeholder engagement, labor standards, working conditions, employee and community relations, social equity, gender balance, human rights, good governance, and anti-corruption measures.[4] Once a firm dives into the CSR pool, it can quickly become overwhelmed as it tries to establish itself as a good corporate citizen, while at the same time wanting to prove that it is genuine in its beliefs and its efforts.

Shades of grey permeate the CSR landscape. Is it a duty or an obligation? Is it something a firm 'should' do or rather something it 'should want' to do? Milton Friedman's narrowly focused approach to CSR contends that the only 'social' responsibilities of business organizations are to make profits and to obey laws. By this view, the free and competitive marketplace will 'moralize' corporate behavior. In essence, morality, responsibility, and conscience reside with the invisible hand of the free market system, not in the hands of organizations within the system, much less managers within organizations.[5]

An alternative view to the invisible hand of the *marketplace* is the hand of *government*, where corporations pursue objectives that are rational and purely economic. John Kenneth Galbraith was the leading proponent of this view, believing that the regulatory hands of the law and the political process turn these objectives to the common good. Neither the invisible hand nor the hand of government arguments entrust corporate leaders with stewardship over CSR values. Under the 'hand of management' argument, it is the thought process of an organization that guides the CSR process.[6] A corporation *can* have a conscience; the trick is to make *having* a conscience and *wanting* to do things 'the right way' standard operating procedure.

Whether you subscribe to the market, government or hand of management view, CSR has become such a phenomenon and such an integral aspect of doing global business that any firm that is active globally really must have a CSR

---

[3] B. Kytle and J.G. Ruggie, *CSR as Risk Management*, Harvard University, JFK School of Government, Working Paper 10, March 2005, 9.

[4] UNIDO, *What is CSR?*

[5] K.E. Goodpaster and J.B. Matthews, "Can a Corporation Have a Conscience?," *Harvard Business Review*, January 1982, accessed November 25, 2015, https://hbr.org/1982/01/can-a-corporation-have-a-conscience.

[6] Goodpaster and Matthews, "Can a Corporation Have a Conscience?"

philosophy and plan of action—for its *own* benefit and protection. Regardless of a firm's motivations for jumping into the CSR pool, the task of any business operating internationally that wishes to embrace CSR is to maintain acceptable levels of profitability while at the same time displaying a genuine concern for the people and environment associated with international operations, and simultaneously making that approach a hallmark of corporate culture.

Governing Principles

There are four basic principles that govern CSR—Accountability, Fairness, Responsibility and Transparency—all of which can impact firm performance. These principles should ideally also be elements of corporate philosophy outside the CSR realm, in order to increase shareholder value and satisfaction. CSR is ultimately focused on creating balance between the economic and social goals of a company, including the efficient use of resources, accountability in the use of power, and the behavior of the corporation in its social environment. While the definition and measurement of good corporate governance remains a subject of debate, we can say that good corporate governance includes the following aspirations:

- Creating sustainable value.
- Achieving firm objectives.
- Increasing shareholder satisfaction.
- Efficient and effective organizational management.
- Enhancing credibility.
- Ensuring efficient risk management.
- Creating an early warning system to address all forms of risk.
- Maintaining a balance between economic and social benefit.
- Ensuring efficient use of resources.
- Distributing responsibility equitably.
- Producing all necessary information for stakeholders.
- Keeping the board independent from management.
- Facilitating sustainable performance.[7]

The CSR 'debate' is focused on whether a company's sole purpose is to maximize shareholder value or to consider the needs of a broader range of

---

[7] S.P. Harish and S.P. Santosh, "Affects of Globalization and Limitations of CSR," *IOSR Journal of Humanities and Social Science*, vol. 14, issue 4 (September–October 2013), 46.

stakeholders in the decision-making process. There is, of course, more than one way to get to the finish line, but what is shared by most companies in the CSR space is a desire to keep all stakeholders happy. That is sometimes easier said than done, but if a company can be seen to at least be making an honest attempt to stay engaged in the CSR process and demonstrate the above-referenced attributes, that is half the battle.

CSR and Globalization

In the 1990s, social and environmental issues were considered the domain of social activists. Today, these issues are in the mainstream—among the most critical factors shaping corporate strategy and government policy—and consistent with not only the globalization process, but, specifically, the globalization of laws and reporting practices. Common standards now exist for governance, compliance, and regulations across a broad range of sectors and industries. That is good for CSR, because the globalization process brings with it some norms and standards that propel business toward CSR, rather than away from it.

Leading the way toward CSR are a host of initiatives developed by multilateral institutions and designed to encourage and enforce good behavior by businesses in the global community. Among the most significant of these are:

1. *Global Reporting Initiative (GRI)*: The GRI was established to improve sustainability reporting practices while achieving credibility, comparability, timeliness, and the verifiability of report information. In essence, the Initiative promotes globally accepted sustainability reporting guidelines.
2. *Global Sullivan Principles (GSP)*: The GSP constitute a voluntary code of conduct seeking to enhance human rights, social justice, protection of the environment, and economic opportunities for workers of all nations.
3. *ILO Tripartite Declaration*: The International Labor Organization's (ILO's) Declaration encourages the positive contribution of multinational enterprises (MNEs) toward economic and social progress by respecting international standards, obeying national laws, and honoring voluntary commitments that harmonize their operations with the social objectives of the countries in which they operate.
4. *OECD Guidelines for MNEs*: The Guidelines are the longest-standing initiative for the promotion of high corporate standards (since 1976) and contain voluntary principles for business conduct in such areas as human rights, supply chain management, labor relations, the environment, com-

petition and consumer welfare. Their aim is to promote the positive contribution of MNEs to economic, environmental and social progress.

5. *Principles for Responsible Investment (PRI)*: The PRI is a voluntary initiative that strives to act on common ground vis-à-vis the goals of institutional investors and the sustainable development objectives of the UN. While observance of the PRI is voluntary for companies, its member governments make a formal commitment to promote their observance among their own MNEs. The adhering countries comprise nearly 90 percent of the world's FDI and are home to most of the world's MNEs.

6. *Sustainable Development Goals (SDG)*: The SDG replaced the Millennium Development Goals with a post-2015 global development agenda that includes integration with the UN's evolving development agenda, the involvement of all relevant stakeholders in the development process, and the implementation of the agendas of all major summits related to society, economy, and the environment.

7. *UN Global Compact*: The Compact is a voluntary initiative based on ten core principles related to anti-corruption, the environment, human rights, and labor standards. The Compact's principles are derived from the Universal Declaration on Human Rights, the ILO's Declaration on Fundamental Principles and Rights at Work, the Rio Declaration on Environment and Development, and the UN's Convention Against Corruption.

8. *UN Norms*: Officially entitled "The UN Norms on the Responsibilities of Transnational Corporations and other Business Enterprises with regard to Human Rights", the Norms seek to establish a comprehensive legal framework for companies' human rights responsibilities. The Norms seek to standardize existing standards regarding human rights and labor standards, and address such issues as corruption, environmental sustainability, security and worker's rights.[8]

These have been developed over decades and form the backbone of a formidable framework for the coexistence of business, government, and social actors. While a variety of companies do a good job of adhering to these principles and guidelines, others do not. Considerable progress has been made on the environmental front—in large part the result of widespread awareness and a realization that we really are 'all in this together'—but in such areas as cor-

---

[8] Adapted from "CSR: Impact of Globalization and International Business, Bond University," *Corporate Governance eJournal*, April 2007, 9–12.

ruption, human rights, corporate transparency and labor standards, progress has been more constrained.

CSR remains largely a voluntary process that some governments promote more strongly than others. A gap remains with regard to the legal accountability for CSR practices, especially in relation to MNE operations outside their home jurisdiction. Depending on the country and industry, there remains great resistance to the notion of holding companies legally responsible for any actions contrary to the collective objectives of CSR global initiatives. This is unlikely to change in the near or even medium term, as a result of the diversity of companies, governments and legal regimes. It is and will remain up to companies to decide to become part of the CSR community, and to adhere to global frameworks and voluntary commitments.

The Social Risk Landscape

The risk mosaic that companies face today is naturally very different than was the case ten or twenty years ago. Technology and political risk came into vogue in the 1990s, risks associated with globalization and terrorism gathered steam in the last decade, and the perils associated with climate change and cyber risk are coming increasingly to the forefront of today's business culture. So is 'social' risk, broadly defined as *when an empowered stakeholder takes up a social issue and applies pressure on a corporation so that the company will change its policies or approaches in the marketplace.*[9]

More than one stakeholder may apply pressure on a company at the same time, with potentially wide-ranging implications. The stakeholder can be employees, suppliers, investors, NGOs, or even employees who can have either a direct or an indirect impact on a company's management, strategy, operations, finance, marketing, public relations and even its board of directors. Whether or not the company has a well-deserved reputation for violating broadly accepted standards of CSR, pressure may be applied. If a firm uses labor from developing countries, engages in mining or other environmentally sensitive practices, or produces food products, it is already likely to be on an NGO's list.

Merely having existing global reach, or pursuing customers in other countries, may be sufficient to trigger a backlash. Social actors may simply be

[9] B. Kytle and J.G. Ruggie, *CSR as Risk Management*, Harvard University, JFK School of Government, Working Paper 10, March 2005, 6.

looking for ways to enhance their own influence, fill governance gaps or compensate for government failures: Here are a few examples:

- Starbucks was targeted by the U.S. Organic Consumers Association because it was the only large coffee company in the world that proclaimed itself to be socially responsible.
- Activist groups have targeted Nike in the past because it had large profit margins (and could presumably pay higher wages), had been one of the first such companies to go global, and because it was a market leader.
- Former McDonald's CEO Jack Greenberg said that he believed his company had been targeted because activists were looking to publicize their point of view, and brands like McDonald's are highly visible.[10]
- AIDS activists chose Coca-Cola for specific embarrassment in 2002 at the Barcelona AIDS conference not because the company had a particular connection to HIV/AIDS, but because it was a prominent global brand and had large distribution networks throughout Africa (Coca-Cola subsequently agreed to provide treatment not only to its own employees, but also to the staff of its independently owned African bottlers).[11]

This kind of pressure works, which is why it continues to be applied by social actors. Let us remember, however, that the objective of the anti-globalization movement is to instill change in the existing global economic model. It does not seek to change one company at a time. *Individual companies are simply points of entry into the larger debate.*

So, social risk may have nothing to do with perceived right or wrong, but is roughly analogous to a host of other risks faced by companies. Whether a company it targeted by a social actor may be little more than fate, or arriving at its 'turn', in the eyes of a social actor. By the same token, whether an organization is targeted may also have to do with perceived vulnerability and/ or perceived payoff.

When in 2013 the Breast Cancer Fund accused Revlon of using chemicals linked to cancer in its cosmetics, the company called the charges false and defamatory, demanded a retraction and threatened to sue. In the end, Revlon for the first time published an ingredients policy and it is now seen as one of the industry's most comprehensive cosmetic safety policies. The Fund perceived Revlon to be vulnerable, and judged that if it could force the company to make a change, it would have a high perceived payoff. Revlon maintained

[10] Globe Envision, *Corporate Responsibility and Globalization*, January 25, 2007, p. 2. www.globalenvision. org/library/8/1433.
[11] Globe Envision, *Corporate Responsibility and Globalization*, 8.

that such a change was already underway and credited the Fund with helping them realize that they had done a poor job of messaging that change.[12]

The stakes can be high. For example, environmental groups had been pursuing Indonesia-based Asia Pulp and Paper (AP&P) for more than a decade about its alleged damage to Indonesia's forests. In 2013 the company committed to a non-deforestation policy, and to the restoration of a million hectares of natural forests in the country. Once AP&P had agreed to its deforestation policy, Indonesia's second-largest paper company agreed to do the same. Then Cargill and the Singapore-based Wilmar, the world's largest palm oil trader, also agreed to do the same.[13] When actors see this type of impact, they are emboldened to maintain the pressure.

Managing Strategic Partnerships

The potential impact of social risks on a company's operations, profitability, and reputation can be so large as to warrant thinking about them as an integral and standardized component of the larger risk management process. *CSR is really concerned primarily with managing stakeholder relationships*, which is best thought of as a two-way street: the company engages in a partnership with stakeholders in which they are made to feel that they are involved in the corporate decision-making process. In return, the company learns what issues and concerns it should be sensitive to in the planning process, and responsive to in the operational process.

Stakeholder relationships should be thought of as a form of intelligence gathering that can prevent problems from arising and also greatly enhance the chances of arriving at solutions for those problems that do arise. Over time, integrated information flows between stakeholders and companies can form a systemic foundation of knowledge. Among the important questions that can be answered by engaging with stakeholders in this way are:

- What is working in the current approach and what is not working?
- What is the scope of the problem?
- Who has an interest in solving problems?
- What can be accomplished by engaging others in the problem-solving process?[14]

---

[12] M. Gunther, "Under Pressure: Campaigns That Persuaded Companies to Change the World," *The Guardian*, February 9, 2015, accessed November 25, 2015, http://www.theguardian.com/sustainable-business/2015/feb/09/corporate-ngo-campaign-environment-climate-change.

[13] Gunther, "Under Pressure."

[14] Kytle and Ruggie, *CSR as Risk Management*, 11.

Some companies use multi-stakeholder initiatives to maximize the gains of generating strategic intelligence this way, whether structured around specific industries or global issues. The UN Global Compact is perhaps the largest corporate citizenship initiative in the world, with more than 8,000 company members in more than 160 countries that address social issues, the environment, governance, sustainable development and supply chains.

When businesses and social actors work together, they can achieve far more than they could by operating in isolation. The 2014 Corporate/NGO Partnerships Barometer indicated that 87 percent of corporate respondents felt having corporate/NGO partnerships improved their business understanding of social and environmental issues; 59 percent stated that their business practices had been changed for the better as a result.[15]

In today's risk management arena, there is probably more risk associated with *not* generating strategic intelligence this way than there is in doing so. How else can an organization truly understand the landscape and the constantly evolving issues associated with social risks? For example, there is far more upside than downside risk to be gained by getting ahead of the dynamics of the minimum wage debate in the U.S., or fair trade, or the genetically modified food phenomenon. Can your firm afford *not* to be involved in the decision-making process on these, and other similar issues? Those companies that innovate, reach out to their partners, are transparent, and craft mutually beneficial solutions are far likelier to have a good story to tell than those that do not.

CSR and Shareholder Value

In spite of the growth in CSR awareness among global companies, and the broad-based acceptance that CSR is a valuable addition to the corporate risk management process, many firms continue to struggle to integrate CSR into their strategy, planning, risk management, and operational processes. They also struggle to effectively communicate CSR initiatives to what may be skeptical investors, employees, customers, and social actors. The debate over whether CSR adds to or takes away from shareholder value is at the heart of the issue.

In most corporations, the two officers most accountable for shareholder value are CEOs and CFOs, the majority of whom (at least, in Fortune 500 companies) share a common background, being male, having studied business

---

[15] *C&E Corporate-NGO Partnerships Barometer 2015*, candeadvisory.com, accessed November 25, 2015, http://www.candeadvisory.com/barometer.

administration, economics, engineering or accounting as undergraduates, and having also acquired MBAs. The most common paths to the CEO role were through departments concerned with finance, marketing, or operations, while nearly half of the CFOs had previously held accounting positions. The reasonable conclusion, therefore, is that the two executives most responsible for determining shareholder value in these companies came from both educational and functional backgrounds built on quantitative and financial analysis, with an emphasis on measurable results.

Comparable information on CSR executives is scarce, partly because there is no common title for the role and partly because in most companies the role itself is relatively new. Based on the information that *is* available,[16] however, it can be concluded that approximately half of the positions are filled by women and their educational backgrounds are, in general, focused more on social sciences than on business. This implies that, in terms of a frame of reference, CSR executives are more likely to be qualitatively oriented rather than quantitatively oriented in their thinking processes.

Traditional approaches to determining shareholder value are relatively unambiguous. Securities and Exchange Commission (SEC), Generally Accepted Accounting Principles (GAAP), and International Accounting Standards (IAS) accounting rules define what financial information is disclosed and how, and comparable rules apply globally. Revenue, profit, and market capitalization are universally understood concepts. That is not the case with CSR, since there is no singular, widely understood and accepted definition for how CSR contributes to shareholder value. How does one quantify compliance with environmental laws, reducing a carbon footprint, or a campaign to reduce employee obesity in the first place? Wikipedia lists 14 accounting, auditing, and reporting methodologies that are used globally,[17] each of which measures something different and uses its own methodology.

Yet it should be possible to reconcile CSR practices with accepted reporting standards. By focusing on those CSR practices that have a clear connection with financial performance, the primacy of shareholder value should be maintained without any sacrifice of analytical rigor. Basic finance theory states

---

[16] A. Longworth et al. "The Sustainability Executive: Profile and Progress," PWC, October 2012, accessed November 25, 2015, http://www.pwc.com/us/en/corporate-sustainability-climate-change/publications/sustainability-executive-profile-and-progress.html; and J. Davies and the Greenbiz Group, *State of the Profession 2013*, January 2013, accessed November 25, 2015, http://info.greenbiz.com/rs/greenbizgroup/images/State%20of%20the%20Profession%202013.pdf?mkt_tok=3RkMMJWWfF9wsRonu6TAZKXonjHpfsX74%2BkqX6axlMI%2F0ER3fOvrPUfGjI4ATMFnI%2BSLDwEYGJlv6SgFSLHEMa5qw7gMXRQ%3D.

[17] "Corporate Social Responsibility," Wikipedia, accessed November 25, 2015, https://en.wikipedia.org/wiki/Corporate_social_responsibility.

that a company's share price is the present value of future cash flows. CSR activities create shareholder value if they increase profits or reduce the risk of cash flows. Since it is clear that CSR can improve financial performance by reducing costs, increasing revenues and reducing risk, the challenge is to find a way to quantify those benefits in a manner consistent with firm standards.

Some activists have argued that adopting CSR standards enables companies to build brand value by associating their brands with emotions, ideas, and beliefs that have inherent appeal to consumers. They believe that building brand value with CSR initiatives is less expensive than trying to accomplish the same thing via advertising and public relations. Some in the corporate world disagree, noting there is little evidence that short-term CSR initiatives increase sales. Depending on what study one refers to, there is evidence of both an *enhancement* to the bottom line in the short term, and a *reduction*, but the longer a company practices CSR, the more positive the impact on net income tends to be.[18]

What Consumers Really Want

Some of the largest companies in the world have been very vocal in their support for aspects of CSR, and their bottom lines have benefitted as a result. For example, in 2005 General Electric (GE) launched its "Ecomagination" initiative and it made a lot of money in the process. Between 2005 and 2014, the company reaped $160 billion from the program, with the revenues attributable to the program growing twice as fast as total company sales.[19] So 'going green' can add to the bottom line—and that is fine. Businesses would certainly not be incentivized to make CSR part of their corporate culture if doing so took *away* from the bottom line.

Yet CSR isn't simply about 'doing good' and making money while doing it—it can also be concerned with protecting your firm's reputation. Given the seemingly endless corporate scandals around the world related to bribery, corruption, financial shenanigans, and other forms of malfeasance, establishing and maintaining trust and a good reputation is as important as demonstrating ongoing profitability. Consumers want to do business with firms that stand for the things they stand for, can provide integrity, and have a track record

---

[18] Adapted from R. T. Bliss, "Shareholder Value and CSR: Friends or Foes?," CFO.com, February 9, 2015, accessed November 25, 2015, http://ww2.cfo.com/risk-management/2015/02/shareholder-value-csr-friends-foes/.

[19] A. Winston, "GE is Avoiding Hard Choices About Ecomagination," *Harvard Business Review*, August 1, 2014, accessed November 25, 2015, https://hbr.org/2014/08/ges-failure-of-ecomagination.

of consistency. For those that cannot comply, an army of non-governmental organizations (NGOs) is waiting to pounce on them.

Companies are now constantly being watched by NGOs and other third-party actors, and a growing host of rankings pressure firms to report non-financial performance. Although being on an NGO's radar does not necessarily mean profits will be negatively impacted by bad headlines, the lasting impact on a brand's reputation can transcend profits. In recent decades some of the biggest names in business have been singled out for alleged CSR abuses:

1. The 1996 *Life* magazine photo of a 12-year-old Pakistan boy sewing together a Nike soccer ball pushed Nike to raise wages and improve conditions for its assembly workers.
2. In 2007 Human Rights Watch published a detailed report claiming that Walmart stood out for the magnitude and aggressiveness of its anti-union actions.
3. In 2010 the International Labor Rights Forum named Chiquita and RJ Reynolds as among the worst companies of the year for the labor rights practices.
4. In 2013, Apple's manufacturing partner in China, Foxconn, was caught employing under-aged workers and disregarding workplace safety, which resulted in injuries and death.
5. The NGO Earthworks waged an aggressive campaign to get Macy's to sign a pledge to commit to responsible sourcing for its jewelry.[20]

Increasingly, the business world is taking notice. In 2014 Nielsen conducted a global poll[21] of 30,000 consumers in 60 countries to determine how passionate consumers were about sustainable practices related to purchase considerations, which consumer segments were most supportive of ecological or other socially responsible efforts, and the social issues/causes that were attracting the most concern. More than half (55 percent) of global respondents in the survey said they were willing to pay extra for products and services from companies that were committed to positive social and environmental impact—an increase from 50 percent for 2012.

The findings reveal that two-thirds of the "sustainable mainstream" population will choose products from sustainable sources over other conventional

---

[20] *10 Huge U.S. Brands Who Profit From What Americans Would Call Slave Labor*, criminaljusticedegreesguide.com, accessed November 25, 2015, http://www.criminaljusticedegreesguide.com/features/10-huge-u-s-brands-who-profit-from-what-americans-would-call-slave-labor.html.

[21] Nielsen, *Doing Well by Doing Good*, June 2014, accessed November 25, 2015, http://www.nielsen.com/content/dam/nielsenglobal/apac/docs/reports/2014/Nielsen-Global-Corporate-Social-Responsibility-Report-June-2014.pdf.

products. They will buy as many eco-friendly products as they can, and have personally changed their purchasing behavior to minimize their effect on global warming. The referenced respondents said *they will be more likely to buy products repeatedly from a company if they know the company is mindful of its impact on the environment and society.*

For many consumers making a firm's social commitment clearly visible in product packaging is the difference between a purchase and a pass. In fact, for more than half of global respondents in Nielsen's survey (52 percent overall, and nearly two-thirds in Asia, Latin America and the Middle East), their purchasing decisions were dependent on the packaging. These respondents said they checked the labeling first before buying to ensure the brand is committed to achieving a positive social and environmental impact. These are compelling numbers.

CSR pulls its weight beyond manufacturing and consumer goods. A growing number of the largest financial institutions—including Goldman Sachs and Union Bank of Switzerland—now integrate governance, social, and environmental issues into their equity research. Even private equity is responding to public pressure by agreeing to voluntary codes of transparency. Companies are also increasingly finding that they must create a CSR campaign to attract and retain staff, since many employees prefer to work for socially conscious enterprises.

In response, many firms have taken it upon themselves not only to gravitate toward the CSR world, but to reach out to NGOs and governments to create codes of conduct and commit themselves to increased operational transparency. While some of this can be characterized as defensive strategies, companies realize that there can clearly be benefits for those that successfully stay ahead of the curve. 'Doing well by doing good' has become fashionable, prompting more and more companies to want to make CSR part of their corporate DNA.

The business of trying to be good is confronting executives with difficult questions. How can CSR performance be measured? Is it in a company's long-term interest to be cooperating with NGOs, and even competitors, to craft and practice CSR policies? Can a green strategy really result in competitive advantage? And how does the rise of companies in emerging markets change the rules of the game? If CSR is not taken seriously and done well, it can be perceived as insincere and can backfire on a company. If it *is* done well, however, it may well be perceived as a virtue, and just good business.[22]

---

[22] "Just Good Business," economist.com, January 17, 2008, accessed November 25, 2015, http://www.economist.com/node/10491077.

Challenges Companies Can Bring Upon Themselves

It is clear that CSR can bring a company many benefits, while at the same time making it a good global citizen. Of course, this comes at a price. The way a company may do business after becoming that good citizen can have a dramatic impact on its operations, reputation, and character. Not every company that dives into the CSR pool ends up with a good story to tell. As with any corporate transformation, it is difficult to envision what the final product will look like and what the implications are. What follows are a few examples that serve to illustrate some of the challenges.

GE

Earlier in the chapter we mentioned how profitable GE's Ecomagination initiative had been since it began in 2005. After nine years, GE ended up taking Ecomagination into some complicated territory. This included an open innovation program that encouraged ideas to reduce greenhouse gases from Canadian oil sands. In doing so, GE risked what had become a valuable business and brand asset and associated it with a sector not exactly considered consistent with 'green' initiatives.

At the time of its launch in 2005, Ecomagination's DNA was not well defined. It was difficult to tell whether GE's original intent was that of a marketing campaign, corporate strategy, or a product. In time it transpired that it was all three of these things. GE had purposely kept the program broad in scope, including green products such as wind turbines as well as energy-efficient motors. As mentioned previously, between 2005 and 2014 the company earned some $160 billion from the program—revenues which grew twice as fast as total company sales. By any measure, it was successful, particularly because consumers began to think of GE not just as an industrial and financial giant.

That impression began to change in 2014, when the company made two announcements that took the risky concept of mixing 'green' with 'brown' to a whole new level. It committed $10 billion toward research and development for "clean-tech" technology by 2020, signaling the goal of developing alternative water technologies for fracking natural gas as a top priority. Fracking is a controversial process, but one that did receive support from some environmentalists, as long as there was no significant methane leakage or no water supplies were contaminated during extraction.

That same year GE launched the third installment of its Ecomagination Innovation Challenge, an open innovation competition seeking ideas to help "power the grid" and "power the home". In support of its initiative, GE and its venture capital partners invested $140 million into some promising clean-tech businesses. However, this third Challenge was different, given its focus on the Canadian oil sands.

Oil sands require a lot of energy to actually melt and process the tar, so while reducing production emissions is a worthy goal, the negative impacts of using oil remain by continuing its use as a primary energy source. Exploiting oil sands leaves issues such as the objective of moving the world toward a more climate-friendly energy source on the back burner. Cleaning up oil sands is kind of like putting a filter on a cigarette—it's a better alternative, but will still kill you in the end.

Given shifting regulatory and social attitudes toward carbon, supporting oil sands is probably not the best idea, particularly if your intent as a company is to portray yourself as environmentally friendly. Using the valuable and well-deserved public relations capital associated with Ecomagination to promote the exploitation of tar sands stands a very good chance of tarnishing the brand. From a purely green perspective, the oil sands industry moves the world in the wrong direction on carbon, and sucks up intellectual and fiscal capital that could be better allocated.

If a major company like GE wants to make some of its product offerings as environmentally sound as possible and portray itself as green, then, ideally, any green initiatives should be considered to be part of business as usual. Of course, it would be disingenuous to imagine that an industrial, energy or natural resource company would be thought of 'primarily' as a green company, but for such a company to make a genuine effort to become more green should certainly be commended. GE should rightly be applauded for its efforts. It should not matter that the company makes a lot of money in the process. But it does raise question about where a company should draw the line between portraying itself as green and actually being green.[23]

Pepsi

Early on in Pepsi CEO's Indra Nooyi's tenure she recognized that consumers would continue to pressure the company to make its products healthier, and she argued that a shift toward healthier products would be good for society as

[23] Adapted from A. Winston, "GE is Avoiding Hard Choices about Ecomagination," hbr.org, August 1, 2014, accessed November 25, 2015, https://hbr.org/2014/08/ges-failure-of-ecomagination.

well as for Pepsi's bottom line. Nooyi subsequently acquired healthier brands like Tropicana and Quaker Oats, created Pepsi Next (a lower-calorie version of the flagship brand), and even hired a former official from the WHO to oversee the reforms.

Nooyi initially won widespread acclaim for her efforts, but Pepsi's investors have long been skeptical. During her tenure, there was a doubling in the stock price of arch-rival Coca-Cola, while Pepsi's stock stagnated and Pepsi lost its number-two position in the cola market to Diet Coke. Investors believed that Nooyi's socially responsible vision was a bad business strategy that had diverted resources from Pepsi's successful (yet unhealthy) core brands.

While hundreds of millions of new consumers in emerging markets sought Pepsi's existing products, and with more dollars allocated to promoting Pepsi's 'healthier' products, the company's marketing budget was spread thin, and some consumers bought Coke instead. As a result of the reduced profits, Pepsi announced management changes and appeared to signal that it would step back from Nooyi's "performance with a purpose" business strategy. This case study shows that 'doing good' does not necessarily mean doing well financially, even if your firm happens to already be a leader in its space.[24]

There is theory and there is practice. It does little good to 'go green' only to find that your stock price tanks as a result. A company can either make a decision to 'do good' and go for it full throttle, or it can dip its toe in the water and see how it feels. Neither approach will insulate a company from either the business cycle or unforeseen events. As noted above, there is no doubt that consumers prefer products and companies associated with 'doing good', and embracing CSR can do wonders for the bottom line. Some companies that are exclusively focused on fair trade and green products do very well indeed. Others just cannot seem to get it right.

The only CSR initiatives likely to survive are activities that a firm should be doing anyway to increase profits. When Walmart requires its suppliers to be more energy-efficient, the company lowers its costs. When in 2014 CVS refused to sell cigarettes, it declared an entirely new business strategy. It just so happens that the company was rewarded richly for doing so, with its stock rising from $80 to $110 in a period of just eight months. Starbucks' focus on placing social impact ahead of short-term profits certainly didn't hurt its bottom line, either; between 2009 and 2015 its stock rose from less than $5 to more than $60.

That said, investors' ruthless focus on quarterly results gives companies little incentive to continue their focus on social impact when their profits are

---

[24] Adapted from A. Chatterji, "When Corporations Fail at Doing Good," www.newyorker.com, August 29, 2013, accessed November 25, 2015, http://www.newyorker.com/business/currency/when-corporations-fail-at-doing-good.

at risk. In the end, CSR is a business decision. If it is pursued on the basis of sound assumptions, it will be a net positive to the bottom line. If not, it will be treated just like any other business decision that did not turn out well.

Promoting Social Responsibility in the Developing World

While project financing is important, projects undertaken in the developing world that are managed by socially responsible sponsors, financiers, and insurers prove to be the most profitable. The importance of project finance in promoting infrastructure development in the developing world is well known. Without the billions of dollars of support generated for infrastructure projects through the use of project finance, hundreds of millions of poor people in the developing world would not have access to basic needs such as electricity, clean water, and sewage treatment.

Typical project financing involves the issuance of a non-recourse loan, wherein the sponsor has no obligation to make payments on the project loan if the revenues generated by the project are insufficient to cover the principal and interest payments. Lenders seek to minimize the risks associated with making non-recourse loans by requiring indirect credit supports in the form of guarantees, warranties, and other covenants from the sponsor, its affiliates, or other third parties involved with the project. Political risk insurance (PRI), which protects a sponsor or lender against non-commercial risks (such as expropriation, currency inconvertibility/non-transfer, and political violence, or breach of contract), is often utilized to remove country risk from the equation. Project sponsors and lenders thereby assume the commercial risks associated with a given project.

The Project Finance Challenge

Project financiers have had to balance their desire to participate in sound, profitable business ventures with the needs and capabilities of the people and governments of developing countries, as well as the interests of NGOs. This has been a slippery slope for many in the project finance business. Project sponsors, financiers, and insurers alike have had to find balance between the many competing forces that impact the construction and operation of infrastructure projects in the developing world. Their need and desire to adhere to strict credit, accounting, design, construction, and operational standards has often conflicted with equally important objectives such as strict environmental compliance, greater socioeconomic benefits for workers, and the rights of indigenous peoples.

Not long ago, project sponsors, lenders, and insurers did not pay sufficient attention to the latter issues. When non-recourse project finance first emerged in the early 1990s the emphasis tended to be on how to get an important project in a difficult country funded, rather than how to do so in a manner consistent with the objectives of all the involved parties. What project financiers often failed to appreciate was that it was in their interest to create a win/win environment with the governments and people of the developing world.

Now that project finance has become a standardized means of promoting infrastructure development, the project finance industry as a whole (which includes the sponsors and the providers of PRI) has come to realize that it is very much in its interest that:

- The people who work at each project, as well as local inhabitants, have a sense of participation in and belonging to each project.
- The long-term interests of projects are served by meeting the long-term interests of the governments and people of the countries where these projects are located.
- A fair and competitive long-term price should be charged for services provided.
- A uniform, conservative environmental standard should be used (World Bank standards are commonly used).
- It contributes to the long-term peace and stability of the country and region where a project is located.

Businesses are increasingly recognizing that only in stable operating environments are projects most likely to earn an acceptable rate of return. What is perhaps most important is that a project not contribute to or accentuate perceived imbalances between ethnic groups, social classes, or geographical sub-regions. Particularly in the areas where conflict exists, extra attention needs to be paid to ensuring transparency in all aspects of project implementation.

The Importance of CSR

CSR has been an integral part of the planning and implementation of infrastructure projects since the 1980s. Over this period project sponsors, financiers, and insurers have learned some tough lessons about the dangers of not paying sufficient attention to these issues.

One good example of the possible consequences of not paying enough attention to the social and environmental issues associated with owning and operating a mine is Bougainville in Papua New Guinea (PNG). In 1988 a small group of villagers blew up some of the mine's installations, coming in the wake of demands for compensation for loss of land and resources to the project, and the alleged pollution of the local river system. Refusal by the mine owner and the PNG government to address the demands prompted escalating guerilla action against the mine and its employees. The company closed the mine down the following year, and it has remained closed. Thousands of people died in an ensuing civil war, and litigation against the mine and its owners continues to this day.

NGOs have also learned some lessons. One of the best examples is the Freeport mine in West Papua (formerly, Irian Jaya), Indonesia. An NGO sought to have Freeport's PRI canceled for alleged violations of the environmental conditions set out in the insurance provided by the Overseas Private Investment Corporation (OPIC, a U.S. government agency) and the Multilateral Investment Guarantee Agency (MIGA, a member of the World Bank Group). Because covenants of the insurance appeared to have been breached by the company, OPIC canceled the coverage.

Freeport took OPIC to court, had the insurance reinstated, and then itself canceled OPIC's and MIGA's insurance, meaning that the NGO's objective of stricter environmental compliance backfired. When the insurance was canceled, Freeport was no longer obligated to adhere to strict, internationally accepted environmental regulations.

Many mining companies have subsequently made great strides in taking the initiative to avoid conflict with NGOs and indigenous peoples by ensuring that they are inclusive in the planning, construction, and operation of mines. Before Rio Tinto built the Lihir gold mine in PNG, it spent a great deal of time establishing strong relationships with tribal leaders, acquired an understanding of the concerns of local inhabitants, and negotiating a sensible agreement with the government. They went so far as to remunerate islanders for every coconut tree that was cut down, and they studied 100 years of islander family histories to establish property rights based on written history. The banks that financed the mine, and the PRI providers who insured it, recognized the benefits of a well-thought-out approach to making an investment in a country as culturally complex as PNG.

These examples demonstrate that it is in the interest of all parties to collaborate in achieving mutually satisfactory guidelines for the construction and operation of infrastructure projects in the developing world. It is only through cooperation and collaboration that the objectives of all parties can be reached.

Public/Private Sector Collaboration

Multilateral Development Banks (MDBs)—such as the ADB and the World Bank—have a unique role to play in this regard. Since their work is by nature oriented toward the promotion of development and the alleviation of poverty, the concept of social responsibility strikes a familiar chord. MDBs have strict covenants governing all aspects of their participation in project finance. From a "no child labor" policy to a requirement for total transparency to no tolerance for bribery and corruption, MDBs have played a pivotal role in ensuring that social responsibility is a common theme in the projects in which they become engaged.

Many private sector banks and insurers have attained a surprising degree of convergence with the MDBs in the general area of social responsibility. In addition to the commonality that has been achieved in applying World Bank environmental standards to projects that are financed and insured in the private sector, the type of due diligence now applied by most private sector institutions has achieved a remarkable degree of similarity with those of the MDBs.

While public sector insurers have, for example, addressed a host of developmental and social responsibility issues in the natural course of conducting their project risk analyses, many private banks and insurance companies now routinely address the same issues during their due diligence process. In determining whether a project is worth supporting, for example, it is now common to assess the degree to which substantial tax revenues are generated, local workers are hired, and technology transfer is present.

Banks and insurance companies face a host of performance, reputation, and ethics issues when engaging in project finance, particularly now that there is such an emphasis on the whole concept of "corporate governance." PRI providers, in particular, have moved from making "traditional" assessments of country risk based largely on the social context in which a project operates to better understand the degree to which infrastructure projects impact the wider geographical area and contribute to indigenous and cross-border conflict. This is particularly important because so much foreign direct investment in the developing world exists in areas that are inherently politically unstable. Promoting CSR is therefore an underlying, albeit indirect, objective of PRI providers.

Information Sharing

Despite the great progress made in the convergence of the interests of public and private PRI providers, there remains a need for even greater collaboration with respect to information gathering and sharing. MDBs and export credit agencies (ECAs) have a distinct advantage in collecting project-related information because they have access to sources private sector entities do not have.

For example, MIGA has access to IMF and World Bank Group data, and OPIC can access any number of information sources from within the U.S. government. Similarly, information gathered from local sources by private sector institutions could prove to be extremely valuable to MDBs. All such institutions could benefit from greater adherence to widely acknowledged guiding principles, such as the OECD's Guidelines for Multinational Enterprises, or those of the United Nation's Global Compact Conflict Dialogue. Enhanced information sharing, and the establishment of greater commonalities in project assessment and operational safeguards, will contribute to a greater degree of conflict avoidance in the project finance process.

The Path Forward

Private sector sponsors, financiers, and insurers of project finance-related infrastructure projects should be given credit for having moved solidly in the direction of CSR. It is clearly in the interests of all parties involved in the development process that the maximum amount of attention be paid to promoting social responsibility, and to minimizing the potential for conflict. Much remains to be done, however.

Greater information sharing is one important aspect to enhancing the risk assessment, which is key to being able to better predict where problems are likely to arise. Generating accurate risk assessments is critical to increasing the flow of foreign direct investment to the most difficult, conflict-ridden areas of the globe. The problem is that the utter unpredictability of political events makes the creation of more accurate risk analyses even more difficult to produce. Where and when will the next terrorist attack occur? What will its impact be on the foreign investment climate? Will a host government's response to terrorist attacks create an investment climate that is less conducive to attracting foreign investment? These are the types of questions political risk analysts now face.

MDBs can play a better, more effective role in supporting access to project finance during periods of crisis by improving the finance methodologies they use so that they can be introduced quickly and economically into new markets, even before a crisis starts. They can focus on filling market gaps that might appear, so that capital flows from private banks may remain open longer.

MDBs can also consider using the least amount of intervention possible in times of crisis, giving priority to financing tools that help private-to-private flows first, leaving the public-to-public foreign exchange loans as a last resort.

MDB intervention can and should occur in times of crisis, but only when the ordinary functioning of capital markets fails, so as to avoid creating a future financial burden in crisis-ridden countries. The loans provided by MDBs must eventually be repaid.

Finally, increased adherence to CSR principles among all stakeholders in the project finance business will certainly minimize the extent to which such business aggravates conflict-prone locations. Increasingly, project financiers are insisting that adherence to such guidelines be a prior condition to the issuing of loans. As the example of Rio Tinto in PNG illustrates, future conflict between project and local stakeholders can be anticipated and even neutralized by thoughtful planning. In the future, the hope is that it will be those projects managed by socially responsible sponsors, financiers, and insurers will prove to be the most profitable.[25]

Conclusion

CSR was initially considered by some to be idealistic, something beyond the realm of core corporate strategy, or something that might be nice to have in the mix at some point in the future. It has now become an integral part of risk management, strategic planning, and corporate culture. It is no longer a question of simply cutting an annual check to a CSR-friendly NGO or being able to say that a corporation has a CSR mission statement. Now CSR really has become an important part of being considered part of the responsible global business community.

What global trends are likeliest to influence the development and acceptance of CSR in the future? The following are some trends to watch:

1. The business world and consumers continue to embrace 'disruptive' companies.
2. There is a growing global commitment to sustainability.
3. The lines between 'for profit' and 'not for profit' companies continues to blur with the emergence of the 'sharing economy'.
4. A growing proportion of start-ups adapt to CSR from the outset, which is helping to transform the global business landscape.

---

[25] Adapted from D. Wagner, *Project Financiers' and Insurers' Roles in Promoting Social Responsibility in the Developing World*, International Risk Management Institute, January 2004, accessed November 25, 2015, https://www.irmi.com/articles/expert-commentary/promoting-social-responsibility-in-the-developing-world.

5. More consumers want to do business with companies that are perceived to be making a difference in the world. Those numbers can only grow.
6. There is an increasing focus on income inequality, which can only enhance the core principles of CSR.
7. Publishing CSR initiatives in corporate publications is quickly becoming a business requirement, rather than an optional extra.
8. Companies and non-profits alike are leveraging big data for the social good.

Those companies that can integrate these trends into their DNA will not only stay ahead of the curve, but will also be able to profit from the ongoing shifts in the global business paradigm. If your firm has not yet embraced CSR, it is time to do so.

11

Country Risk Management

Defining Country Risk

How political power is exercised often determines whether government action—or inaction—threatens a firm's value. For example, a dramatic political event may pose little risk to a multinational enterprise, while subtle policy changes can have a substantial impact on a firm's performance. A student-led protest for political change may not change the investment climate at all, while a change in local tax law can erode a firm's profits very quickly. It is usually the task of a local manager, corporate risk manager, treasurer or CFO to identify whether a government action poses a threat to a firm's financial well-being, and act upon it.

To do so, risk managers and decision makers must first be able to identify and measure what country risk is before being able to make the right decisions as a result of it, however. The first distinction that must be made is between *firm-specific political risks and country-specific political risks.* Firm-specific political risks are risks directed at a particular company and are, by nature, discriminatory. For instance, the risk that a government will nullify its contract with a given firm, or that a terrorist group will target the firm's physical operations, are firm-specific. By contrast, country-specific political risks are not directed at a firm, but are country- or region-specific, and may affect a firm's performance. Examples include a government's decision to forbid currency transfers or the outbreak of a civil war within a host country.

© The Author(s) 2016
D. Wagner, D. Disparte, *Global Risk Agility and Decision Making*,
DOI 10.1057/978-1-349-94860-4_11

Table 11.1 Firm-specific versus country-level risks

	Government risks	Instability risks
Firm-specific risks	Discriminatory regulations "Creeping" expropriation Breach of contract	Sabotage Kidnappings Firm-specific boycotts
Country-level risks	Mass nationalizations Regulatory changes Currency inconvertibility	Mass labor strikes Urban rioting Civil wars

Source: D. Wagner, *Defining Political Risk*, International Risk Management Institute, October 2000, accessed November 25, 2015, https://www.irmi.com/articles/expert-commentary/defining-political-risk

Firms may be able to reduce both the likelihood and impact of firm-specific risks by incorporating strong arbitration language into a contract or by enhancing on-site security to protect against terrorist attacks. By contrast, firms usually have little control over the impact of country-level political risks on their operations. The only sure way to avoid country-level political risks is to stop operating in the country in question.

There is a second key distinction to be made between types of political risk: *government risks* and *instability risks*. Government risks are those that arise from the actions of a governmental authority, whether that authority is used legally or not. A legitimately enacted tax hike, or an extortion ring that is allowed to operate and is led by a local police chief, may both be considered government risks. Indeed, many government risks—particularly firm-specific ones—contain an ambiguous mixture of legal and illegal elements. Instability risks, on the other hand, arise from political power struggles. These conflicts could be between members of a government fighting over succession, or mass riots in response to deteriorating social conditions. These parameters are outlined in Table 11.1.

Country vs Sovereign vs Political Risk

It is important to distinguish between some commonly used and often misunderstood terminology in the country risk arena. Too often, risk practitioners tend to use the terms 'country risk', 'sovereign risk' and 'political risk' interchangeably, when, in fact, they mean different things. Although the term "country risk" is widely and generically used to refer to the risks assumed by operating in another country, country risk is really a misnomer in this context. Even the term "political risk" does not fully encompass the scope of risks a company encounters when operating in a foreign country. The following definitions are useful in considering the differences between the three:

- *Country Risk*: This term broadly refers to the likelihood that a sovereign state may be unable or unwilling to fulfill its obligations towards one or more lenders.[1] It involves an assessment of economic performance in the context of a country's demand for external financing and judgments about the prospect for changes in financial returns.
- *Sovereign Risk*: The risk that a foreign central bank will alter its foreign-exchange regulations thereby significantly reducing or completely nullifying the value of foreign-exchange contracts.[2] It also refers to the risk of government default on a loan made to it or guaranteed by it.
- *Political Risk*: Those political and social developments that can have an impact upon the value or repatriation of foreign investment, or on the repayment of cross-border lending, which may originate either within a host country, the home country, or the international arena.[3] This includes arbitrary or discriminatory actions taken by governments, political groups, or individuals that have an adverse impact on trade or investment transactions.

Based on these definitions, Country and Sovereign Risk are more related to payment risk, while Political Risk is more related to trade and investment risk. For our purpose, country risk shall be defined as: "*changes in a business environment that may adversely affect operating profits or the value of assets in a specific country.*"[4]

To help distinguish among between, sovereign, and political risk, and to emphasize the point that each transaction has a unique risk profile, it is more useful to think of country risk in terms of "*Transactional Risk*", which can be defined as *the country, sovereign, political, economic, financial, technical, environmental, developmental, and socio-cultural risk that an organization assumes in every international transaction it engages in.* The idea here is that every trade, investment and lending transaction is unique, given the time, place, and nature of the transaction. By definition, each such transaction will have some

---

[1] T. Krayenbuehl, *Country Risk: Assessment and Monitoring* (Cambridge: Woodhead-Faulkner 1985).

[2] *Definition of Sovereign Risk*, Thefreedictionary.com, accessed November 25, 2015, http://financial-dictionary.thefreedictionary.com/sovereign+risk.

[3] J.D. Simon, "Political Risk Analysis for International Banks and Multinational Enterprises," in R.L. Solberg (ed.), *Country Risk Analysis: A Handbook* (London: Routledge 1992), 118.

[4] *Definition of Country Risk*, Wikipedia.com, accessed November 25, 2015, http://en.wikipedia.org/wiki/Country_risk.

characteristic, or set of characteristics, that distinguishes it from every other transaction. The ability to understand how all of these variables interact to form a 'risk mosaic' that defines the scope and nature of risk is essential. This concept will be explored at greater length in Chapter 12.[5]

Effective Country Risk Management in the New Normal

Managing cross-border risk is more important today than in recent memory for a simple reason: the rules of engagement for conducting international business have changed—the risks associated with international transactions are high and risk aversion is high, but the margin for errors is low.

It is only natural, after recovering from global economic trauma, that international businesses would think more carefully about assuming and managing cross-border risk, but doing so has become more difficult. One of the things that has changed over the past several years is that the 'new normal' includes a paradigm shift: the rule book has changed as a result of a combination of two decades of modern globalization and a decoupling in growth patterns between the developed and developing worlds, which implies a change in risk profile between the two. The developed world is in some respects riskier than the developing world. Just consider that the Great Recession was born in the U.S. and Europe, and that Western countries are in the crosshairs of global terrorism, with the U.S. being the perpetual number one target.

What has become a clear trend over the past couple of decades is that the largest developing countries have collectively galloped ahead of the developed countries, with growth rates in the mid-to-high single digits not being uncommon, while North America and Europe continue to have comparatively anemic growth rates. The temptation among many international companies will be to trade and invest in developing countries as a result, perhaps without fully considering the implications of doing so from a country risk perspective. The need to ensure proper cross-border due diligence was always present, but the manner in which many businesses traded or invested internationally before the crisis did not require the same degree of due diligence as is required today.

To the extent that international companies devote any resources at all to understanding cross-border trade and investment climates (and the vast majority do not), they tend to be overreliant on externally generated country

---

[5] Adapted from International Risk Management Institute, *Defining Political Risk*.

risk analyses, which are more often than not produced generically and are not entirely appropriate for specific transactions. This is perhaps the most common mistake made by risk managers. They believe that because they may have information about the general political and economic profile of a country, they have a true handle on the nature of the risks associated with doing business there. What about gauging legal and regulatory risk, the country's friendliness toward foreign trade and investment, and other companies' experience there?

Too often, companies get caught in an 'investment trap': they commit long-term resources to a country only to find that the bill of goods they were sold—or thought they understood—turned out to be something completely different. There are plenty of stories out there about companies whose investments turned into disaster because the regulatory environment changed, a legal issue arose, international sanctions impacted their ability to operate, or they just selected the wrong joint-venture partner. After the investment has been made, it is usually too late to withdraw without incurring large losses and experiencing reputational risk once the story hits the press.

A risk manager may have the right information, but based on a short-term assessment of the risks. The long-term view may be completely different, but in the absence of knowing what questions to ask, and having clear lines of communication, the right information may not be taken into consideration.

The simple way to limit the possibility that unforeseen events will occur is to establish clear reporting lines and do your homework—really do your homework—and either hire one or more individuals in your company to focus full time on managing these risks and/or hire an external firm to create a customized risk profile for each and every investment your company plans to make. The expense involved pays for itself many times over when a problem is uncovered and avoided, yet many companies are happy to invest millions of dollars to make cross-border investments without doing their homework.[6]

The Boardroom Vacuum

The following real-world example will illustrate just how important it is to ensure that every organization operating across borders has a well-developed capability to understand and manage cross-border risk. At a board meeting

---

[6] Adapted from D. Wagner, *Managing Political Risk in the New Normal*, International Risk Management Institute, January 21, 2011, accessed November 25, 2015, https://www.irmi.com/articles/expert-commentary/managing-political-risk-in-the-new-normal.

of a top 20 multinational corporation, the question of whether or not to invest $50 million in a project in a Middle Eastern country was discussed. The president of the company's subsidiary who was seeking the board's approval insisted that the country was a safe place to invest because of its recent history of economic and political stability. Satisfied with the president's assurances and facts, the board approved the investment—a decision they came to regret.

As it turns out, the country in question was not as stable as it was portrayed and the company's investment became tied up in costly legal limbo, which had far-reaching and potentially damaging implications for the company's brand and reputation. To make matters worse, the interests of some of the corporate actors involved were not directly aligned with those of the company. The corporate sales team and underwriters promoting the transaction, for instance, were incentivized to sell the deal internally so they could meet their production targets and receive their annual bonuses. While the underwriters sought the views of the country risk manager charged with vetting the transaction, as they were supposed to do, a large portion of the analysis produced by the risk manager was deleted by the underwriters from the final transaction justification.

As a result of the absence of checks and balances in the risk management system, neither the country risk manager nor the corporate risk manager had any way of knowing that the underwriters had manipulated the risk analysis. The corporate risk manager sent it to the division president for approval before sending it on to the board of directors. Everyone up the chain of command believed all necessary approvals had been obtained in the manner previously mandated by the board and senior management.

This is just one example where the due diligence process in place failed to alert decision makers to the risks associated with their international business operations. As is illustrated by this example, companies often rely exclusively on their own managers and risk management processes, which they believe are bulletproof, but which may in fact be riddled with holes, inconsistencies, and contradictions. Clearly, the company's risk management function and final version of the country risk analysis were faulty. Without more comprehensive data or insight of its own, the company's senior management and board were too reliant on the deal team's version of the risk assessment to make an informed decision and fulfill their duty to protect the interests of the company and shareholders.

If the board had been better educated about the political, economic, and social, situation in that country, it might have been able to detect and respond to the errors in the assessment it received. The board may then have forced the company to conduct more thorough due diligence before requesting a vote, rejected the request outright, or made the approval conditional on receipt of the company's plans to mitigate and address the potential risks.

This example is far from unique. Over the last few years, a number of high-profile international investments have faced serious unanticipated obstacles, for example, a Middle Eastern port operator's management of U.S. ports was derailed by political opposition, an Australian mining company's executives were jailed in China accused of engaging in industrial espionage, and the Mexican government nullified a lucrative contract with a Chinese company to build a high speed rail project. We will discuss more about the need for decision makers to become better informed in Chapter 13.[7]

How the Arab Awakening Impacted Country Risk Analysis

Political change in the Middle East and North Africa (MENA) has had, and continues to have, a profound impact on many countries in the region and beyond, pummeling many of the most established governments in the world. One of the unintended consequences of this change has been to prompt some country risk analysts to reevaluate how they analyze country risk. In the rarified atmosphere of country risk analysis, this is a useful exercise, but for individuals and organizations that already think about the world in an esoteric fashion, the challenges are unique.

Country risk analysis probably sounds to a layperson as if it is the domain of number crunchers, political scientists, and intelligence agencies. In fact, this is the case, but how the numbers and theories are used to arrive at a meaningful conclusion varies widely, depending on the individual or organization doing the analysis. For example, many banks tend to be heavily skewed toward plugging numbers into algorithms or spreadsheets, political scientists often apply political theory to the behavior of nation-states, and intelligence officers often use information to draw conclusions about the likely behavior of leaders, political parties, militaries, and other state and non-state actors.

The job of a country risk analyst can be overwhelming, given the amounts of information that must be absorbed and synthesized into easily digestible text. When one considers that country risk analysis is not simply about politics, but also encompasses economics, socio-cultural dynamics, and history, there is a lot for an analyst to contemplate in drawing conclusions. This must,

---

[7] Adapted from D. Wagner, "The Boardroom Vacuum," *Risk Management Magazine*, December 2009, accessed November 25, 2015, http://cf.rims.org/Magazine/PrintTemplate.cfm?AID=4020.

of course, all be applied to each country at a given point in time. Due consideration must be given to scenarios, unexpected events, and short- and long-term trends.

Given this, what are some of the lessons a country risk analyst may learn from the ongoing upheaval in MENA? First, even though political change may be expected in any country at some point in time as a result of fundamental socioeconomic disparities, high levels of corruption, rumblings in the military, or the actions of a neighboring government, the ability to anticipate when and how such change may occur presents a daunting task.

In the case of Tunisia—which was the source of the spark that ignited the flame of political change throughout the region—the fact that so many people were disenfranchised within society, and that President Ben Ali had been in power (and had been abusing his power) for so long, prompted large segments of society to coalesce together to promote rapid political change. In the absence of the poor, middle class, business community and elements of the military not coalescing together at the same time, the 'spark' of the self-immolation of the fruit vendor would in all likelihood not have created a flame. So one lesson here is that social, political and economic disenfranchisement can reach a boiling point in an instant, given the right circumstances.

Second, although an analyst may expect that political change is likely to erupt from a larger, more important country in a given region, the very fact that a country is larger and more important may prevent it from embracing political change. Very few would have predicted that Tunisia was to become the venue for the spark that turned in to the flame in North Africa. Egypt was a likelier candidate, but Egypt's geopolitical importance, its prominence with respect to U.S. foreign policy and military aid, and the size and strength of its military made it a less likely candidate for radical political change. The regional implications for radical political change in Egypt made it harder to achieve as a catalyst, but once change had begun elsewhere in the region, it was certainly ripe to participate.

The great irony (and tragedy) of the Arab Awakening is that, after all that has happened, so little has changed. The fundamental structure of the governing systems that were in place prior to 2011 continue unabated. The monarchies remain in place in Bahrain, Qatar, Oman, Kuwait, Morocco, Jordan, and Saudi Arabia. The 'democratically' elected strong men remain in place in Algeria and Syria (at a terrible price). Only Tunisia successfully transitioned from dictatorship to democracy; Libya and Yemen replaced their dictators, but the result was chaos.

The aspirations of so many people throughout the region have been thwarted. Frustration levels are higher now than they were before Ben Ali,

Mubarak, and Saleh departed the scene, because so little has changed, and so little appears likely to change in the near or medium-term. In the case of Egypt, that frustration level reached a boiling point, as many citizens realized that the net result of their efforts was another military-led government protecting the status quo. So, herein lies yet another lesson—what may appear to be fundamental change may not really be change at all, or even change for the worse.

Fourth, in today's linked-in, globalized and interconnected world, political change that may previously have been limited in regional impact has great potential to explode in scope. Not that profound political change cannot happen in remote corners of the world that are less linked in or globalized—of course it can and does—but what might have been limited to the departure of former President Ben Ali in Tunisia became a tour de force for the entire region, with implications for the world. A corollary here is that, by definition, it will become more difficult for country risk analysts to accurately predict what happens in the future, in the short and long term. While few analysts would have believed that dictators would fall in succession in MENA, few also would have imagined that the result—in Iraq, Libya, Syria and Yemen—would be anarchy and chaos. No one can really know with any certainty what will come next.

The job of country risk analysts has therefore become more complicated. Many previously accepted assumptions about the way the world works have been shattered as a result of what has happened in MENA over the past few years. Many unexpected events will no doubt continue to occur, impacting the foreign policies of most of the world's major governments. In order for country risk analysts to stay ahead of the game, they must excel in being able to use the past and present to try to predict the future. In that regard, their job is really no different than it was before—just a bit more complicated.

The ability to effectively manage cross-border risk used to be considered something an international business either may not need or could not afford, but the risk management landscape has experienced a transformation since the new century began. In our multidimensional, constantly changing and leaderless world (the G-Zero world[8]), traders, investors, and lenders no longer have the luxury of assuming everything will work out, or maintain the mistaken belief that the horror stories that have befallen other organizations will not happen to them.

---

[8] *A G-zero World*, www.foreignaffairs.com, accessed November 25, 2015, https://www.foreignaffairs.com/articles/2011-01-31/g-zero-world.

Conventional wisdom used to dictate that because everyone else is investing in a given country, it must be the right place to invest, the idea being that strength lay in numbers, and that, surely, not everyone could be wrong. Yet the 'herd' mentality and chasing the 'hot dollar' has gotten many companies into a lot of trouble. While global investors know that Mr Putin's Russia is fraught with risks and that Brazil's economy consistently fails to perform anywhere close to its potential, they continue to invest in such destinations, believing that eventually, things will get better, even though they must know that is often not the case.[9]

Is Country Risk Really Rising?

It has become fashionable for political and economic pundits to declare in unison that the world is today a dangerous place, and that the risks associated with conducting cross-border business are increasing. Is it? Is cross-border risk really more dangerous to today than it was in, say, 1989 or 2001? Do trade and investment climates become inherently riskier in times of economic crisis?

Today's risk-averse business climate is not the result of a single event—such as the cataclysmic collapse of the former Soviet Union or 9/11—both of which significantly altered the long-term global trade and investment landscape. Rather, what got us here is a combination of greed, inadequate regulation of the banking industry, improper regulatory enforcement, inexplicable comfortability with obscene amounts of debt on an institutional and individual level, and short memories. The world is now adapting to an evolutionary change in the international banking and credit systems that should result in an improved cross-border trade and investment environment in the long term.

In 1989 or 2001 the world was neither as globalized nor as interconnected as it is today, and trade and investment landscapes were more easily defined and categorized. Trade and investment decisions then were based on less information and less sophisticated means of managing risk. Today, cross-border traders and investors benefit from a more level playing field with respect to access to information, more open markets, and a more competitive landscape. More countries than ever before want to attract FDI, enhance international trade, and be members of the global 'club'. To do so, they must maintain a competitive footing and constantly reinforce their comparative attractiveness as trade and investment destinations. That makes the global trade and invest-

---

[9] Adapted from D. Wagner, *How Political Change in MENA is Affecting Country Risk Analysis*, International Risk Management Institute, April 21, 2011, accessed November 25, 2015, https://www.irmi.com/articles/expert-commentary/how-political-change-in-the-middle-east-and-north-africa-is-affecting-country-risk-analysis.

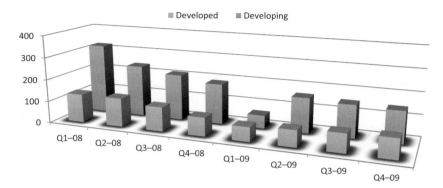

Graph 11.1   Global Net FDI Flows: 2008–2009 (Billions of U.S. Dollars) (*Global Net FDI Flows: 2008–2009*, United Nations Conference on Trade and Development, World Investment Report 2010, 2010, accessed November 25, 2015, http://unctad.org/en/Docs/wir2010ch1_en.pdf)

ment climate less risky than in recent history. As will be demonstrated below, protectionism remained largely muted throughout the Great Recession and country risk in general remained stable during the crisis.

Statistics compiled by the World Bank[10] show that net FDI flows contracted by approximately 40 percent in 2009, representing the sharpest decline in 20 years, but this was much less than the net decline in private bank lending, which plummeted 134 percent in 2009 (because banks were calling in loans). FDI began to improve in the second quarter of 2009 among both developed and developing countries. FDI into developed countries fell further than into developing countries from 2008 through 2009, but proportionately, developing countries made up more ground after Q1 2009 than did developed countries (see Graph 11.1). If the collective view of foreign investors was that country risk was rising during the period, the FDI statistics would not have demonstrated such strength following the peak of the crisis among either developed or developing countries.

Trade Protectionism Largely Absent

According to the World Trade Organization (WTO) and World Bank,[11] in contrast to the Great Depression, overt acts of trade protectionism were largely absent from the global trade arena during the Recession, but the num-

---

[10] *Prospects Weekly: Protectionism Muted, FDI Plummets in 2009, Global Oil Demand Now Rising*, blogs. worldbank.org, accessed November 25, 2015, http://blogs.worldbank.org/prospects/prospects-weekly-protectionism-muted-fdi-plummets-in-2009-global-oil-demand-now-rising.

[11] Worldbank.org, Prospects Weekly.

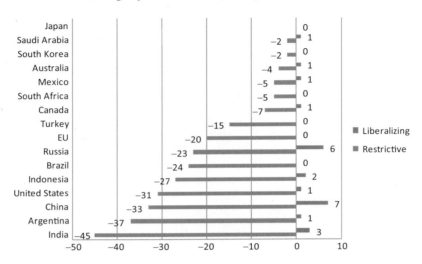

Graph 11.2 Trade Measures taken by the G20: October 2008–February 2010 (*OECD and the G20: Monitoring Investment and Trade Measures*, www.oecd.org, accessed November 25, 2015, http://www.oecd.org/g20/topics/trade-and-investment/g20.htm)

ber of restrictive trade actions taken on the part of governments exceeded those of liberalized trade actions by 10 to 1. This is not surprising, as countries naturally seek to protect domestic industries in times of crisis. The top five countries restricting trade transactions were (in order) India, Argentina, China, the U.S. and Indonesia (see Graph 11.2).

In spite of this, global trade volumes *rose* by 21 percent year-on-year in January 2010, in terms of both volume and value. Interestingly, during the period October 2008 to February 2010, the number of anti-dumping investigations initiated by G20 governments fell by 21 percent. Given the number of restrictive actions taken by governments during the period, new anti-dumping investigations *should* have risen considerably, but did not. So, does this point to rising country risk? Again, the answer appears to be no. Having avoided tit-for-tat protectionist measures among the world's major economies, and having seen an impressive rebound in trade during the height of the Recession, country risk remained stable, even if the *perception* may have been higher.

An Evolving Perception of Risk

What all this means is that our *perceptions* of risk must change. Simple categorization of countries into 'good' or 'bad', 'rich' or 'poor', and 'risky' or 'not risky' no longer captures the scope of risk companies face when investing in today's evolving mosaic of investment climates. Greece has clearly perceived as riskier than a whole host of developing countries since 2009, but that was not the case in 2008. Egypt and Syria were considered bastions of stability until 2011. The BRICS countries were all thought to have one trajectory—up—until they no longer did. Rather than simply saying the world is a riskier place, it is more accurate to say that depending on where a company invests, in what sector, and who with, a developed country can easily be riskier than a developing country. For example, the country that has been one of the boldest in taxing mining company profits is not a corrupt, poor, developing country, but Australia.

Long gone are the days when the West calls the shots and the rest of the world snaps to attention. Gone also is the time when so many of the good ideas, best risk management practices, and acceptable standards of behavior are automatically derived from the developed world. Countries such as Brazil, China, and India are showing dramatic progress in establishing improved governance, business practices, and advances in technological prowess. If the global economy is akin to a business cycle, then the developed countries are mature markets in the process of gradual decline, while the most dynamic economies of the emerging world have yet to hit their prime.

Country risk management is a function of where one invests, in what sectors, with whom, and in what manner. In that regard, risks associated with doing business in other countries is indeed rising, but largely in the *developed* world, where the price being paid and the ongoing risk of contagion from some of the mistakes that were made over the past couple of decades will linger, and where the monetary resources and political will required to make some of the structural changes necessary to ensure long-term growth are absent. Country risk can be characterized as falling in many parts of the emerging world, where opportunity abounds, governments continue to liberalize foreign trade and investment regimes, and trade and investment volumes continue to outpace that of the developed world.[12]

What the Euro Crisis Implies About Managing Country Risk

Having the right risk management tools at your disposal—whether it is knowledge, experience, or an insurance product—is essential to effectively managing cross-border risk. Yet doing so involves much more than simply

---

[12] Adapted from D. Wagner, *Is Country Risk Really Rising?* International Risk Management Institute, July 2010, accessed November 25, 2015, http://www.irmi.com/articles/expert-commentary/is-country-risk-really-rising.

collecting data, which is relatively easy given the wealth of information at our disposal. Interpreting data is the real challenge, as is the ability to distinguish perception or risk from reality. A good risk analyst must be able to tell the difference.

Data can only tell you so much. *Gut instinct* is often as important as all the risk indicators one can identify—sometimes even more so. Including intuition and experience in the analytical process is extremely important, as an integral part of analyzing cross-border risk. An overreliance on numbers and quantitative methodologies can lead to trouble, as they may prevent an analyst from thinking in sufficiently broad terms to adequately capture the true nature of risk. Ultimately, each risk must pass our own 'smell' test, based on our own inherent experience, knowledge, and biases, for country risk analysis cannot be a purely objective undertaking.

Country risk management is all about casting aside long-held assumptions, and applying insight and foresight to imagine what a country and the world will look like in the future. Consider the question of the Euro Crisis. Logic tells us that continuing to throw good money down what appears to be a black hole may make little sense, yet, as the 2015 bailout of Greece demonstrated once again, the Eurocrats continue to do just that. Rather than focusing on creating and implementing policies that will generate long-term growth, policymakers are stubbornly continuing down a path that has proven to be unsuccessful, and they will likely continue to be so.

If we look forward five to ten years to imagine what Europe will look like if the same policies are pursued, what can we expect? More than likely, we can expect a prolonged crisis, with even worse unemployment, higher budget deficits, greater uncertainty, and less confidence about what comes next. Incorporating some of the concepts noted above, this implies a greater likelihood of social tension, economic hardship, and rising political risk. We do not need confirmation of continually declining GDP figures and rising unemployment to tell us that the business climate may suffer for many years to come. Common sense tells us that.

Putting the pieces of the puzzle together, it is not difficult to come to the conclusion that the business climate will continue to suffer in Europe for many years. This, in turn, has profound implications for countries that trade and invest in and with Europe. Many developing countries have paid a high price for the European imbroglio, but have few resources or options in which to counter it. This implies a distressed trade and investment climate for much of the world for as a long as the Euro Crisis continues.

To manage risk in this environment, risk managers, strategic planners, and senior management in corporations throughout the world must not only think outside the box—they must *turn the pyramid upside down* and look at the world in a whole new way. Managing country risk in this kind of environment requires a range of skills that are possessed by few companies. So companies that have muddled along in a unidimensional manner attempting to address risk are now forced adopt a multidimensional approach to managing cross-border risk. Some will make the transition successfully; others will not.[13]

Rising Social and Political Risks

While income inequality has been falling globally since the 1980s, it has actually been rising in most of Europe. The Gini Coefficient, a widely used measure of income inequality, has risen since that time not only in poorer Eastern Europe, but among Europe's wealthiest societies. Austria, Finland, Germany, Luxembourg, Norway, and Sweden are among the Western European countries where inequality has risen consistently over the past several decades. The fact that this has been occurring for so long points to a deep malaise in European societies that the Euro Crisis has only served to exacerbate, and has raised a host of serious issues European law makers must address beyond short-term economic and fiscal concerns.

According to Eurostat, the percentage of the population categorized as 'at risk' of poverty *after* the receipt of social transfer payments in Europe overall was 25 percent in 2013.[14] Unemployment rates have risen dramatically in Europe since 2008 and reached a new high of 24 percent in 2013, while youth unemployment in Greece and Spain exceeded 50 percent.[15] Among the long-term impacts of this unbelievably high unemployment is a brain drain among the young, corresponding with an older population over time.

In societies where governments have made a social compact to provide a broad range of long-term entitlements, this spells looming disaster. Eurostat notes that since 1960, the percentage of citizens over 65 years of age has tripled—from 10 percent to 30 percent. Elderly benefits account for nearly 40 percent of average social spending among the EU27 governments, with healthcare taking nearly 30 percent of the pie. The traditional tax base of these

---

[13] Adapted from D. Wagner, *What the Euro Crisis Implies about Managing Country Risk*, International Risk Management Institute, August 17, 2012, accessed November 25, 2015, https://www.irmi.com/articles/expert-commentary/euro-crisis-and-country-risk.

[14] *Living Conditions in Europe*, Eurostat, ec.europa.eu, 2014, accessed November 25, 2015, http://ec.europa.eu/eurostat/documents/3217494/6303711/KS-DZ-14-001-EN-N.pdf/d867b24b-da98-427d-bca2-d8bc212ff7a8.

[15] Eurostat, *Unemployment Statistics*, ec.europa.eu, September 2015, accessed November 25, 2015, http://ec.europa.eu/eurostat/statistics-explained/index.php/Unemployment_statistics.

countries is declining while the number of individuals dependent on entitlements is rising.

Also according to Eurostat, between 1995 and 2010, total taxes among all European governments dropped an average of 2 percent, with some countries losing as much as 5–6 percent of their tax base during that time. Overall, European tax rates on corporations declined more than 8 percent, and labor taxes were down more than 3 percent, between 2000 and 2010. European governments responded by raising Value Added Tax rates, environment and other forms of taxes in an effort to counter the revenue decline.

The corresponding gradual erosion of standards of living in Europe has manifested itself in a number of other ways, including increasing crime rates, tougher immigration laws, and a backlash against minorities and immigrants. The popularity of radical right parties has risen dramatically across Europe since the 1980s, even in countries which have not been the worst affected by the economic crisis. For example, in Belgium far right parties took 1 percent of the popular vote between 1980 and 1984, compared with 14 percent between 2005 and 2009.[16] During the same period, in Norway the figures were 4.5 percent compared with 22.5 percent, and in Switzerland 3.8 percent compared with 30 percent.

A greater percentage of the population throughout Europe is becoming increasingly disillusioned with what the post-war social, financial, and political compact has produced, and are seeking more radical alternatives. Previously 'fringe' political movements have benefitted while the political center is increasingly being eroded by candidates and parties catering to people's worst fears and darkest visions. Europe's middle class is eroding slowly. People are losing hope. Their sense of security is disappearing. The notion that a better future can be achieved through a painful present is increasingly being challenged.

In Europe, new political divisions are emerging. It is no longer simply a question of left versus right or nationalism versus xenophobia. Today, the choice is increasingly between Old World and New World, and a future which envisions participation in the EU versus one that does not. Voters are likely to support candidates who can demonstrate political sobriety, realism, common sense, and a break from the past. As the Crisis endures and the solutions pursued continue to fail to produce lasting and meaningful change, there is a rising risk of civil disobedience, street demonstrations, and political violence.

---

[16] Friedrich-Ebert-Stiftung Forum, *Right Wing Extremism in Europe*, 2013, accessed November 25, 2015, http://library.fes.de/pdf-files/dialog/10031.pdf.

Many analysts continue to focus on the economic and fiscal manifestations of the Euro Crisis, while the social and political implications—which are every bit as important—seem to garner less attention, yet are having a profound impact on when, how, or if the Crisis gets resolved. The Crisis is multifaceted and more than likely has another five to ten years to play out. Europe must address its long-term political, economic, and social demons in a realistic and common sense fashion. It will do Europe little good in the long term to forcefully address its fiscal and economic challenges without simultaneously addressing its daunting social and political problems. The risks of not doing so are rising.

In our debt- and conflict-ridden world, with increasing forms of risk from unexpected places, country risk can only continue to rise in the near and medium-term. The ability to look at the present and future with an open, creative and foresightful mind is essential. If your company is not up to the task, there is a good chance future profitability will suffer. The best advice is to modify your corporate risk management policies to anticipate and account for future challenges to growth and profitability.[17]

Common Sense Political Forecasting

As global politics continue to gyrate, the pace of change poses ever greater challenges to accurately predicting the future. Many forecasters have gotten into the habit of declaring—*ex post facto*—that they 'got it' right, when in fact they failed to accurately predict the course of events. Given the multifaceted nature of the world today, predicting the future is quickly becoming a fool's game, and calls for a new paradigm.

Today's oracles confront a dizzying array of information in an attempt to make sense of the future. Looking for precedents in unprecedented times exposes the shortcomings of managing risk by looking in the rear view mirror. It has gotten to the point where there is so much information out there, it is impossible to consume it all. And yet most of this information is little more than a regurgitation of news, and dated analysis—which is often passé as soon as it is produced. The average consumer of such information rarely recognizes the difference because they do not have the luxury of spending enough time consuming the information.

---

[17] Adapted from D. Wagner, "Europe's Rising Social and Political Risks," *The Huffington Post*, March 11, 2013, accessed November 25, 2015, http://www.huffingtonpost.com/daniel-wagner/europes-rising-social-and_b_2853191.html.

It is increasingly becoming the case, given how complex the world has become, that having real insight into the future calls for little more than an appreciation for history, some 'street smarts', and good old-fashioned common sense. This seems to have gotten lost in the blizzard of quantitative indicators and seemingly endless stream of pontifications available in the information markets today. Many of these 'wise men' wish us to believe that they possess some kind of magical insight into the future, when in fact, many of them don't know any more than the rest of us—they may just package it well.

The truth is, no person or entity can possibly know everything that needs to be known to accurately predict the future. Even the intelligence agencies often get it badly wrong. So, how can the rest of us hope to get it right? Part of the answer surely resides in the rise of social media, and the 'nuggets' of information that is the result of free-flowing thought by ordinary people. A billion imaginations bloom on Facebook, LinkedIn and Twitter, and perceptions can quickly become reality. With mobile density nearing 90 percent globally, this new paradigm of near perfect interconnectedness is the new form of intelligence that, if properly harnessed, can hold an important key to remaining a step ahead of the competition.

Equally importantly, we should all be paying even more attention to the lessons of history, which is the best source of information and guide to the future. History teaches us that we will often end up where we started. For example, in Iraq, the mistake of creating artificial country boundaries that tried to forcefully integrate Kurds, Sunnis, and Shia was bound to fail eventually. In Malaysia, granting preferential rights to Malays at the expense of the Chinese was a policy that could not possibly exist without a backlash. And in South Africa, white majority rule could never last.

What do all these examples have in common? They are joined by human nature, and the notion that people will eventually rise up when a situation is grossly unfair. We do not need an expensive information provider to tell us that. In our view, all we really need to know is the difference between right and wrong, and the nature of human nature, to have real insight into what direction events will ultimately take a country.

Traditionally, the interconnection between politics and economics is what has driven markets, but today, the impact of development, environmental issues, and socio-cultural dynamics must be included in the mix. Taking a static view toward forecasting the direction of markets that does not include this range of variables will likely lead to failure. Each transaction is unique and each risk profile is specific, so using a broad brush to 'paint' a risk landscape will rarely yield the desired result.

Simple use of intuition and the 'smell test' may be more accurate indicators of risk than all the quantitative methods so often used to interpret the world today, which fail to introduce texture and common sense into the equation. So, study history and listen to your gut. Your chances of being right are as good, if not better, than an overemphasis on numbers—and a lot less expensive.[18]

Why Every Manager Needs to Be a Country Risk Manager

While most risk managers and decision makers would probably say that they do not have any particular expertise in assessing and monitoring country risk, the truth is that they do so every day in an international business, whether they realize it or not. Since most businesses (apart from some financial institutions and natural resource companies) tend *not* to have country risk specialists on their staff, it is incumbent on risk managers and decision makers alike to *become* country risk managers.

This may appear to be a daunting task—*and it is*. Country risk management is an art, not a science, and the average corporate manager will have had no formal training in the subject. Managing country risk is all about creating and understanding a mosaic of risk factors that form a unique risk profile for each transaction, whether it involves trading, investing, or lending.

Country risk management isn't merely about considering obvious variables—such as the health of a country's economy and its political stability—but, in addition, such things as the regulatory environment, developmental issues, environmental concerns, and socio-cultural considerations. Since each transaction has a unique risk profile, in order to be effective as a country risk manager, the ability to identify what truly distinguishes one transaction from another, and their implied short- and long-term risks, is critical to making the decision to invest, and stay invested. Doing so requires ongoing monitoring, which can be rather time consuming.

The number of companies—including some of the largest and best-known names in business—that have no formal methodology or staff for assessing and monitoring cross-border risk is surprising. Their belief is that they are either so large, so well known, or have such a long operating history in a

---

[18] Adapted from D. Disparte and D. Wagner, *Common Sense Political Forecasting*, International Risk Management Institute, February 24, 2012, accessed November 25, 2015, https://www.irmi.com/articles/expert-commentary/common-sense-political-forecasting.

country that all they may need to do when a problem arises is send the CEO in to the country to 'solve' the problem. That is an obviously flawed approach and can have the opposite effect.

It is therefore incumbent upon every company doing business abroad to implement a meaningful approach to cross-border risk management. Smaller companies may have no choice but to rely on third-party-produced reports. The problem with doing so is that these reports are often out of date as soon as they are published, and in general do not make recommendations about whether or how to proceed with a transaction.

Making an attempt to understand an investment or trade climate is better than no effort at all, but, by the same token, some companies become deluded into believing that because they may have a person or 'system' in place to monitor country risk, they are bullet proof. Too often, gaps and inconsistencies in the risk management process, the nature of the internal reporting system, or the manner in which senior management delivers information to the board, can itself can be a significant problem. It is recommended, therefore, that organizations with significant international exposures devote the resources necessary to adequately address cross-border risk, which will only become a more important component of the risk management landscape with time.

Much of what determines whether a country risk manager can do a good job will ultimately be outside his or her control. No one, no matter how experienced, can know or anticipate precisely where a country is heading all of the time. All we can do is make educated guesses based on what history teaches us, and what we have learned in the process.

That said, to be effective, a country risk manager must have the tools necessary to do the job and have the backing of senior management to both integrate the country risk function into the decision-making process, and take that process seriously. Whether that occurs may also be outside the risk manager's control.

In the end, the ability to anticipate what the future will bring using a combination of knowledge, insight, and a healthy sixth sense can make all the difference. Listening to your gut and sense of smell are, in the end, as important as all the other tools at one's disposal. A good risk manager knows when to listen to it.[19]

---

[19] Adapted from D. Wagner, "The Need for Country Risk Management," *Risk Management Magazine*, June 2, 2015, accessed November 25, 2015, http://www.rmmagazine.com/2015/06/02/the-need-for-country-risk-management/.

Part III

Effective Decision Making

12

Transactional Risk Management

Recall that in Chapter 11 we first introduced the concept of "Transactional" Risk Management (TRM), which is defined as *the country, sovereign, political, economic, financial, technical, environmental, developmental, and socio-cultural risk that an organization assumes in every international transaction in which it engages.* While every trade, investment, and lending transaction is unique, given the time, place, and nature of the transaction, by definition, each such transaction will have some characteristic, or set of characteristics, that distinguishes it from every other transaction. The ability to understand how all of these variables interact to form a 'risk mosaic' that defines the scope and nature of risk is essential.

In considering how best to integrate TRM into your organization's risk management lexicon, bear in mind the following basic guidelines:

1. *Know your data sources.* Data may be obtained from any number of sources, but it is important to be able to trust the information sources being used. Where did the source you may wish to use obtain its information? Too often, external information providers or consultants may not reveal where they got their information, which may raise question about its validity and accuracy.

2. *Question official statistics.* Analysts tend to be overreliant on statistics generated by governments or multilateral organizations. While being important sources of information to include in the analytical process, it should be remembered that governments often manipulate statistics for political pur-

© The Author(s) 2016
D. Wagner, D. Disparte, *Global Risk Agility and Decision Making,*
DOI 10.1057/978-1-349-94860-4_12

poses, and that international organizations often rely on such information to produce their own analysis and projections.

3. *Benefit from the power of observation.* There is no better way to arrive at a conclusion about the nature of country risk than to see it for yourself. If you have the opportunity to visit the country and/or project site, it is often the best means of arriving at a conclusion about the nature of the risks associated with a given transaction. One's own experience and interpretive abilities add a lot of legitimacy to the analytical process.

4. *Qualitative risk assessment can ultimately undo all quantitative approaches.* An overreliance on quantitative measures can be dangerous. Adding texture to analyses by refusing to categorize them as either black or white, or good or bad, can be the single most effective addition to the analytical process. Human behavior (an unpredictable variable) can only be assessed through the incorporation of qualitative measurement.

5. *Politics can defy logic.* The political process is generally not logical, so applying logic to the country risk management process makes little sense.

Your ability to integrate these concepts into your firm's overall risk management structure will be critical to introducing agility to the risk management process. In the remainder of the chapter we will further explore some of the challenges decision makers face in crafting effective risk management programs, and what is involved in embracing TRM. Much of the discussion will be framed around themes related to managing country risk.

Elements of an Effective Risk Management Process

In order to be in a position to integrate TRM into the overall risk management process, an organization must first possess a sound risk management foundation. Sound risk management processes include certain basic elements that result in the creation of an environment conducive to effectively managing risk[1]:

- Effective oversight by a board of directors.
- Adequate risk management policies and procedures.
- Adequate internal controls and an audit function.
- An accurate system for reporting country exposures.

---

[1] *Country Risk Management,* Comptroller of the Currency, Comptroller Handbook, October 2001, 3.

- An effective process for analyzing country risk.
- A country risk rating system.
- Established country exposure limits.
- Regular monitoring of country conditions.
- Periodic stress testing of foreign exposures.

Within this context, it is important to establish clear tolerance limits, delineate clear lines of responsibility and accountability for decisions made, and identify in advance desirable and undesirable types of business. Policies, standards, and practices should be clearly communicated to the affected staff and offices, and should then be enforced. Reports on a quarterly basis should be imposed—or the reports should be even more frequent if foreign exchange exposure is likely to impact a given investment.

It is also naturally important that analyses be adequately documented and conclusions communicated in a way that gives decision makers an accurate basis on which to gauge exposure levels, and that sufficient resources be devoted to the task of assessing risk. Communication methods should clearly convey the level of risk and urgency. Some banks have centralized the analytical process and engage in periodic assessments of risk on a more regionalized basis (as opposed to strictly on a country-specific basis).

Best practices dictate that a number of actions should be taken to create conditions conducive to a transactional risk management program. Among them are:

- The transactional risk management function should be centralized.
- Transactional risk guidelines should be established and widely disseminated.
- Country/sector limits should be established.
- A system to better delineate the severity of perceived risks should be established.
- Quarterly transactional risk reporting should be implemented.
- A firm should make maximal use of internal information capabilities while incorporating a wide array of external information sources into analyses.

Much can be learned at a corporate level by the approach and experience of international banks in addressing transactional risk. Nearly all banks have developed formal programs to manage transactional risk, and most of these are centralized so as to establish and maintain control over an entire network of operations. Almost all banks assign formal country ratings, most of which cover a broad definition of risk. Ratings are typically assigned to all types of credit and investment risk, including local currency lending.

Transactional risk ratings establish a ceiling that also applies to credit risk ratings. Most banks do not generally have formal regional limits to lending, but some banks monitor exposures for a given region informally, and most have specific country limits. Many banks apply a single country rating to all types of exposure, while distinguishing between foreign and local currency funding. Formal exposure limits tend to be set annually and managed through the use of aggregate country exposures.

The ability to obtain primary knowledge through inputs from local offices, as well as by regular visits on the part of in-country risk officers, cannot be overemphasized. Best practice should encourage in-house assessments before relying on external sources of information in order to build internal rating applications. This should not come at the expense of utilizing external sources of information, however. Too often, organizations over-rely on internal sources of information and in the end self-impose 'blinders' on the risk management process.

In most organizations, the country risk function operates autonomously, as there tend to be diverging interests between the operating side of the business and risk management. It is important, therefore, for senior management to effectively oversee interaction between the two sides. The risk assessment decision chain should be transparent and independent of compromise by business unit practices.

Even if there is a rating guide at one's disposal, it is best to utilize information from a variety of sources, identify the central themes that keep reappearing, and make a judgment about the nature of the risk. This should not be done in a vacuum, however. The underwriting and pricing process should be viewed as collaborative, seeking the affirmation of others in the decision-making chain who actually have the ability to contribute meaningfully to the risk management process.

Another common issue is that the lines of communication between risk management personnel, risk management and decision makers, or between decision-makers are either bypassed, convoluted, or just plain wrong. Some examples are:

- Risk management is given only cursory participation in the transaction approval process.
- Sales teams bypass risk management entirely, or ignore risk management's recommendations, because they fear a transaction will be canceled as a result of unacceptably high levels of risk.

- A CEO delivers a presentation to a board of directors that is false, but he/she believes it to be true, because the risk manager and his/her staff said it was.

A risk manager may have the right information, but it may be based on a short-term assessment of the risks. The long-term view may be completely different. In the absence of knowing what questions to ask and having clear lines of communication, the right information may not be taken into consideration. A board of directors often has no idea what questions they should be asking of corporate decision makers. Executive education on the subject of risk management generally, and also on specific elements of it, can be invaluable to decision makers up and down the chain, and could save an organization millions of dollars by having a better ability to avoid costly mistakes.

Embracing Best Board Practices

Too many boards are composed of individuals who do not necessarily have the skills and experience to make meaningful contributions to the decision-making process. It is also regrettably the case that too many boards are composed on individuals that were hand-picked by the CEO without a transparent vetting process and act as a rubber stamp in the decision-making process. This is clearly not in the best interests of either the organization or the decision maker, and it certainly can prevent adopting a rigorous TRM process.

Boards have a responsibility to implement optimal oversight of the risk management process. There are a number of things boards can do to help ensure that best practices are adopted and embraced, so that in the process, they are in a position to make a valuable contribution to the risk governance and culture of an organization. To do so, decision makers and boards should together consider implementing the following best practices:

1. Require management to establish a top-down ERM program that addresses key risks across the company and elevates risk discussions to the strategic level.
2. Create a heat map that identifies and collates exposures across the company, reveals linkages between exposures, and identifies fundamental risk drivers.
3. Require an in-depth, prioritized analysis of the top three-to-five risks that can make or break the business (the company's 'big bets' and key exposures).

4. Require an integrated, multi-factor scenario analysis that includes assumptions about a range of economic and business-specific drivers.
5. Establish a board-level risk review process that includes management insightful risk reports.
6. Establish a clear understanding of risk capacity based on metrics that management can measure and track.
7. Produce a strategy statement that clarifies risk appetite, risk ownership, and the strategy to be used for the company's key risks.
8. Require management to formally integrate risk thinking into core management processes (such as strategic planning, capital allocation, and financing).
9. Clarify risk governance and risk-related committee structures at the board level, and review board composition to ensure effective risk oversight.
10. Define a clear interaction model between the board and management to ensure an effective risk dialogue.
11. Require that management conduct a diagnostic of the organization's risk culture and formulate an approach to address gaps.
12. Review top management's compensation structure to ensure performance is also measured in light of the risks taken.

These measures are not easy to complete, particularly if an organization has not established a robust foundation upon which to achieve them. Yet doing so will make achieving any risk-related goal easier, and will certainly make managing transactional risk easier to achieve. At the very minimum, board members should take the time necessary to learn what they need to know about a firm's established risk management practices. This will greatly assist the board in crafting the right types of questions to ask, and answer.

Board members should identify key participants in the organization's risk management structure, seek them out, and interact with them directly. This will help provide an environment conducive to mapping out a plan of action for identifying critical issues, and provide a pathway to possible solutions. There can also be great value in seeking an independent third party's review of a firm's risk management practices—not only to provide practical insights, but also to learn how an organization's practices compare with that of other in the same peer group, and beyond.

Decision makers and management should work together to determine how best to integrate risk management into the strategic planning process, since the two are inextricably linked. Ideally, the board should be involved in such things as helping to assess both the upside and the downside risks of investment proposals. The larger point is that the board should not simply be thought of as a mechanism for getting investment proposals approved, but should be part of the larger decision-making process.[2]

Having the Right Tools and Orientation

Having the right risk management tools at your disposal—whether it is an insurance product or a hedging instrument—is the best means of managing risk. Yet assessing cross-border risk involves much more than simply collecting data. Getting good data is relatively easy, given the wealth of information at our disposal. *Interpreting* data is the real challenge. To do so well, risk analyses should be:

- *Consistent*: Made using rigorous frameworks that allow for valid cross-country comparisons;
- *Concise*: With the conclusion easy to understand, but with sufficient detail to make it meaningful;
- *Informative*: Giving the end user the rationale behind any assessment without any "black boxes" that are difficult to understand; and
- *Decisive*: Having a clearly defined position on prevailing country conditions and future implications.

The risk manager or decision maker's ultimate challenge when assessing transactional risk is to determine whether an event poses a current or future threat to a firm's financial performance. A mass demonstration in a stable *developing* country may be less significant to a firm's performance than one occurring in an unstable *developed* country. Similarly, a worker strike for higher wages is very different from a nationwide strike intended to overthrow an incumbent government.

The nature of transactional risk varies most fundamentally by the category of investor (direct or portfolio) because their exposures to cross-border risk may differ. In general, portfolio investors are more likely to be affected by country-specific risks, such as a sudden hike in interest rates or an unanticipated currency devaluation, while direct investors tend to be affected more by firm-specific risks. It is therefore necessary to focus on those dynamics that affect the overall business environment in a host country.

---

[2] McKinsey Global Institute, *A Board Perspective on Enterprise Risk Management*, February 2010, accessed November 26, 2015, www.mckinsey.com/~/media/mckinsey/dotcom/client_service/risk/working%20 papers/18_a_board-perspective_on_enterprise_risk_management.ashx

Effective TRM requires the ability to distinguish emerging events and trends that pose true risks—such as a well-defined threat to corporate performance—from events that are merely dramatic. When assessing political stability, the focus should be on the legitimacy of state authority, the ability of that authority to impose and enforce decrees, the level of corruption that pervades the system of authority, and the degree of political fractionalization that is present. Where economic policy is concerned, the focus would be more along the lines of the degree of government participation in an economy, the government's external debt burden, and the degree to which interest groups can successfully obstruct the decision-making process.

Although there are a number of ways to protect a firm against country-specific risks, *proper planning* and *due diligence* are the most important. Too many businesses begin operations in an unfamiliar country without having taken the time and devoted the resources necessary to ensure a better-than-average chance of success. Developing solid relations with relevant governing authorities is often the preferred approach, but this may not always be possible, or even desirable.

Another important component of creating a friendly investment environment is to establish a good relationship with the local workforce. Too often, foreign businesses are perceived as having uncaring managers who do not appreciate their workers—something which can have dire consequences. One of the best ways to protect a firm's assets is to generate a loyal workforce. Management can be replaced much more easily than can a workforce, and it is becoming increasingly common for host governments to remove foreign managers and replace them with local managers who will operate in accordance with government objectives.

It is easy to lose track of the bigger political picture once an operation is established. After an operating environment has been changed, it is often too late to do anything about it. It is important, therefore, to remain engaged with one's local embassy, chambers of commerce and other support networks and information collection sources. A collective voice is more powerful than that of an individual firm, even if a firm has a solid relationship with governing authorities.

Engaging Successfully with the World

In considering how to 'engage' with the global marketplace, a variety of types of global organizations have found it difficult to adapt to the rapid pace of change. A 'global' firm may end up having a more difficult time adapting to

change by virtue of its size and how quickly it moves in sync with its environment. Here, firms that are more locally focused can have a distinct advantage,
but there is no single most effective organizational model for how companies
should adapt to change.

Those firms that are, for example, in the natural resource space have little
choice but to adapt quickly, for the regulatory, environmental, and security
environments in which they operate are often in a rapid state of metamorphosis. By definition, the set of challenges and opportunities that each firm faces
will vary depending on the nature of their operating environment and business
model. Many firms find themselves with conflicting demands—having to meet
local needs on the ground while maintaining practices that are consistent with
global trends or requirements. Also, firms that have been accustomed to growing organically may be at odds with those that have adopted more standardized business models. Firms that have grown through acquisitions may have a
completely different set of needs than those that have made no acquisitions.

Making the most of a firm's local assets can prove to be a challenge. Apart
from the linguistic and cultural dimension, transferring knowledge from a local
environment to headquarters, and being able to successfully integrate it into a
meaningful translation can be difficult indeed. The same is, of course, true in
reverse. If corporate strategy can be communicated effectively across a global
firm, headquarters need to be nimble and sensitive enough to understand the
nuances of the feedback that may result. Is the organization well enough developed culturally to keep from being biased in delivering and interpreting two-
way information? Sometimes unintended biases are introduced in firms that
may not wish to create a 'single' corporate culture, but do so anyway.

Those firms that are bound to be more successful in the global marketplace
will seek to integrate the experiences and lessons to be learned from operating internationally into their risk management strategy, their strategic planning
processes, but, equally importantly, into their corporate culture. There is an
inherent disconnect between firms wanting to be global but incorporating the
type of top-down management structure that acts more like a bulldozer rather
than a construction crane. If a firm wants to be able to adapt to global change, it
will recognize that it stands to benefit from being flexible, creative, and nimble.

Myth Busting

Risk managers and decision makers may *think* they can either decide to manage risks at home or abroad, but if your firm is international, there is little distinction to be made. Domestic operations are part of the global risk mosaic.

Table 12.1 Doing business rankings of four selected countries (2015)

Variable	Japan	Russia	Georgia	Thailand
Overall ranking	29	62	15	26
Starting a business	83	34	5	75
Construction permits	83	156	3	6
Getting electricity	28	143	37	12
Registering property	73	12	1	28
Getting credit	71	61	7	89
Protecting minority investors	35	100	43	25
Paying taxes	122	49	38	62
Trading across borders	20	155	33	36
Enforcing contracts	26	14	23	25
Resolving insolvency	2	65	122	45

Source: *Doing Business*, World Bank Group, 2015, accessed November 26, 2015, http://www.doingbusiness.org

Both domestic and international operations imply risk, of course, but perceived lower risks associated with simply staying at home can be deceptive. For example, if your organization is based in a developed country, staying at home implies enhanced regulatory and governance risk—something that is often not nearly as onerous in a developing country. On the other hand, operating in a developing country can imply having to worry about things you may not generally have to worry much about in a developed country—such as reliable electricity and rule of law. Or perhaps not.

Every year the World Bank publishes its "Doing Business"[3] report, ranking nations according to the ease with which they allow commercial enterprises to be undertaken. You may be surprised about how some developed countries stack up against a variety of developing countries in ten basic criteria the Bank uses to determine the perceived desirability of doing business across the globe (in 189 countries). Table 12.1 presents the rankings of two 'developed' and two 'developing' countries: Japan, Russia, Georgia, and Thailand. The lower the score, the better the ranking.

The first thing one notices is that although Japan and Russia are recognized as being among the most developed countries in the world, their overall rankings are both below that of Georgia and Thailand (Russia being well below). Japan's best ranking is a number 2 (resolving insolvencies) and its worst is 122 (paying taxes). Six of its scores are ranked under 35 (in the top 19 percent overall). Compare that with Russia. Its best score is a 12 (registering property), it has only three of the ten rankings under 35, and three rankings are among the worst (construction permits, getting electricity and trading across

---

[3] World Bank Group, *Doing Business*, 2015, accessed November 26, 2015, http://www.doingbusiness.org.

borders). So, Japan seems like a decent place to do business, while Russia appears to be a risky proposition, based on the rankings.

Compare that with the two developing countries selected. In Georgia, four of the ten rankings are in the top seven, and all but one (resolving insolvency) are in the top 43. In Thailand's case, two of the rankings are in the top 12, seven are in the top 36, and its worst score is an 89 (getting credit). If, as a manager, you knew no better, you would probably automatically assume that doing business in Japan is easier than doing business in Georgia, or that doing business in Russia is better than doing business in Thailand. Based on these rankings, you would have been wrong.

This example serves once again to emphasize the need to turn the pyramid upside down and assume that your preconceived notions of fact versus fiction may in fact be exactly wrong. Georgia is one of the best smaller countries in which to do business, yet many outside of Central Asia would never even consider doing business there.

Necessary Transformation

Clearly, in the absence of embracing new modes of thinking about the world, the way it works today, how it may look five or ten years from now, and casting aside conventional approaches to risk management, any organization that has international operations can easily get into trouble. The forces of change have taken firm hold and are exposing global organizations to a plethora of multidirectional risks.

Businesses that are rising to the challenge and making fundamental changes to the manner in which they think about risk are rapidly finding that they must confront a range of issues for the first time. Many businesses have not previously had to address such issues as government austerity, interactive supply chains, non-state actors, or global market shifts. Some businesses are creating new business models that integrate digital technology, revamp resource acquisition, and penetrate completely new marketplaces, while others are busy bringing existing overseas operations back home to benefit from preferential logistics and cost differentials. The important point is that they are running toward transformation, rather than away from it.

Doing so exposes companies to new forms of risk, particularly because while they are changing, their competition and the marketplace in which they operate both continue to evolve. The interplay of businesses and marketplaces is itself creating complex linkages that are difficult to detect, and whose path is difficult to predict. All this is happening at the same time that investors are making more and more demands on companies with less tolerance for mistakes, clients demand more for less, and there are more hoops to jump through with respect to such issues as fair labor practices, environmental compliance, and local sourcing.

In this type of ecosystem, in which markets and businesses jointly function in what can be described as a global web of complex and interconnected challenges, risks can and do metastasize rapidly, having a knock-on effect. Earlier we discussed the impact of Ebola initially on West Africa, and then on other areas of the globe. Consider the impact of how Hurricane Sandy in 2012 shut down Wall Street and interrupted global financial activities, or how the pursuit of the Sustainable Development Goals will impact global standards for doing business internationally. It is the cascading impact of events and a changing landscape that poses one of the greatest operational challenges for global business.

The more businesses dive into the global marketplace, the more risks they encounter and must manage. Small and medium-sized businesses sometimes encounter the greatest challenges, because they are often the least prepared to address risks in the political, economic, regulatory, compliance, and sociocultural arenas. For them, especially, but for all global businesses in practice, the ability to ask on an ongoing basis questions such as "What could go wrong?" and "What may be wrong with the way we're tackling this problem?" is essential not only to staying ahead of the curve, but to simply surviving. Smaller firms, in particular, can easily be wiped out by a global-sized problem.

That said, it is clear that many business transformation attempts are not well-timed, well-executed, or, ultimately, successful. There are inherent risks involved in agreeing to pursue a transformation process while up to 70 percent of business transformations actually fail.[4] Among the most common reasons given for these failures are the high costs implied in pursuing change, the amount of time change can take to be achieved, a failure to communicate effectively within the organization about the change, unclear goals, and a lack of agility. Those companies that eventually succeed in executing a transformation have clear targets and a structure that can accommodate change, they use limited resources wisely, and they find efficient ways to engage their employees.

---

[4] Morgan Franklin Consulting, "Top 8 Reasons Why Transformations Fail," *Core Confi dence* , volume 1 (2014), accessed November 26, 2015, http://www.morganfranklin.com/core-confidence/article/top-8-reasons-why-transformations-fail.

The ability to avoid critical mistakes along the way can make a big difference between maintaining momentum or losing it, and making it to the finish line. Among the things that matter are:

1. Creating a vision to direct the change effort.
2. Establishing a sense of urgency.
3. Making realistic and accurate assessments of the competitive landscape.
4. Honestly assessing significant opportunities and potential crises.
5. Assembling a group with the power to lead the change effort.
6. Ensuring that the group work together as a team.
7. Effectively communicating the new vision and strategies.
8. Empowering others to act on the vision.
9. Removing obstacles to change.
10. Creating an enabling environment for short-term wins.
11. Consolidating improvements and using them to spur more change.
12. Institutionalizing the new normal.[5]

To increase their chance of success, some companies are using tools that make it easier to identify linkages between risks, so that they do not cascade into a crisis. This involves scenario planning based on potential events in the future, and hypothesizing about the range of potential outcomes. Early warning systems can be used to enable systematic monitoring of changes to an operating environment. Producing risk appetite statements can be important in ensuring that decision making is aligned with strategy. Stress testing, and reverse stress testing, can also prove to be valuable in measuring the impact of changes in variables have on overall risk.

Using the Right Tools/Asking the Right Questions

One of the challenges decision makers and risk analysts face in crafting a methodology that will gain wide acceptance in an organization is a tendency for existing risk management practices and biases to inhibit the development of tools that do not reproduce defects inherent in the system. It is almost as if there is a built-in tendency for existing systems to have a reproduction function in its DNA. Among the most common impediments standing in the way of broad acceptance of new and better ways of doing things are:

---

[5] J. P. Kotter, "Leading Change: Why Transformation Efforts Fail," www.hbr.org, January 2007, accessed November 26, 2015, https://hbr.org/2007/01/leading-change-why-transformation-efforts-fail.

- A breakdown in communication between analysts, risk management, and senior management.
- Unresolved conflicts between staff and management.
- A tendency to overlook or ignore defects or problems.
- Killing the messenger instead of rewarding the messenger.
- Covering up mistakes.
- Screening information, then denying there is a problem.
- A lack of incentives to find and correct a problem.
- A tendency to accept the most popular or favorable point of view—and not to rock the boat.
- Lack of flexibility and innovation.
- Loss of institutional memory.

Any tools that are adopted, or systems that are put in place, must be simple and easy to use, so as to encourage their broad acceptance and avoid the replication of common pitfalls, or, in extremis, of making things worse. There is an inherent trade-off between the relative breadth of an analytical tool and the amount of detail it can provide. Analytical tools are used to make judgments, but they are not omnipotent, nor should they be used as an alternative to other forms of due diligence. Most tools fail to capture subtleties that can be important in the risk analysis process, such as the economic impact of political change.

Broad indicators are more likely to produce greater variance and error, while more narrowly focused indicators may be more accurate, but for a limited set of problems. So the objective in crafting any such tool is to include enough variables so as to be able to read between the lines and derive sufficient nuance to be able to arrive at a meaningful conclusion. It is also important to be clear about underlying assumptions, rationales, and purpose of creating and using analytical tools, which can introduce bias into the analytical process.

Some of the best thinkers about risk management disagree about relative risks, weights, and measures, for there are many ways to approach a given problem or challenge. The key is to craft a risk management program that meets your organization's specific needs while utilizing information you have confidence in. One of the best ways to achieve that is to use information from a plethora of sources, toss out the 'fringe' viewpoints, and concentrate on what they all seem to be saying.

A good place to start is to create a common platform for classifying all the countries where you may be doing business. One easy way to do that is simply to categorize them by number and color, based on a composite of information sources. You may wish to subscribe to several information providers, create a

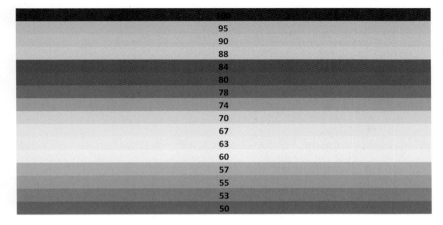

100
95
90
88
84
80
78
74
70
67
63
60
57
55
53
50

Illustration 12.1 Country classification by number and color (*Source: D. Wagner, Managing Country Risk* (New York: CRC Press, 2012), 82)

weight for their rankings (which allows for comparison), and assign a number, corresponding to a color for them. One option is to use green/higher numbers to identify countries of lower risk, and red/lower numbers to show higher risk (see Illustration 12.1). Using a color-coded system also helps communicate the level of risk in a universally understood manner.

A simple risk tolerance color-coding identification approach is:

- *Green*: High grade—no restrictions (excellent recovery potential).
- *Blue*: Investment grade—no restrictions (good recovery potential).
- *Yellow*: Speculative—restricted (moderate recovery potential).
- *Orange*: Low grade—heavily restricted (poor recovery potential).
- *Red*: Prohibited (no recovery potential).

To determine whether a country may need to be listed as 'red', consider the following parameters:

- Is the country known to be fully compliant with international law?
- Are there any outstanding embargos imposed by the UN or individual states?
- Is there a history of embargos?
- Has the country ever been labeled a state sponsor of terrorism?
- Does it produce chemical, biological, or nuclear weapons?
- Does it produce military- or defense-related items, and does it sell such items to any country that could result in its being labeled either an illegal arms shipper or a transit point for illegal arms?

Table 12.2 Two grade-based rating options

Letter Option A	Letter Option B
Aaa	AAA
Aa	AA
A	A
Baa	BBB
Ba	BB
B	B
Caa	CCC
Ca	CC
C	C

Source: Wagner, D. "Managing Country
Risk," (New York: CRC Press, 2012), 82

- Have any financial restrictions been placed on the country?
- Is travel prohibited to or from that country, by whom, and why?

You may wish to establish cut-off points for country approvals, wherein, for example, any country ranked above 90 would gain automatic managerial approval and any country below 50 would automatically fall into the 'forbidden zone'. Countries could also simply be assigned a color or number, instead of both. Some examples of how countries might be categorized by color are:

- *Green*: Australia, Canada, the United Kingdom.
- *Blue*: Chile, Malaysia, South Africa.
- *Yellow*: Costa Rica, Jordan, Mauritius.
- *Orange*: Bangladesh, Ecuador, Ivory Coast.
- *Red*: Iran, North Korea, Venezuela.

Three other simple qualitative approaches are:

1. Using two options side by side for a simple grade-based rating scheme (Table 12.2) (which depicts lower risk in the "A" category and higher risk in the "C" category);
2. Combining coding with expected impact (Table 12.3); and
3. Combining letter ratings with levels of risk (Table 12.4).

There are numerous quantitative variations on a theme that can add depth to an analytical framework. Naturally, the objective will determine which variable is being measured, and how. Here are some topics, along with subcategories, that may be used:

Table 12.3 Color coding and expected impact

Risk Rating	Level of Risk	Operational Impact	Physical Asset Impact	Financial Impact
1	Negligible	Little likelihood of interruption	No meaningful likelihood of loss or damage to property	Low chance
2	Minor	Interruption, but little impact	May or may not require and insurance claim	Impact will not affect financial performance
3	Moderate	Results in partial shutdown	Requires an insurance claim to repair damage	Financial performance impacted
4	Significant	Long-term shutdown	Damage may or may not be repairable	Severe impact on the bottom line
5	Extreme	Permanent cessation of operations	Damage not repairable. Asset must be abandoned.	Could result in significant impact on corporate performance

Source: Wagner, D. "Managing Country Risk," (New York: CRC Press, 2012), 85

Table 12.4 Letter ratings with levels of risk

Letter Rating	Level of Risk
A	Industrialized economy with negligible risk of political instability or liquidity crisis. A strong ability to withstand economic crisis.
B	A developed economy with low political or economic risk.
C	A country with structural weakness and moderate risk of either political or economic distress.
D	A developing economy with a range of structural weaknesses and a high risk of political or economic distress.
E	A failed or failing state prone to political and economic instability. No ability to withstand economic distress.

Source: Wagner, D. "Managing Country Risk," (New York: CRC Press, 2012), 85

Economic Growth

- GDP growth vs oil demand growth
- Growth in industrial production
- Foreign capital as a share of total investment
- Household and corporate liquidity
- Urban vs rural population growth
- Population growth by age
- Projected labor force growth

Economic Health

- Gross domestic savings as a percentage of GDP
- Expenditure on health and education as a percentage of GDP
- Military expenditures as a percentage of GDP
- Labor force as a percentage of the total population
- Government/provincial/municipality fiscal deficit
- Government expenditure by municipality, province, or state
- Source and allocation of funds by financial industry
- A country's share of global projection for a given commodity or product
- Destination of and dependence on exports

Power Sector

- Sources of demand for power (by economic sector)
- Amounts of energy used to generate GDP growth
- Power generation capacity growth and utilization hours
- Power demand forecast and generation mix
- Variation in energy prices by province or city

In developing a methodology best suited for your organization or purpose, it is useful to first determine the risk attributes/indicators that should be included. One easy way to do this is to identify which indicators are likely to apply to the majority of countries which will be included in the framework. This can be as general or specific, and as broad or as narrow, as necessary. Graph 12.1 includes both broad and narrow topics that can be used to create a broad analytical framework.

Let us examine some of the subtopics that might be included in each (what follows is meant to be suggestive, rather than comprehensive in nature). The topics a specific firm may wish to include will be dependent on the scope and nature of its international operations; however, much of what is listed here will apply to many different types of trade and investment transactions:

Graph 12.1 Examples of indicators for inclusion in an analytical methodology (*Source:* D. Wagner, *Managing Country Risk* (New York: CRC Press, 2012), 87)

1. Economic Environment
   - Degree of economic openness
   - Monetary stability
   - External financing needs
   - Fiscal imbalance
2. Trade and Investment Climate
   - Import/export restrictions
   - Capital controls
   - Protectionism
   - Embargos
   - Shifting regulatory climate
3. Financial Considerations
   - Availability of consumer credit
   - Excessive growth in national debt
   - Ability to freely convert/transfer currency
4. Local Environment
   - Natural disaster-prone
   - Endemic corruption
   - Labor unrest
   - Existence of civil society
5. Developmental Issues
   - Adequacy of infrastructure
   - Degree of poverty
   - Sufficiency of income levels
   - Progress toward middle-income or developed country status

6. Cultural Issues
   - Orientation toward western values
   - Work ethic
   - History of ethnic conflict
   - Degree of homogeneousness of population
7. Geo-strategic Considerations
   - Position/perceived value on the global political stage
   - Relationship/centrality to existing or future regional conflicts
   - Perceived military value in a regional conflict
8. Social Environment
   - Friendliness toward people from other cultures
   - International orientation
   - Active civil society influence
9. Political Stability
   - Degree of political competition
   - Regularity of elections
   - Political legitimacy
   - Frequency of changes in government
10. Personnel Risks
    - Mobility of labor force
    - Employment regulations
    - Staffing regulations
    - Labor relations
    - Legal environment (enforceability of contracts, rule of law)
11. Currency Risks
    - Nature of conversion/transfer regime
    - Existence of wide fluctuations in currency value
    - Vulnerability to exogenous shocks
12. Legal/Regulatory Risks
    - Rule of law
    - Sanctity of contracts
    - Central bank independence
    - Judiciary independence
    - Presence of nepotism/corruption
13. Asset Risks
    - Local ownership requirements
    - Control issues (designated board seats)
    - Joint-venture partner
    - Intellectual property protection

14. Supply/Delivery Risks
   - Vulnerability of critical raw materials
   - History of supply interruption
   - Reliability of transportation into and out of the country
15. Operational Risks
   - Domestic participation requirements
   - Denial of permits or licenses
   - Production quotas
16. Security Risks
   - Existence of civil war, insurrection, protracted conflict
   - History of cross-border conflict
   - Vulnerability of location of specific project/transaction
   - High-profile nature of project/transaction
   - Security protocol/procedures at project site[6]

Risk Manager Beware

This above list gives a sense of the incredibly diverse array of issues to be addressed as a transactional risk manager, and some of the tools at your disposal. Of course, this is just a small sampling, but it provides enough texture to drive home the point that there is plenty of room for error. Accurately assessing risk is an art, not a science. Quantitative measures can only go so far toward telling the whole story. The importance of integrating qualitative risk measures cannot be overstated.

It is not in the risk manager's purview to simply head for the exit from a country experiencing challenges, whether temporary or permanent. A pragmatic risk manager would simply say that many of the implied challenges are the cost of doing business, and that a firm's pricing model and risk management strategy should adequately reflect such heightened exposures. Some companies do indeed operate on this basis, making a long-term commitment in the belief that conditions will improve with time. Sometimes they do, but often, they do not.

Many risk managers tend to worry less about macro-level risk and more about 'spectacular' events, such as high-profile kidnappings and acts of terrorism. The reality, however, is that all the variables that erode economic and political conditions ultimately fall under the remit of a risk manager—whether they may be mitigated or not. The challenge is to be able to decipher

---

[6] Wagner, *Managing Country Risk*, 88–89.

risks that are manageable from those that are not, and to do an effective job of taking counter-measures.

A transactional risk manager can tell the difference between routine headlines and game-changing events. Just as insurance underwriters—the final repository of risk—are trained to be able to tell the difference, it is incumbent upon decision makers and risk managers to be able to do the same. Since retreat is usually not an option, risks should be managed proactively, and mitigating measures should be utilized early and reviewed often. In the end, there is no substitute for anticipating your firm's risk management needs, and being able to address them in a meaningful fashion.

13

Anticipatory Risk Management

We started this book with a discussion about some of the most important challenges facing the world, such as climate change, global pandemics, and the rise of Islamic extremism. What makes these issues important from a risk management perspective is that we recognize that our collective vulnerability to them is permanent, and that although our ability to 'manage' them may be limited, we have the ability to address them in some way *today*, even if the manner in which we do so may only result in partial success over time.

Anticipatory Risk Management (ARM) may be defined as a permanent orientation toward understanding and embracing future risk, generating mitigation strategies to address those risks, and integrating them into the planning, operation, and risk management functions of a business. Businesses are all too often focused on the short term (the competition, consumers, and profits). ARM requires that the orientation toward the medium- and long-term future must be permanent, that the strategies to mitigate them be realistic and achievable, and that their integration into the core functions of a business be seamless. Doing so is no easy task, but given the evolving nature of the risk landscape, it is no longer a luxury to have—it is essential to firm survival.

© The Author(s) 2016
D. Wagner, D. Disparte, *Global Risk Agility and Decision Making*,
DOI 10.1057/978-1-349-94860-4_13

Self-induced 'Grey Swans'

A grey swan event is an occurrence that is possible, known about, and potentially significant, but considered unlikely to happen.[1] Large-impact, low-likelihood events (grey swan events) are happening with greater frequency as a result of the increased interconnectivity of supply chains, financial markets, and information sources. A good example of a grey swan event was the 1999 earthquake in Taiwan, which resulted in the near doubling of semiconductor prices overnight as a result of Taiwan being a primary supplier of semiconductors. An earthquake does not normally have such 'ripple' effects around the globe. Other examples include the collapse of Long-term Capital Management in 1998 (which sent shudders through the global financial system) and 9/11 in 2001 (which, in combination with the Iraq War in response, completely altered the global terrorism risk landscape).

The Taiwan and 9/11 events prompted a fundamental re-think among some companies about how they plan for business interruption and supply chain disruption. Many companies did not have plans in place to enable them to continue operating when such disruptions occur, in part because there were divergent views between risk managers and decision makers about what should be their primary areas of concern. CEOs may view corporate tax rates or the ability to hire key personnel as being more important than data privacy or regulatory compliance. This is why it is incumbent upon decision makers to ensure that there is synchronicity between departments with competing agendas with respect to risk management.

Grey swans can also be self-induced, if an organization and/or its decision makers fail to account for the unforeseen—something that *should* or *could* have been foreseen, but was not. In the two following *true* examples, the companies involved 'could' have foreseen the problems that soon materialized before them—if only they had done a little more research, put on a broader thinking cap, and thought more about things that could go wrong. The first example is in Palestine, an investment destination a typical business person from outside the region may feel leery about, but would probably never imagine an investment could go so terribly wrong so quickly. The second is in South Korea, a country normally associated with efficiency and predictability.

In the Palestine example, this astonishing chain of events—all linked to a fast food restaurant—could have been avoided relatively easily. A transactional risk manager practicing ARM would never have allowed the investment

---

[1] Definition of 'Grey Swan', www.macmillandictionary.com, 2015, accessed November 26, 2015, http://www.macmillandictionary.com/us/dictionary/american/grey-swan.

to go forward without making at least one site visit, and conducting extensive on-the-ground due diligence. He/she would also not have simply taken the word of the business partner about the precise location of the venture. This is a perfect example of what can happen when companies fail to think and operate in a multidimensional manner.

A Fast Food Chain in Palestine[2]

In 1999, Company A asked its Israeli franchise holder to remove the company's brand name from a three-month old hamburger stand in a new mall in Maale Adumim, the largest settlement in the West Bank. The Israeli company that owned the stand refused to do so, defending its right to offer kosher food at the food court in the disputed territory. What Company A thought of as a commercial issue with a franchisee, quickly became a political confrontation, and a public relations nightmare.

At the beginning of the summer, Arab-American groups protested the establishment of an Israeli-owned fast food company in the West Bank. They led demonstrations at Company A's restaurants in the U.S. and started a boycott of the company for what they saw as a business act legitimizing Israeli settlement in land claimed by the Palestinians. Their cause was picked up the Arab League, which threatened to impose an all-Arab boycott of Company A's restaurants. When the company decided to cancel its franchise contract, Jewish groups accused the company of capitulating to Arab pressure. Jewish groups in the U.S. promised their own boycotts if the company did not back down.

For its part, Company A claimed to have been duped by its Israeli partner, which said the address of Maale Adumim was in Israel, rather than the West Bank or Occupied Territories. Company A did not perform a site visit or do on the ground due diligence before approving the application to open the restaurant; a company inspector was sent from Dubai to take a look at the premises, unaware that the restaurant was located in the West Bank.

After the owner refused to shut the restaurant down, Company A took its action. When it decided to cancel its franchise contract, Jewish groups in the U.S., led by the Anti-Defamation League, vowed to take its complaint against Company A to the U.S. Attorney General and Congress. Christian organizations, which vehemently supported Israeli sovereignty over the West Bank, threatened to join the boycott—all because of a failure to communicate effectively and perform proper due diligence.

---

[2] Wagner, *Managing Country Risk*, 144–145.

In the following Korean example, the U.S. company in question was large, globally experienced, and had the resources to devote to 'getting it right' at the outset, but failed to anticipate the impact nationalism can have on investment climates and foreign investors. As previously noted, nationalism can become overwhelming in terms of the legal and regulatory environment, as well as impacting public opinion. This is a perfect case of having checked all of the prescribed boxes, but not taking into consideration a variable that is there for all to see, if one is looking for it.

A Technology Company in South Korea[3]

In 1997 one of Korea's leading software developers was one step from bankruptcy. Poor management, rampant software piracy and Asia's financial crisis had all taken their toll on the firm. A U.S. technology company (Company B) proposed to save the company from bankruptcy by investing $20 million in return for the local company ceasing production of its highly popular Korea-language word processing software (which accounted for 80 percent of the local market) and to start selling Company B's version of the software (which then commanded just 15 percent of the market).

When the Korean press got hold of the story, the Korean public—well known for their fervent nationalism—was outraged. Rather than proceed with the alliance with Company B, the local company instead chose to accept a $10 million investment from Korean computer-related businesses and associations that had led a nationwide campaign to save the local company—which many Koreans regarded as a national technological treasure—from foreign control.

In the year that followed, the local company's market capitalization increased five-fold. Ironically, both the local and U.S. companies experienced increased revenues in part because of the country's robust economic recovery, but also because the Korean software piracy rate fell 10 percent as a result of the debacle, thanks at least in part to the campaign to save the local firm. In the meantime, the battle between the two companies continued, with the local firm filing dumping charges against Company B, accusing it of trying to gain market share illegally.

---

[3] Wagner, *Managing Country Risk*, 147.

As these examples serve to illustrate, some grey swan events can be avoided simply by doing your homework and taking the time necessary to complete appropriate due diligence before making an investment. Sometimes 'the basics' get lost in the shuffle. A lot of risk management procedures tend to focus on short-term risks, which may be insufficient to address grey swan risks. Businesses need to institutionalize longer-term risk management processes to integrate creative thought, rigorous due diligence, scenario planning, and common sense into the mix.

Managing the 'grey swan effect' requires businesses to proactively engage multiple layers that impact their operations, such as consumers, suppliers, and regulators. Businesses should identify the sources of grey swans through an early warning system and establish a mechanism to evaluate the implications of these events and prepare adequate responses to deflect downside risk and leverage upside opportunity. Depending on the nature of a grey swan event, businesses should ideally have in place both an 'anticipate and respond' and an 'absorb and rebound' framework.[4]

Transitioning Away from Enterprise Risk Management

It is clear that in a grey swan-filled world, risk management frameworks have two choices: evolve, or become irrelevant. Enterprise Risk Management (ERM) was originally developed to enable companies to respond more meaningfully to an increasingly complex risk environment, with a focus on asserting stronger control over financial and operational risks. However, because it approached risk from a conventional perspective, presuming that events can be predicted and controlled with a reasonable degree of certainty, ERM is quickly losing its usefulness in a world filled with uncertainty and unpredictability.

ERM can actually interfere with organizational risk agility, because it can encourage a 'process-led' approach to risk management that can lead decision makers to separate risk management from business decisions and deflect responsibility for risk management issues exclusively on to risk managers, rather than on others in the organization who share in creating and ultimately managing risk. This type of myopic approach moves an organization away from focusing on the risks associated with interconnectedness and can increase the propensity for contagion.

---

[4] PWC, *Grey Swans: Transformation of Risk in an Interconnected World*, 2013, 18.

Thinking about risk management as a business enabler means embedding risk assumptions within a company's overall growth plans and strategy, so that they overlap, rather than operating in separate silos. The most risk-resilient companies already do this. Those that do not find themselves more vulnerable to things like product failures, supply chain interruptions, and difficulty with entry into new markets. Companies with risk resilience view risk taking as a deliberate choice, rather than a by-product of a given strategy.

In order to create conditions conducive to transitioning from ERM to ARM, the first step is to develop a corporate culture that can become multidimensional in focus, with a culture that naturally gravitates to a more organic form of risk awareness—one that actually presumes some events may be unpredictable, uncontrollable, and not managed in a traditional way. Clearly defining risk appetite, mapping the boundaries of acceptability, and creating a new risk orientation aligned with corporate strategy is ultimately the path to achieving ARM.

That said, less than a third of private U.S. companies appear to recognize the need to overhaul their approach to managing risk—even in the face of uncertainty and events that have disrupted their business—and even fewer have made changes directly in response to external shocks. A 2012 survey[5] noted that although some of these companies had experienced a direct financial impact associated with major disruptive events (such as the Arab Awakening, the European debt crisis, or the Fukushima nuclear crisis in Japan), most of them did not revise their approach to risk management.

Another survey[6] found only a minority of private companies were confident that they would be effective in key aspects of risk management, and that most companies were doubtful that they could keep pace with the evolving risk environment. Most of the companies that were revisiting their approach to risk felt the need to focus more on the broad range of emerging risks, which is a move in the right direction. While companies clearly cannot detect, analyze, and monitor all forms of risk all of the time, doing what they can to identify and manage those risks that have had or will have an impact on their ability to operate makes real sense.

A culture of risk resilience requires a risk management platform be built that:

- Encourages collaboration horizontally and vertically across a business;
- Incorporates risk considerations in the strategic planning process;

---

[5] PWC, *15th Annual Global CEO Survey*, 2012 (interview with more than 1,200 CEOs from 60 countries), accessed November 26, 2015, https://www.pwc.com/gx/en/ceo-survey/pdf/15th-global-pwc-ceo-survey.pdf.

[6] PWC, *Trendsetter Barometer*, 2012, accessed November 26, 2015, http://www.pwc.com/us/en/press-releases/2012/private-company-optimism-and-revenue.html.

- Establishes boundaries in advance;
- Looks at outcomes first and identifies possible causes;
- Prepares for the consequences of assuming risk before it may become necessary;
- Engages in routine and rigorous scenario planning;
- Develops recovery strategies for mission-critical processes; and
- Designs and tests resiliency strategies to ensure that crucial information flows during a crisis.[7]

Those companies focused on the long haul make resilience the centerpiece of their risk management strategy, recognizing that it is the outcome of any given trigger that the firm will ultimately have to address. Risk resilience is similar to peeling an onion to get to the core of the problem in advance. By definition, a consequence-focused approach requires drawing on resources from across an organization—exactly the type of thinking that is consistent with ARM.

The Importance of Having an Informed Board

As company operations and holdings continue to expand into all corners of the globe, decision makers too often pay too little attention to specific international risks that may turn out to be of crucial importance. Many boards of directors are particularly guilty of this, as a result of a lack of direct insight into a particular country or region of the world, an inability to discern fact from fiction, or not knowing the right questions to ask of corporate management. Since boards are the final stop in the risk management chain of command, they should actually be the *most* informed, but can sometimes end up negating all of the thought and effort that went into getting an item on their agenda in the first place.

How can boards make better, more informed decisions? One place to start is in the composition of the board itself. Too often, board members are selected from a small group of high-profile, well-connected, and prestigious individuals who may not have relevant experience in foreign investments or operations, and who may not want to appear ignorant about a subject matter being discussed, so they may fail to contribute meaningfully (or at all) to board discussions.

A case in point is the former board of directors of Lehman Brothers. Lehman's board consisted of ten high-profile individuals, many of whom had

---

[7] PWC, *Risk Resilience: Reckoning with a New Era of Threats*, 2013, 9.

served for more than ten years before the firm's collapse. Nine of the board members were retired, four were more than 75 years old, and only two had any experience of financial services. Dina Merrill, the 83-year-old daughter of E.F. Hutton, and a former actress, served on Lehman's board for 18 years, until 2006 (two years before Lehman's collapse). It is hard to imagine what value she added to the board; she was presumably a net negative.[8]

Similarly, during the tenure of Hank Greenberg, AIG's board was comprised mostly of loyal friends and colleagues of Greenberg, former politicians and government officials specifically chosen to add prestige to the board. A board with this kind of composition is neither likely to challenge the CEO nor have the knowledge necessary to ask the kind of penetrating questions that would add value to the decision-making process.

The CEO of AIG Financial Products (the division of the company that imploded, and took the company along with it), Joe Cassano, apparently lacked the mathematical skills to truly understand the business his division was in. Enron's board was selected for their connections to Enron, rather than their skillset. And prior the infamous Texas City explosion, the British Petroleum director with board responsibility for all BP refineries—including safety—had no refining experience prior to his appointment.[9]

In case you may think reputational risk is insufficient to bring down a company, consider Arthur Andersen, whose leadership clearly did not understand or take action to protect the company's reputation by its association with Enron. Similarly, Northern Rock's management failed to appreciate the importance of maintaining a bank's reputation for paying depositors on demand. Firestone's management team failed to appreciate how the company's reputational capital had been eroded as a result of its first tire recall crisis in 1978. It took a huge hit when the company had a second major recall crisis in 2000. Management and the board did not prioritize the safeguarding of its reputation as a trusted tire manufacturer.[10]

When selecting board members company management clearly needs to emphasize experience, knowledge, relevance, and potential addition to value. That said, it is difficult, if not impossible, to find individuals who have direct and timely experience in every country that may be an investment target for a large corporation. Another solution is for boards of directors to press compa-

---

[8] M.T. Williams, *Uncontrolled Risk* (New York: McGraw-Hill 2010), 189.
[9] Association of Insurance and Risk Managers (AIRMIC), *A Study of Major Risks: Their Origins, Impact and Implications*, Cass Business School, London, England 2011, 8.
[10] AIRMIC, *Major Risks*, 9.

nies to regularly update their own risk management procedures and insist on instituting appropriate checks and balances.

Given the competing interests that may influence an internal risk management team, a better solution may be to look outside of the company's risk management function and insist that either the company hire an independent third-party assessor or, ideally, they do it themselves. A qualified third party can conduct regular risk management audits that test and stress the system, provide insights into the target country that incorporate political, economic, and social risk, and thus can provide board members with unbiased information, empowering them to ask the right questions.

Especially when considering international business operations, the underwriting and pricing process should be viewed as collaborative, seeking the affirmation of risk specialists in the decision-making chain. Unfortunately, even when a problem is identified, boards are sometimes reluctant to confront management. Candor often gets lost in the politeness of board proceedings, and, all too often, boards are focused on building consensus, which inhibits due diligence and proper risk management. By remaining polite and silent, boards can do more than contribute to monetary losses and they may unwittingly cause reputational risk, often with long-lasting and severe consequences. If a company consistently performs well—above expectations, year after year, and in spite of business cycles or the state of the global economy— the board should be asking why, instead of merely applauding it.

When gathering and managing information, it is best to utilize information from a variety of sources, identify the central themes that keep reoccurring, and make a judgment about the nature of the risk. This should not be done in a vacuum, however. Therefore, board members must exercise their responsibilities with renewed vigor and with a solid base of knowledge and insight. If that had been the case with the companies described above, their business outcomes would have been very different.[11]

If risk managers are to be able to support boards effectively on risk-related issues, the purpose and scope of risk management within an organization needs to be reconsidered from the board level down, in order to capture those risks that have not yet been identified. Risk professionals within a company need to expand their skillsets so that they are able to identify, analyze, and discuss risks emerging from the organization's ethos, culture, and strategy, as well as their leadership's activities and behavior, if necessary. And finally, top-level

---

[11] Adapted from D. Wagner, *Anticipating Country Risk*, International Risk Management Institute, June 2014, accessed November 26, 2015, https://www.irmi.com/articles/expert-commentary/anticipating-country-risk.

decision makers must themselves recognize the importance of enhancing and understanding the risk management process, so that they may be able to fully appreciate the data that will be presented to them.[12]

The Moral Compass Imperative

Having a defined business culture is naturally important, and boards are critical to both establishing that culture and ensuring that it is implemented effectively, but it isn't sufficient to have a culture in place if an organization fails to have a moral compass to go along with it. The reason is that being able to distinguish right from wrong in a business setting—not as it relates to short-term profit-oriented decision making, but as it relates to 'doing the right thing'—can make the difference between avoiding problems in the first place and being able to extricate an organization from troubled waters quickly and with as little pain as possible.

The experience of Arthur Andersen (AA) is instructive. When Arthur E. Andersen first founded AA, he is said to have cemented his reputation when he told a local railway chief that there was not enough money in Chicago to persuade him to agree to enhance reported profits by using 'creative' accounting. He lost the account, and the railroad firm went bankrupt soon thereafter. Mr. Andersen had a clear moral compass. By the 1980s, the firm named after him was adopting the 'Big Five' auditors' new business model, which was to grow business by selling consulting services on the back of an audit relationship. AA did well in doing so by embracing the "2X" model—generate twice as much consultancy as audit revenue. Those who succeeded in doing that were rewarded, while those who did not faced sanctions.

Through its work for Enron, AA earned $25 million in audit fees and $27 million in consultancy fees in the year 2000. Over the years AA had been involved in 'creative' accounting measures, such as aggressive revenue recognition and mark-to-market accounting, in addition to the creation of special purpose vehicles, which were used for dubious purposes. The firm was sufficiently concerned that in 2001 it gathered together 14 partners to discuss whether the firm had retained sufficient independence from Enron. Noting that revenues could reach $100 million from Enron, they decided to keep Enron's account, even though Enron was in big trouble and heading for bankruptcy later that year. Mr Andersen would not have been pleased.

---

[12] AIRMIC, *Major Risks*, 19.

As news of the SEC investigation into Enron spread to Andersen, the Houston practice manager told the audit team that when AA had recently been investigated by the SEC. AA learned that most of the SEC's information against AA had come from AA's own files. He said that while they could not destroy documents once a lawsuit had been filed, they could legally be destroyed prior to a suit being filed. Over the coming days, AA's paper shredders were working overtime, all over the world. This loss of moral compass was an important factor in AA's ultimate collapse.[13]

There are plenty of other examples to cite as evidence of a lack of a moral compass and resultant damage to a company's reputation, such as when Coca-Cola knowingly sold processed tap water as 'spring' water in its 'Dasani' brand, or when Northern Rock permitted employees to be pressured into underreporting mortgage arrears, or when Shell had understated oil reserves.[14] Once a pristine reputation has been tarnished in the public's mind, it can almost never be restored. Any decision maker should be able to 'anticipate' that risk. The pursuit of short-term profit can never be worth the long-term risk of loss of trust from consumers.

Making the Right Choices

Decision makers and risk managers cannot effectively anticipate risk if they are unable to apply a moral compass in their organization with a free flow of information that clearly communicates boundaries and expectations. In the absence of having the ability to do so, the 'unknown unknowns' simply multiply, or decisions are made with rose-tinted glasses on, often yielding an undesirable result. So establishing a culture that encourages information flow throughout the organization, encourages a 'learning' culture, and rewards employees for asking good questions and expecting good answers is an essential part of ARM.

Some companies make straightforward, sensible decisions that end up making it easy to operate, and reduce the chance of creating a self-induced crisis. Yet for some inexplicable reason, other companies do exactly the opposite, which increases their chance of having a serious problem at some point in the future. One excellent example of this is the design and initial manufacturing process of the Airbus A380 aircraft. The aircraft's major components are built in France, Germany, Spain, and the UK, with subassembly completed

---

[13] C. Barney, *The Enlightened Organization* (Philadelphia: Kogan Page Limited 2014), 10.
[14] AIRMIC, *Major Risks*, 13.

in numerous other places around the world. All parts have to be shipped to Toulouse, France for assembly—nearly 100 miles from the nearest body of water (clearly, that decision was made for political reasons, and not because it made any real logistical sense).

As if establishing that approach to aircraft manufacturing was not bad enough, Airbus decided at the same time to take considerable risk to use non-standardized technology in the design and construction of the aircraft. When major assembly parts were matched for final assembly it was discovered that the wiring harnesses did not match, and middle managers did not inform senior managers for six months. The parts had to be discarded and required to a new design, which cost the company as much as €5 billion. Senior company leaders, and their political sponsors (each leader being from a different country), soon became embroiled in a series of internal disputes, which became much more complicated than they needed to be as a result of the company's culture and the structure of its leadership.[15]

By contrast, Coca-Cola had clearly benefitted from lessons learned from its 1999 experience in Belgium, when a tainted batch of its soft drinks resulted in the illness of dozens of children and prompted a government recall of all of its products that eventually spread throughout Europe. This experience helped the company more effectively respond to the Dasani water debacle. The same can be said about any number of banks that put in place safeguards to detect rogue trading activity following the collapse of Barings Bank. Of course, that did not prevent rogue traders from emerging—they were simply detected and stopped more swiftly.[16]

To appreciate the value of ARM one need look no further than how some of the world's best-known organizations have incurred, managed, and suffered from unanticipated crises. Those firms that tend to be the worst affected have underlying weaknesses that make them particularly prone to incurring a crisis in the first place, but practicing ARM can certainly serve to reduce the likelihood that an event may occur, and limit its negative impact.

Every organization has weaknesses that can, if left unchecked, potentially threaten firm survival, but the good news is that they can be identified and managed. Among them are such issues as poor leadership, defects in corporate culture, excessively complex operations, ineffective staffing policies, and inappropriately designed employee incentives. Many of these 'internal' risk factors are choices—a company can 'decide' not to streamline its operational processes or choose not to properly screen employees before they are hired. A

---

[15] AIRMIC, *Major Risks*, 11.
[16] AIRMIC, *Major Risks*, 15–16.

combination of corporate culture and leadership decisions (or lack of action) can, by themselves, seal a company's fate.

Being proactive about making the right choices seems to be little more than common sense. If so, then why do so many companies not do so? Part of the reason is a simple failure to think strategically, establish a risk appetite, and put in place mechanisms to enable a firm to enforce them. Of course, it sounds easy on paper, but doing so in a meaningful way requires enormous resources and determination. And that assumes that the company is heading in the right direction, when it may not be.

Business Continuity Management

Given how unpredictable events can and do have a global impact, the ability of businesses to endure shocks and be able to continue to operate has become imperative. Companies need to have the resiliency to endure a wide range of potential interruptions to their normal flow of business, and be able to spring back quickly after a crisis strikes. For this reason, Business Continuity Management (BCM) has become a key component of ARM.

BCM is a management process that identifies risk, threats, and vulnerabilities that could impact an entity's continued operations, and provides a framework for building organizational resilience and the capability for an effective response. Its objective is to make a firm more resilient to potential threats and to allow it to resume or continue operations under adverse or abnormal conditions. This is accomplished by applying resilience strategies to reduce the likelihood and impact of a threat, and the development of plans to respond and recover from threats that cannot be controlled or mitigated.[17]

While BCM is not a new concept, it deserves a fresh look, given that many companies have no BCM process in place. Some boards of directors are looking for high-resiliency strategies, given how recent disruptions may have highlighted their companies' vulnerabilities. Some existing BCM disaster plans may be focused on IT concerns, leaving other aspects of business operations, facilities, and people exposed. Furthermore, since some businesses increasingly migrate toward external service providers, BCM needs to account for the possibility that those entities' ability to provide service may be interrupted at some point in time.

---

[17] DRI International, *Certification: Professional Practices*, www.drii.org, accessed November 26, 2015, https://www.drii.org/certification/professionalprac.php.

Other factors have combined to change operating realities from just a few years ago. More employees work away from the office, prompting the question: can the wireless equipment needed to keep a business functioning do so without network connectivity for an extended period of time? The firm running your data center may have itself outsourced its operations to a third-party provider. What happens if that provider is knocked out? Cloud-based IT systems introduce another layer of dependency on a firm. Does it have back-up systems, or is there an alternative provider you may rely on when necessary? Some BCM programs are limited in scope to data backup and storage, with no similar program in place to support physical assets and facilities.

While it is not possible to prepare for every conceivable risk, efforts are being made to think in a new way about forms of risk that did not previously encompass BCM protocols. A new generation of BCM standards are being developed to address a whole range of issues that were previously unaccounted for, such as public health risks and biodefense, so that businesses can respond efficiently to natural and man-made threats. Bio-situational awareness is a new paradigm designed to address the outbreak of a biological event as it develops over time. Although this may seem inappropriate for your business, soon it will become standard protocol. The same used to be said about terrorism risk.[18]

Crisis Management: Managing the Media

An important part of ARM involves avoiding a crisis, but after a crisis has occurred, managing it before it gets into the media and spins outside corporate leaders' control can also be extremely important. Simply waiting for the media to get hold of a story that will be damaging to your company and its reputation is a bad strategy. Crises are, by their nature, often messy, and can unfold at a pace that makes it difficult to create a thoughtfully crafted response. Whether crises erupt as a result of forecast or unforeseen events, they always pose a test of leadership skill. If handled well, the response to a crisis can not only enhance a firm's reputation, but prompt needed internal dialogue and operational change.

If stakeholders know that a company is aware that there is a problem, this may be enough to maintain goodwill until the problem can either be addressed or resolved, but how and when to engage the media can be difficult.

---

[18] PWC, *10 Minutes on Business Continuity*, accessed November 26, 2015, http://www.pwc.com/us/en/10minutes/business-continuity-management.html.

Sometimes not talking to the media at all can be a perfectly good strategy, but more often than not, it is necessary to interact with the media swiftly. The most common business response to a crisis is to say nothing and hope the problem goes away, or to hope that the public is not paying close attention. In the absence of a full picture of what happened and why, this tactic rarely does anything other than allow the court of public opinion to reach a verdict. A lack of information usually fuels anxiety rather than defusing it.

Managing crises is ultimately all about managing risk—to individuals, products, institutions, and reputations—and the need to communicate with those who are affected by a crisis. Minimizing the *perception* of risk is where proper messaging comes in. If a leader says the wrong thing to the media, it is hard to correct, but there are things that can be done to minimize the damage after a mistake is made. While a crisis can, of course, occur at any time or place, certain issues which have a propensity to become a crisis and spin out of control, for example:

- Labor disputes
- Management changes
- Litigation
- Product liability
- Natural disasters
- Alleged financial impropriety
- Hostile takeovers
- Fraud

Anything out of the ordinary can lead to a crisis—even something as simple as a violation of business conduct. One good example is the former president of Hewlett-Packard, Mark Hurd, who was relieved of his duties in 2010 when it was revealed he had falsified documents to conceal a relationship with a former contractor, and helped her to get paid for work she did not do. An investigation found that Hurd falsified expense reports and other financial documents to conceal the relationship. The company said it found that although its sexual harassment policy was not violated, its "standards of business conduct" had been. The resulting media storm saw the shares of its stock tumble within minutes, wiping out billions of dollars in market capitalization. Investors and shareholders wondered whether his impropriety said something more fundamental about HP's business practices. It did not.

The impact of a crisis can depend on many things: the depth of the problem, the length of media coverage, management (or lack of management) of the media, the damage actually incurred, and the communities or populations affected. The larger the crisis, the longer the public remembers. If a firm refuses to take responsibility and it is clear that the company is at least partly at fault, the media will do what it does best. If it appears that a firm is doing what is in its own best interest instead of what is in the best interests of those impacted, or that they were too slow in reacting to a problem, they are sure to be attacked by the media. This occurs even in a place like Japan, not known for the veracity of its media. The Japanese media were relentless in their pursuit of both the Japanese government and Tepco over the Fukushima nuclear disaster.[19]

One excellent example of how a crisis can be badly mishandled in the media is BP's Deepwater Horizon crisis. In April 2010, an explosion at the Deepwater Horizon oilrig, situated in the Gulf of Mexico, caused the largest maritime disaster oilspill in U.S. history, and resulted in considerable reputation and financial losses to BP. A number of lessons can be learned from the aftermath of this event, among them:

- As is the case with many companies, BP seemed to ignore the role of crisis communication, before, during, and after the crisis. Specifically, BP ignored crisis communication during the first hours of the crisis, which had remarkable implications for the company's image and brand.
- Stakeholders want to feel informed, safe, and connected when a crisis occurs. Open, timely, and trustworthy reporting as well as regular dialogue and communication with all stakeholders should be ensured before, during, and after the crisis. The company should be particularly honest about what it knows and does not know, which gives it more credibility.
- Crisis communication should be viewed as a proactive function rather than as primarily reactive. In doing so, crisis communication will help crisis managers to anticipate possible crises, reduce its likelihood, be more able to manage and resolve a crisis when it happens, prepare key stakeholders for the crisis, and build company credibility before the crisis occurs.[20]

Whether there is lasting impact resulting from a crisis can be a function of helping to define *perception* versus *reality*. Crisis management consists, to a large degree, of managing the communication process effectively. Every organization can prepare itself for crises that will erupt eventually. If they do so

---

[19] Adapted from D. Wagner, *Managing the Media in Times of Crisis*, www.smashwords.com, 2012, accessed November 26, 2015, https://www.smashwords.com/books/view/252158.

[20] D. DeWolf and M. Mejri, "Crisis Communications Failures: The BP Case Study," *International Journal of Advances in Management and Economics*, March–April 2013, Volume 2, 54–55.

in advance, and methodically, they will be much better off than attempting a haphazard or incomplete response. Although much of this will appear to be common sense, here are some basic steps to take in order to become more effective in crisis communications:

- *Anticipate crises*: They will happen, so you had better get prepared.
- *Identify your crisis communications team*: If you don't have a specialist on staff, at least retain a third party who can step in at a moment's notice.
- *Identify and train spokespersons*: If you opt to handle it in-house, identify the right person(s) and get them trained up in advance.
- *Establish notification and monitoring systems*: To be effective, the situation must be able to be monitored effectively.
- *Know your stakeholders*: Know in advance who you wish to target your messaging to, and identify in advance the best means of doing so.
- *Assess the crisis situation*: Once a crisis has occurred, have a methodology in place for understanding the scope of the crisis, and what it likely means.
- *Have key messages ready*: Run through some scenarios so that you can develop key talking points in advance.
- *Post-crisis analysis*: Take a hard look at what could have been done differently and better, next time.

It is easy for a management team in an otherwise healthy organization to brush off the notion that a crisis will happen to them. That is undoubtedly what any number of other companies believed before they were in the headlines. All the other preparation an organization may do to manage risk can be undone in no time at all if a crisis is not handled effectively. If your firm does not have a crisis communication plan in place, make it a priority.

Staying Ahead of the Curve

In business, as in life, much of what determines what we are able to do is completely outside our control. We can neither know if, when or how a problem will arise—or when, if or how we may be able to control it when it does arise. We can be good students of history, however, recognizing what our individual and collective experiences have taught us, which will help guide us to the answers we need. That said, in the absence of having the resources necessary to get the job done, all the wisdom in the world may not help. The problems generated by man-made risk, and its collision with natural risk, require us to integrate wisdom with adequate resources, a forward thinking disposition, a willingness to listen to our gut instincts, and a good dose of common sense.

14

Risk Governance

As we have argued throughout the book, risk in the twenty-first century is very much within the span of human control and understanding. With the exception of naturally occurring calamities, even climate change may be reversible through our collective choices and actions—although it is very clear that we are nearing a tipping point. Against this backdrop, risk governance is critically important to attaining risk agility, and improving how firms and systems respond to the challenges that await them.

In order to understand governance in the modern enterprise it is important to consider the various types of organizations that shape the global economy. Although non-state actors have clearly assumed a growing prominence on the global stage, the direction the world will follow remains largely dictated by state actors and their national and supranational governance structures. At the supranational level, international organizations—such as the UN, the EU, the Arab League and the African Union—are among the most prominent examples of where a group of nations have ceded some degree of sovereignty in order to join the comity of nations in pursuit of common goals.

Risk governance has often formed part of the raison d'être for these regional and global organizations. The UN and its predecessor, the League of Nations, were formed to prevent another Great War, and to ensure that, even in the darkest of times, nations continue to pursue dialogue. Governance is at the core of the EU's charter, and is an overriding objective of each of these organizations.

© The Author(s) 2016
D. Wagner, D. Disparte, *Global Risk Agility and Decision Making*,
DOI 10.1057/978-1-349-94860-4_14

The UN and some of its sister institutions in the economic domain, the World Bank and IMF, are central actors providing assurances that the global system will continue to function in times of crisis, while solving some of the world's most intractable challenges, such as poverty alleviation and the promotion of the development process. With greater frequency, these institutions are not only focused on preventing war and improving the lives of people in the developing world, but are also turning their attention and considerable resources to combating risk. From the UN's annual Climate Change Conference to the Global Risk Report, risk and how to manage it has become an increasingly important component of international dialogue.

Herein lies the challenge and one of the principal shortcomings of risk management at the global institutional level—dialogue is what these institutions do best, but implementation is their central weakness. Passivity and inaction are precisely the opposite of risk agility. If these institutions were judged by what they actually did, rather than on what they proposed, nearly all of them would receive a poor grade. Yet they tend to judge themselves based on how many proposals or loan allocations they approve, rather than what they actually achieve. In addition, they often tend to have less than ideal methods of evaluating themselves. As is also sometimes the case in the private sector, entrenched interests drive the bus.

Below the supranational level is the domain of regional institutions, trading blocs and economic unions, which are growing larger and have become even more interconnected. The Trans-Pacific Partnership (TPP) and the Transatlantic Trade and Investment Partnership (TTIP) are two examples where the thrust of globalization and free trade have come to encompass nearly all Pacific Rim and trans-Atlantic trade, which remains the world's largest economic exchange.[1] As many modern trade agreements amplify market access and trade opportunities, as is argued by their proponents, critics often cite these trade deals as examples of *failed* global governance. Not only did much of the negotiations preceding the agreements occur under the cover of darkness with various commercial interests at the table (and *without* other important actors being involved in the negotiation process), their declared objective of lifting developing countries out of economic hardship while maintaining trade competitiveness in advanced markets is the subject of fierce debate.[2] Developing countries argue that both objectives cannot be achieved in tandem.

---

[1] R. Brown, *TPP? TTIP? Key Trade Deal Terms Explained*, The Brookings Institution, May 20, 2015. Accessed November 26, 2015, http://www.brookings.edu/blogs/brookings-now/posts/2015/05/20-trade-terms-explained.

[2] C. Johnston, "Berlin anti-TTIP Trade Deal Protest Attracts Hundreds of Thousands," *The Guardian*, October 10, 2015. Accessed November 26, 2015, http://www.theguardian.com/world/2015/oct/10/berlin-anti-ttip-trade-deal-rally-hundreds-thousands-protesters.

As is the case in an individual firm or household, good governance at the macro level has certain universal characteristics—transparency, trust, and equity are chief among them. On these three points, the global system scores rather poorly. Many of the weightiest security and financial decisions affecting the world—those made in treasuries, the UN Security Council (UNSC), and elite watering holes such as Davos and Aspen—are done in secret and by invitation only to select members that are already *in* 'the club'.

On the subject of trust and equity, few in the world's developing and emerging markets feel that they are adequately represented, given their share of global output and population, at these elite venues. The same is certainly true in terms of the five permanent members of the UNSC, which many have argued is a relic of World War II. That countries such as Brazil, India and Indonesia (with the world's second, third and fifth largest populations, respectively)[3] are not at least permanent *rotating* members of the UNSC smacks of hypocrisy and is the source of much deep-seated resentment of the UN's own approach to governance.

This raises profound global governance challenges, where, much like a banking crisis, we are *privatizing gains* and *socializing losses* from climate change to market access and financial inclusion. No one feels the effects of this marginalization more than the world's developing and frontier nations. This real or perceived inequity is a festering source of instability driving economic migration, refugee crises and deeply ingrained fractures in social order. Indeed, economic inequality is, at its core, what sparked the Arab Spring, with the auto-immolation by Mohammad Bouazizi in Tunisia in 2010.[4] If the world cannot get governance right in its governing institutions, how can it hope to do so on a national, state, or city level?

Governance Structures and Paralysis

If the UNSC and the G7 are, in essence, the world's board of directors, it is no wonder the world is in the state it is in. Just as having Carl Icahn, the famous activist investor, on the board of a company he would like to take over, these global bodies have long been a platform to voice often narrow partisan interests. At the global level, tackling major risks that threaten the planet requires consensus, cooperation and, above all, action. Creating a

---

[3] *The World Population and the Top Ten Countries with the Highest Population*, www.internetworldstats.com, 2015, accessed November 26, 2015, http://www.internetworldstats.com/stats8.htm.

[4] S. Shaikh, "Mohamed Bouazizi: A Fruit Seller's Legacy to the Arab People," CNN, December 17, 2011, accessed November 26, 2015, http://www.cnn.com/2011/12/16/world/meast/bouazizi-arab-spring-tunisia/.

global governance structure that is action-oriented may be nearly impossible with all the entrenched interests and years of ossification that have rendered much of the international system little more than a global talking shop. The net result is paralysis.

Paralysis in the decision-making process is, of course, not solely the domain of international bodies—it impacts governing bodies at all levels. The U.S. is experiencing an intractable partisan era, where Democrats and Republicans risk being excommunicated if they break ranks with their parties in order to do what they were elected to do. In the era of man-made risk, the rise of political risk in America is a preventable loss that has harmed U.S. competitiveness in profound ways.

The U.S. lost its vaulted AAA credit rating in 2011 as a result of political dithering about the country's indebtedness.[5] While sovereign debt is a vitally important issue for all countries, willfully damaging your own credit worthiness seems an odd path to follow in an effort to address a debt crisis. Other Pyrrhic victories included the debt ceiling and 'fiscal cliff' debates, which temporarily shut down the federal government in 2013—a threat that is constantly looming over the horizon.[6] Public servants who should *manage* risk are instead *creating* risk, with real long term harm to their constituents.

The Sad Case of the U.S. Export–Import Bank

Perhaps the most insidious example of the political risk that has crept into U.S. governance is the shutdown of the U.S. Export Import (Ex-Im) Bank. The Ex-Im Bank's mandate from Congress prior to the shutdown expired on June 2015, and it remained closed for months until Congress was pressured by Ex-Im's beneficiaries to reopen the Bank. In the same way that allowing the FDIC to lapse would seriously erode consumer confidence, unilaterally dismantling Ex-Im was bound to have similar consequences for U.S. trade.[7] Small businesses, which made up 90 percent of Ex-Im's customers and com-

[5] W. Brandimarte and D. Bases, "United States Losses Prized AAA Credit Rating from S&P," Reuters, August 7, 2011, accessed November 26, 2015, http://www.reuters.com/article/2011/08/07/us-usa-debt-downgrade-idUSTRE7746VF20110807.

[6] E. Kelly, "Obama Signs Funding Bill Averting Government ShutdownDOUBLEHYPHENFor Now," *USA Today*, September 30, 2015. Accessed November 26, 2015, http://www.usatoday.com/story/news/2015/09/30/senate-approves-funding-bill-avert-government-shutdown/73032366/.

[7] S. Akhtar et al., *Export–import Bank: Reauthorization: Frequently Asked Questions*, Congressional Research Service, 2014, accessed November 26, 2015, http://fas.org/sgp/crs/misc/R43671.pdf.

prised 64 percent of net new private sector jobs,[8] was bound to bear the greatest cost from its demise.

Ex-Im's many congressional critics cited economic favoritism and government welfare for large, blue-chip exporters such as Boeing, Caterpillar, and GE as the reason for the shutdown. While these large firms clearly benefitted from Ex-Im's support, the criticism overlooked the importance of small businesses in the supply chains of large global manufacturers. Joe Kaeser, the CEO of Siemens, a German engineering powerhouse employing around 46,000 people in the U.S., indicated that for every $1 million in revenue the company earned, $300,000 benefitted small businesses. Ex-Im's own targets called for a minimum of 20 percent of its financial and trade credit support to be directed to small and mid-sized businesses.

Putting an organization that paid back $675 million to taxpayers in 2014 in the crosshairs is a dangerous game of brinkmanship.[9, 10] A similar game has been played with the OPIC—one of the few self-sustaining U.S. government organizations, that actually pays U.S. taxpayers back every year—which has been the misguided subject of critics for many years. Another oft-cited criticism is that Ex-Im crowds out private financial institutions and misprices risk by leveraging the Treasury's balance sheet and access to unattainably cheap capital. While this may be true in terms of access to favorable capital, as a self-funded government institution, Ex-Im's return on investment to taxpayers makes it an asset rather than a liability. Adding Ex-Im's low default rate of 0.175 percent and strong capital adequacy covering 17 times the Bank's default rate makes it a benchmark institution, rather than a threat to the private sector. Not crowding out private sector financial institutions is also enshrined in Ex-Im's charter.[11]

So much of Ex-Im's role in the U.S. economy is to fill the void where others dare not tread. More than 90 percent of global consumers are located outside of the U.S. and there are more than 60 export credit agencies in the world, many operating opaquely, jostling to promote their national exports.[12] The decision not to reauthorize Ex-Im was tantamount to unilateral disarmament

---

[8] U.S. Export–Import Bank, *The Facts About EXIM Bank*, accessed November 26, 2015, http://www. exim.gov/about/facts-about-ex-im-bank.

[9] U.S. Export–import Bank, *Fast Facts*.

[10] U.S. Export–import Bank, *Annual Report 2014*, accessed November 26, 2015, http://www.exim.gov/ sites/default/files/reports/annual/EXIM-2014-AR.pdf.

[11] U.S. Export–Import Bank, *The Charter of the Export–Import Bank of the United States*, accessed November 26, 2015, http://www.exim.gov/sites/default/files//newsreleases/Updated_2012_EXIM_ Charter_August_2012_Final.pdf.

[12] E. Gresser, "60 Export Credit Agencies Operate Worldwide," *Progressive Economy*, September 17, 2014, accessed November 26, 2015, http://www.progressive-economy.org/trade_facts/60-export-credit-agencies-operate-worldwide/.

during an era of an enfeebled global economy where the U.S. is one of the few drivers. Christine Lagarde, the head of the IMF, referred to this as an era of "new mediocrity", in which key economic indicators are moving in dangerous lockstep.

With near parity between global GDP growth and the growth in trade, economic headwinds are exacerbated by low productivity and persistently high urban unemployment. Without trade facilitation tools like Ex-Im in the U.S. economic arsenal, U.S. lawmakers have risked putting the global economy in reverse (or, at least, slowing it down) and sending millions of American workers to the unemployment line, while simultaneously harming their global trading partners. The consequences of not reauthorizing Ex-Im were stark, a world where Ex-Im has an open-ended charter is not a zero-sum proposition for U.S. exporters or other nations. In fact, as evidenced by the largely peaceful 70 years since Ex-Im's modern charter, countries that trade together tend to stay together.[13]

Clearly, the U.S. lawmakers who voted in favor of shutting Ex-Im down were not considering the larger picture. They apparently did not realize (or perhaps they did not care) that interfering with the global supply chain has political, economic, and social repercussions. In doing what they did, they failed to consider that lawmakers have a governance obligation to their constituents and their country not to succumb to narrow-minded partisanship, but rather to think about their actions in the context of civil society. That may sound like left-wing gibberish to some, but when put in the context of all the people around the world that their decision impacted (suppliers, manufacturers, and shipping companies), they failed the governance test. One of the key messages of this chapter is that risk management decisions should be undertaken in a broader context, not based simply on short-term gain. U.S. lawmakers came to their senses and re-authorized funding for the Ex-Im Bank in December 2015.

Separation of Powers

Ken Lay and Jeffrey Skilling, the now-infamous leaders of Enron, offer a great example of the depravity of power and greed if it is left unchecked. Like so many risks, the rising smoke of Enron's inglorious demise was missed as an early warning sign. One of the least obvious tell-tale signs that executive pow-

---

[13] D. Disparte, "Ex-Im and the Pax Americana," *Huffington Post*, May 15 2015, accessed November 26, 2015, http://www.huffingtonpost.com/dante-disparte/exim-and-the-pax-american_b_7242152.html.

ers were a part of the problem were the company's annual reports in the years immediately leading up to the firm's collapse. Even a non-financial expert may have caught the full-page photos of Mr. Lay and Mr. Skilling in the company's annual reports the years preceding the crisis. The management team, if they appeared at all, were in mere thumbnail images, making it clear that if you were dealing with Enron as an investor, consumer, or regulator, you were entering the Chairman and CEO's fiefdom.[14, 15]

Although Enron did have a CRO at the time of its scandal, the CRO reported directly to Kenneth Lay, who, as the chairman of the board, was essentially his own overseer. Mr. Lay died before he could be tried in a court of law for his now-legendary financial indiscretions, and in the fallout from the scandal Enron's accounting firm, Arthur Andersen came to personify the very opposite of the professionalism and independence expected of a responsible auditor. The firm fell on its own sword over the Enron scandal, handing the accounting profession a profound setback.

Separating powers and having a system of checks and balances, even if they become paralyzed as those of the U.S. government, is as vitally important for controlling risk as it is for consensus-based decisions. In general, it is wise to be wary of chiefs wearing too many hats. This is not only an early warning sign of potential risk lurking in the recesses of people's minds, but also a sign that there is little or no devolution of power at other organizational levels. This top-down management structure, predominant in large organizations, is the very antithesis of agility.

While a clear separation of power and accountability is generally good, during protracted periods of deadlock, negative consequences can ensue, or opportunities may be lost. On at least one occasion in 2013, the U.S. federal government was shut down because the republicans and democrats could not reach an agreement about the national debt limit and the federal budget. Vitally important national investments in education, infrastructure, and resilience fell by the wayside. At present, political deadlock is eroding U.S. national competitiveness in a way that impacts all Americans, demonstrating the power of man-made risk, even on a national scale, and principally over ideological matters.

The rise of political gridlock is, of course, not unique to the U.S. Europe, at both the individual country level (such as Belgium's protracted inability to form a government) and at the supranational level, has shown governance fail-

---

[14] Enron, *Enron Annual Report 2000*, 2000, accessed November 26, 2015, http://picker.uchicago.edu/Enron/EnronAnnualReport2000.pdf.

[15] Enron, *Enron Annual Report 1999*, 1999, accessed November 26, 2015, http://picker.uchicago.edu/Enron/EnronAnnualReport1999.pdf.

ures in the face of persistent challenges. One particularly profound example of this is the failure on the part of numerous European governments to effectively and efficiently address the flow of refugees from the Middle East. The potential impact on individual countries in Europe, and ancillary impacts on the continent as a whole, may last for many years.

The Effect of Motives, Incentives, and Opacity

One of the most important and least understood drivers of risk of all kinds are peoples' motives, incentive systems, and how they behave under the cover of darkness. When an individual or group of people believe they are unseen or will not be caught, depraved behavior may follow. An unchecked profit motive—from banks, mortgage brokers and homeowners, who used their home equity like a cash machine—was the fuel that ignited the global financial crisis. If greed was the fuse, risky financial products and asset-price bubbles were the fuel. So-called NINJA loans (corresponding to No Income, No Job and No Assets) were popular financial innovations in the lead-up to the crisis. In the hands of frontline mortgage brokers with little or no managerial oversight (or an outright managerial blessing) this 'exotic' category of lending ballooned and ended up imperiling the system.

These and other exotic financial products where packaged alongside 'safer' more traditional mortgage instruments in shared percentages, forming mortgage-backed securities, collateralized debt obligations (CDOs) and other complex financial products. The trade of these instruments as if they were liquid, highly rated asset classes among systemic financial institutions is in no small measure why liquidity dried up as the financial crisis gained speed. Synthetic versions of these products, and subproducts based on further derivations (such as CDO²), continued compounding financial complexity and interconnections in the U.S. economy, while not adequately shifting risk.[16] All of these instruments are meant to improve risk management in the financial system and improve consumer access to the dream of homeownership or commercial property. Left unchecked and in the hands of people with questionable motives and little effective oversight in a highly complex, opaque environment and you have a financial recipe for disaster.

---

[16] V. Acharya, *Systemic Risk and Macro-Prudential Regulation*, NYU Stern/Center for Economic Policy Research/National Bureau of Economic Research, March 2011. Accessed November 26, 2015, http://pages.stern.nyu.edu/~sternfin/vacharya/public_html/ADB%20Systemic%20Risk%20and%20Macroprudential%20Regulation%20-%20Final%20-%20March%202011.pdf.

The Milgram experiment, carried out in 1961, showed the effects of authority on people and proved we are not nearly as virtuous as we might like to think.[17] In the experiment, an overwhelming number of people carried out grim instructions from an authority figure (in this case one who donned a lab coat and clipboard) to press a button, which would be followed by screams of anguish from a separate room. The experiment underscores the perverse effect that a bad system can have on otherwise good people. Paycheck persuasion, especially in the high-stakes world of investment banking and modern finance, can have an equally aberrant impact on otherwise good bankers. Add in hypercompetition for talent among banks and there is a strong inducement to skew both morality and the rules.

While banking offers a deep repository of rogue activities under opacity, such behavior is certainly not exclusive to financial institutions. International Relief & Development (IRD) is an international development NGO carrying out work in developing countries and post-conflict areas that also responds to acute humanitarian crises. Like most NGOs in this line of work, government-backed funding is often the principal source of capital. Since 2007, IRD has received $2.4 billion to administer U.S. Agency for International Development-funded programs, with millions of dollars being diverted to enrich IRD's founders and top management.[18] Other cases abound from virtually all industries, underscoring the importance of new thinking around individual motives, how people perform in teams, and decision making under opacity.

Risk-taking Without Bearing Consequences

In the insurance industry the classic definition of a hazard is a condition that exists and will amplify the likelihood of a particular risk occurring. Most insurance professionals learn about three types of hazards in the course of their studies:

---

[17] N. Nohria, "You're Not As Virtuous As You Think," *Washington Post*, October 2015. Accessed November 26, 2015, https://www.washingtonpost.com/opinions/youre-not-as-virtuous-as-you-think/2015/10/15/fec227c4-66b4-11e5-9ef3-fde182507eac_story.html.

[18] S. Higham, "Longtime USAID Contractor Embroiled in Scandal Fires Top Managers, Others," *Washington Post*, February 20, 2015. Accessed November 26, 2015, https://www.washingtonpost.com/news/federal-eye/wp/2015/02/20/longtime-usaid-contractor-embroiled-in-scandal-fires-top-managers-others/.

1. A *physical* hazard is a risk-enhancing condition that arises from the physical environment or premises. This may occur, for example, in the case of a log cabin in the woods surrounded by tall trees, where the physical hazard of the trees may be exacerbated by a lightning strike.
2. *Morale* hazard is a condition that will enhance risk (or the probability of loss) arising from the attitude or carelessness of an individual, team, or organization. A common example is the pernicious effect on property values and a neighborhood's appeal when one homeowner leaves their property unkempt. The morale hazard may amplify a whole range of risks, from depreciating home values to inviting crime in a dilapidated neighborhood.
3. The final classical hazard—and one that it is vitally important for organizations in all industries to understand—is *moral* hazard. A moral hazard arises from excessive risk-taking behavior where the subject does not bear the consequences of said risk; simply put, risk taking without bearing any consequences. The concept of the personhood of an organization and other legal corporate structures, such as the limited liability company, are examples of how far moral hazard goes.

The global financial crisis was an example of system-wide moral hazard. The government bailout of large financial institutions and their designation as systemically important is an example of privatizing gains while socializing losses. One of the most egregious examples of this moral hazard was AIG's response to being bailed out by the U.S. government with public money. Once the world's largest insurer, AIG was a paragon of resiliency in insurance and financial services.[19] However, one small unit operating as a special product division exposed the firm to a run on complex derivatives and other exotic financial instruments that broke the usual financial soundness of insurance. Stepping outside its 'sweet spot' of insurance ultimately caused the company to collapse, with tens of billions of dollars of obligations falling due to credit default swap holders.

In receiving the $182 billion bailout from the government, AIG's executives proceeded to pay out lavish company bonuses and continue with extravagant company meetings, events,[20] and practices.[21] The 'too-big-to-fail' label or SIFI

---

[19] L. Zuil, "AIG's Title as World's Largest Insurer Gone Forever," *Insurance Journal*, April 29, 2009, accessed November 26, 2015, http://www.insurancejournal.com/news/national/2009/04/29/100066.htm.

[20] W. Greider, "The AIG Scandal," *The Nation*, August, 6, 2010. Accessed November 26, 2015, http://www.thenation.com/article/aig-bailout-scandal/.

[21] Author Daniel Wagner learned, during a visit to AIG Financial Products' headquarters in London shortly after the company had imploded in 2008, that lavish lunches were still being served there. When he inquired what had changed since the implosion, he was told two things: 1. Breakfast was no longer served on the premises, and 2. There was no longer a gym attendant on the premises.

moniker are a dangerous portent that moral hazard has not been removed from the financial system, but rather that it has been amplified. Again, AIG, and the lawsuit brought by its erstwhile CEO and chairman Hank Greenberg, against the U.S. government on the grounds that its bailout eroded shareholder value, drives home the point.[22] So did the pervasive nature of greed on Wall Street and in other financial centers around the world.

Feeling that AIG's share price was driven down by the government bailout (the largest in history), without which AIG would have gone bankrupt, Mr. Greenberg's lawsuit alleged that the firm's enterprise value went down, causing considerable harm to shareholders. The boldness of this case drives home the point that excessive risk-taking behavior apparent in many industries *can* occur without bearing consequences. The AIG executives who were given bonuses were essentially being given bribes to stay with the firm—whether or not they had anything to do with its collapse. By the same token, it can certainly be argued that a whole range of people inside and outside the company bore substantial consequences as a result of the bailout and its subsequent impact on the company and shareholders.

Aside from the political ritual of dragging penitent top executives to Capitol Hill for a mea culpa tour, few bailed-out firms and executives suffered long-term consequences from their actions. Most often, the leadership at the helm either continued in their roles or were sent on their way with massive 'golden parachutes' and nothing more than a perfunctory slap on the wrist. Advocating for a tougher regime of corporate accountability and better balance between risk taking and risk bearing will not only improve the survivorship of these firms (and thus shareholder returns), it will lighten the future bailout burden taxpayers have to shoulder.

This 'public–private' security that is afforded to certain firms in nationally important industries raises questions about the role of regulatory enforcement. The notion that regulatory supervisors can pick winners and losers complicates deeply held free market beliefs in the U.S., Europe, and other advanced economies. It also compounds the effect of systemic moral hazard. Shareholders who invest in publicly listed companies certainly accept the risk that investing can be an 'all or nothing' proposition. However, for certain managers in publicly shielded industries (such as large national banks like Citibank or Chase) the ultimate risk of firm demise is no longer in the range of possibilities, least of all because there is a precedent that SIFIs will be resurrected, especially now that many of them have made so-called living wills

---

[22] S. Ferro, "A Bunch of People Wanted to Punish the Government for Bailing Out AIG," *Business Insider*, June 15, 2015, accessed November 26, 2015, http://www.businessinsider.com/hank-greenberg-aig-lawsuit-win-2015-6.

or resolution plans.[23] Arguably, the system would be safer if 'too-big-to-fail' institutions were divided up into constituent parts that no longer pose such a dire risk to the global economy. This too would send a clear signal to management teams that there is no longer a public springboard to resuscitate dying firms, thus calibrating risk-taking with real consequences.

The competitive advantage derived from an 'undying' class of large systemic firms cannot be overstated. While complaints are raised that the increase in regulatory scrutiny is hampering their competitiveness, SIFIs, like the automotive firms that were bailed out, have emerged not only unscathed from the financial crisis, but actually stronger vis-à-vis their peers. One example is how many of them, after being resurrected with government largesse, were 'forced' to buy the assets of failed firms that were once their rivals. Chase's acquisition of Washington Mutual for $1.9 billion is an example of the fire sale that drove some companies out of existence, while favoring others.[24] Overall, regulators worldwide have great difficulty keeping pace in the game of regulatory catch up. Once they have established norms to address the crisis du jour, industry has moved on, and the issue at hand is no longer capital adequacy, but some new, as yet unseen behavioral risk.

Many firms are aware of this regulatory arbitrage, exploiting it readily, and most often legally. They not only exploit it in single markets, but by virtue of flying 'flags of convenience'; the modern multinational can easily shift operations to less robust jurisdictions. Indeed, many banks in the City of London threatened a wholesale exodus if regulators put caps on pay and bonuses. So, in this system, who is regulating whom? When Adam Smith's invisible hand has been covered in a gentle velvet glove that brings dying firms back to life, a profoundly new risk has been created.

Regulatory arbitrage is taking place worldwide and within countries. While this is not necessarily a bad thing, as with the stock market, market signals often crowd out underpriced assets, eventually correcting their advantage. The challenge lies in the lack of global or national coordination. Even among U.S. states, the competition to increase investment at the expense of a neighbor is often fierce. The former Texas Governor and twice aspiring presidential candidate Rick Perry took out national ad campaigns inviting companies

---

[23] Board of Governors of the Federal Reserve System, U.S. Federal Reserve, *Resolution Plans*, accessed November 26, 2015, http://www.federalreserve.gov/bankinforeg/resolution-plans.htm (Living wills or resolutions plans for 293 financial institutions are available at the Federal Reserve's website).

[24] R. Sidel et al., "WaMu Is Seized, Sold Off to J.P. Morgan, In Largest Failure in U.S. Banking History," *Wall Street Journal*, September 26, 2008, accessed November 26, 2015, http://www.wsj.com/articles/SB122238415586576687.

across the U.S. to domicile in Texas.[25] Governor Perry took aim at everyone from California to New York, often stoking fiery and comical rebuttals from governors across the country.

U.S. companies shelter more than $2.1 trillion in cash overseas,[26] largely due to the fear of double taxation on repatriated earnings and the reality that firms and managers focused on short-term value creation will exploit lower friction environments. In this model few benefit, which speaks to the concept of the modern multinational being a flag-less enterprise. Surely some of these international profits could capitalize a systemic life insurance policy through risk premiums, much like the U.S. FDIC has shored up depositor comfort in the banking system. This systemic life insurance would not necessarily revive a dead company; rather, it would serve to drive orderly resolutions, asset dispositions through transparent auctions, and the general salvage process without public finance. All global SIFIs would have to comply and make contributions to the central risk pool, which can be overseen by an impartial global body such as the Bank of International Settlements or the IMF.

Considering Secondary Impacts and the Value of Thoughtful Planning

In risk management, the notion of contagion is vastly important. It holds that a particular risk (defined as the unforeseen likelihood of an adverse event that is either attritional or acute) will not dissipate entirely once it has manifested itself. Instead it may change hosts and affect another segment of the economy, industry, company or individual. For example, despite the fact that most federal bailouts have been vindicated, even generating a profit, the Troubled Asset Relief Program moved America's sovereign risk clock one notch closer to midnight. Indeed, persistent economic problems in Europe's periphery have thrown good euros after bad in one wave of bailouts after another, with no real change in the underlying crisis. Instead, a deep-seated resentment and a perilous right-wing resurgence have replaced Europe's post-war flirtations

---

[25] J. Thompson, "Texas Governor Rick Perry Spends Thousands on ads to Poach Missouri Businesses," Channel 41 KSHB Kansas City, August 26, 2013. Accessed November 26, 2015, http://www.kshb.com/news/local-news/texas-governor-rick-perry-spends-thousands-on-ads-to-poach-missouri-businesses.

[26] R. Rubin, "U.S. Companies Are Stashing $2.1 Trillion Overseas to Avoid Taxes," Bloomberg, March 4, 2015, accessed November 26, 2015, http://www.bloomberg.com/news/articles/2015-03-04/u-s-companies-are-stashing-2-1-trillion-overseas-to-avoid-taxes.

with solidarity and economic union. For antiagile risk governance, the world is a more dangerous place.

As an example of just how dangerous, the Syrian conflict metastasized to become a regional and European humanitarian crisis, which was having a profound impact on economies and resources throughout the continent. It was predictable, or even perhaps inevitable, that the longer the conflict raged on, the more chance there would be that refugees would seek a better life beyond the camps set up in the countries surrounding Syria. While European governments had literally years to consider their response to such an eventuality, they were taken completely by surprise in 2015 when hundreds of thousands of Syrian (and other Middle Eastern) refugees sought refuge throughout greater Europe. The absence of a future orientation, and an integration into the planning process on this issue for foreign policy and interior ministries across Europe, resulted in a muddled, confusing, and ultimately ineffective response. The possibility of contagion was evidently not taken into serious consideration, even with years of notice.

A forward-thinking approach to a whole range of issues—from financial crises to housing crises to pandemics—appears to be missing at the global, national, state, and city levels across the world. We need only consider how many times there have been financial crises in the past 30 years to recognize that: (a) they are inevitable; and (b) governments and companies have tended not to craft a meaningful response to them until they happen. The housing crisis in London resulting from a shortage of properties for sale and corresponding sky-high prices could surely have been forecast many years in advance, but was allowed to unfold without a near-term solution in sight. And, as has been pointed out previously, the national, regional, and global responses to the 2014 Ebola epidemic were rather late and reactive, rather than early and proactive, as many other responses to other crises have been in the past.

Although all of these examples involve high stakes, there is only a certain amount of predictability that may be introduced into the decision-making process. Previous crises can only provide a limited amount of guidance about what the next one will look like. Yet, we know some things will remain static, such as, there will be another (fill in the blank) crisis, the previous response will have been judged to have been inadequate, more resources will be required to address the problem, and there will be limitations regarding how much money, time and options will be available. That being the case, a sensible approach is to devote more resources to modeling, scenario planning, and considering alternatives not previously considered before a potential problem becomes a problem.

Many of the standard approaches to the risk of loss from an event focus on *direct* impacts (i.e. how much property will be damaged and how many lives will be lost). What is often not accounted for are the *indirect* costs. How long will an airport be closed as a result of a hurricane, who does that impact, and how? What type of impact will a pandemic have on the ability of businesses to operate over time, and how long will it be until children can return to school? How much time will it take for work flows to return to normal following a cyber-attack? These are the types of questions that should be asked and answered—in advance of the occurrence of the event, yet too few governments and businesses do so.

A good example of a thoughtful 'post-mortem' conducted in the aftermath of the 2001 Foot and Mouth Disease outbreak in the United Kingdom, was the independent inquiry conducted by the Royal Society of Edinburgh into the impact of the disease in Scotland, when emphasis was placed on future methods of prevention and control. Among its recommendations on lessons learned and what should be done in the future were that:

- The wider interests of the rural economy and suitable stakeholders should be involved in considering options for the control of the Disease.
- An immediate ban on all livestock movements throughout the country should be imposed as soon as a case of the Disease is confirmed.
- The UK government should press the EU on specific import regulations.
- Comprehensive contingency plans should be devised and scrutinized by a Standing Committee that will coordinate regular and full-scale exercises.
- Resources should be made available as a matter of priority to develop improved tests for the detection of the Disease.
- A Chief Veterinary Officer (for Scotland) should be created with direct responsibility for all aspects of veterinary matters, including the eradication of the Disease.
- A 'Territorial Veterinary Army' should be created to ensure that enough suitably experienced veterinarians are available when needed.[27]

Imagine what kind of risk management plans could be put into place if this type of response was done on a routine basis following each and every catastrophe.

---

[27] The Royal Society of Edinburgh, *Independent Expert Committee Makes Forward-looking Recommendations on Foot & Mouth Disease*, Press Release, 2002, accessed November 26, 2015, https://www.royalsoced.org.uk/209_IndependentexpertcommitteemakesforwardlookingrecommendationsonFootMouthDisease.html.

Part of the challenge in considering a proactive response to large-impact events is that secondary impacts can be greater than the triggering event. A geomagnetic storm may make dozens of satellites inoperable, with knock-on effects that could last for years. The destruction of a power plant could prevent millions of people from receiving electricity for months. Graph 14.1 outlines well both the direct and secondary impacts resulting from a disruption to critical infrastructure. What is noteworthy about the graph is how many forms of disruption can occur from a single event on an ongoing basis. While serving to demonstrate how daunting a task it can be to map out the knock-on impacts of such disruptions, it is critical to the planning process to go through such an exercise, and can make the risk management process much more effective and achievable.

There are significant advantages to identifying secondary impacts well in advance, among them the ability to have a reasonable chance to mitigate their impact, identify both stakeholders and potential victims before the fact, allocate resources before they are actually needed, develop contingency plans, and compensate victims more efficiently.[28] How potential costs and benefits are scrutinized will naturally be different before, during, or following a crisis. It can be difficult to persuade policymakers to allocate funds for what 'might' happen in the future, but simply having the ability to demonstrate that scenarios have been contemplated and well thought out in advance can make a real difference when the time comes.

Planning for Future Shocks to the System

Given that so many of the evolving risks in the twenty-first century are, by their nature, global in orientation, the burden of crafting solutions has naturally fallen upon governments and multilateral organizations. Such entities have a mandate to respond to such risks, and many of their resources and abilities to respond are enshrined in international law. That said, no single organization has either the resources or capability to act on its own to resolve the broad range of threats and challenges that persist. Governments, businesses and non-state actors must find a way to combine their resources to address these challenges if they are to be battled in a truly meaningful manner.

Even when the abilities of all actors are combined, capacity gaps remain. For example, the WHO, which is widely acknowledged as the most able multilat-

---

[28] Organization of Economic Cooperation and Development, *Future Global Shocks: Improving Risk Governance*, 2011, 43, accessed November 26, 2015, http://www.oecd.org/governance/48256382.pdf.

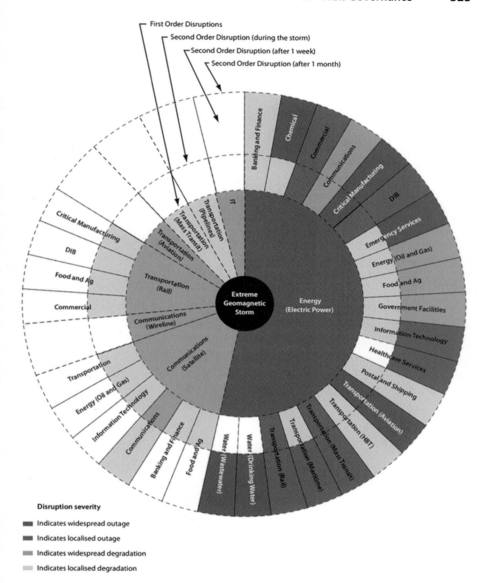

Graph 14.1   Direct and secondary critical infrastructure disruptions (OECD, *Future Global Shocks*, 42)

eral organization in the realm of risk governance, and which counts virtually all of the countries in the world as members, is prevented from being nimble because of its size, and as responsive as it might be by virtue of its multilateral decision-making protocol. The Global Health Institute and London School of Hygiene and Tropical Medicine, which analyzed the response to the Ebola epidemic, concluded that the WHO should be stripped of its role in declaring disease outbreaks to be an international emergency following its catastrophic failure to warn the world of the dangers of Ebola in West Africa.[29] By the same token, an NGO may be restricted by having too narrow a mandate, and a business similarly hobbled by the need to consider shareholder impact before acting in what may otherwise be an 'optimal' manner.

Consider how the major powers reacted to the Great Recession. The G20 sought guidance for what new guidelines would be implemented regarding the manner in which global finance operates (from the IMF and other international institutions)-which the G20 governments participate in, and effectively control-through voting rights and membership. It was essentially giving advice to itself. The Financial Stability Board, which identifies key weaknesses underlying financial turmoil, and which recommends actions to improve markets and institutional resilience, has only 24 member countries, and has no authority to issue binding directives. The absence of an institution that possesses the information, expertise and impartiality to swiftly identify and frankly opine about emerging risks to the global financial system is clearly a major governance gap.[30]

On a broader level, there is no impartial global intermediary that secures data from authoritative sources using universal standards, ensures that such data are disseminated in an impartial manner, and makes recommendations in a genuinely transparent way. What is needed is an organization that combines the comparative advantages of various actors who are in a position to contribute to global problem solving based on widely accepted and realistically achievable standards. Ideally, such an organization would be in a position to produce early warning systems, develop plans for coordinated counter-measures, and prevent contagion.

The challenge is how to get all the parties that are needed to agree on the creation of such an entity to do so while casting aside self-interest and partisanship. That simply is not going to happen. In the absence of that, gov-

---

[29] S. Boseley, "Experts Criticize WHO Delay in Sounding Alarm Over Ebola Outbreak," *The Guardian*, November 22, 2015, accessed November 26, 2015, http://www.theguardian.com/world/2015/nov/22/experts-criticise-world-health-organisation-who-delay-ebola-outbreak.
[30] OECD, *Future Global Shocks*, 106.

ernments and businesses should be focused on designing systems that can withstand foreseeable events, managing shocks as skillfully as conditions and resources will allow, ensuring that the recovery process can be as rapid as possible, and making existing systems as flexible and adaptable as possible. The answer may not only lie in enhancing detection and response capabilities of governments and multilateral institutions, but the ability to utilize military forces from a range of countries, on short notice, to implement policies and contain crises before they metastasize. This again raises the issue of how best to allocate limited resources, while being proactive (rather than reactive) in the process.

Purposely Complex and Ambiguous Risk Governance

Among the threats that have become pervasive in this century are terrorism, cyber risk, climate change, pandemics, and resource nationalism. Each of these has global implications and therefore requires a global approach to solutions. Arriving at those solutions will require a sea change in terms of the inclusiveness of existing multilateral institutions, the manner in which major actors approach problem solving, and an effective marshalling of resources. The bad news is that too little progress is being made, but the good news is that there is a paradigm shift occurring in the world that should set the stage for achieving more meaningful outcomes in the future.

Emerging and developing nations have spoken up loudly about the inequities in the existing global decision-making apparatus, and they are being heard. Even the five permanent members of the UNSC recognize that the current framework for the Council's decision making was literally created in and for another century. It is a matter of time until the leading emerging and developing nations have a stronger voice in that body. Although the U.S., Europe, and Japan continue to lead three of the four largest multilateral development banks (the World Bank, the European Bank for Reconstruction and Development, and the Asian Development Bank), China's new Asian Infrastructure Investment Bank and the BRICS' New Development Bank are both based in China, and represent a fresh new approach to development finance.

New organizations and standards have proliferated at the national, regional, and global levels, which has created both breakthroughs and bottlenecks in terms of how new laws and regulations are crafted and implemented. This represents progress, due in large part to the ongoing pressure being applied to the status quo by state and non-state actors. The longer-term challenge is to ensure that the change that does occur is more in favor of the common good, which is in everyone's long-term interest. Change can only occur while including the existing power structure, because it is needed in order to approve change for the future.

So a balance must be struck between those in favor of maintaining the *ancien régime* and those wishing to disrupt it. This represents a real challenge for businesses, which must embrace change while at the same time not tipping the scales too far in any particular direction. It is a delicate balancing act, requiring the right mix of open-mindedness, adaptability, and accountability. A realistic approach to achieving risk governance standards that are at once bold enough to foster real change, but balanced enough not to rock the boat too much, requires an organizational culture that fosters systematic information sharing and has the ability to evaluate critical information in an efficient, thoughtful, and timely manner.

As it is clear that a meaningful approach to risk governance cannot be achieved in isolation, a diverse range of actors that an organization may not have been accustomed to including in the decision-making process should become part of that process. Given the range of challenges organizations confront, and number of actors on the stage, the solutions are unlikely to be simple and straightforward—rather, by definition, they will be complex and ambiguous. To have the best chance of success, agile risk managers and decision makers should strive to include representative members of civil society in the decision-making process, to ensure a broad range of opinion and obtain broadly based 'buy-in' to whatever decisions are arrived at. Doing so will encourage a more rigorous and thoughtful process.

By definition, risk governance is intended to address risk management in a broad context, which includes as many actors, thought processes, and mechanisms as is practicable. It is not merely the organization that should benefit from this process, it is society at large. How the information is obtained, analyzed, and interpreted will naturally impact the outcome. Since change is the one constant in the management of risk, our ability to embrace change and adapt to it will in the end determine how successful and effective our efforts will be.

In thinking about how to create a framework to achieve this larger objective, some of the basic questions any organization should be asking itself include:

- Who should be involved in creating an enhanced risk management framework?

- What model or framework that already exists should be used from which to build a customized model that ideally meets the needs of the organization?
- What scope of global operations should be included in the model?
- How and from what sources will the information be collected, prioritized, and evaluated?
- What are the likely direct and indirect impacts of the model?
- What are the comparative costs and benefits of pursuing each given course of action?
- What needs to be done to enhance the long-term effectiveness of the chosen course?

When considering the answers to these questions, it should be borne in mind that there will be setbacks and that things will fall through the cracks. Arriving at the best methodology will certainly not be achieved with a single straight line. The long-term objectives should always be kept in mind; focusing on short-term goals will inevitably generate a secondary set of challenges. When mistakes are identified, tackle them, head-on; delaying the inevitable will only make a solution more difficult to achieve (see Graph 14.2).

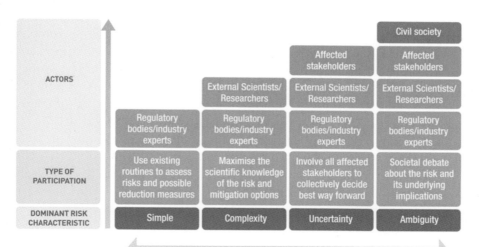

Graph 14.2   Risk governance with stakeholder involvement (*An introduction to the IRGC Risk Governance Framework*, IRGC, October 2007, accessed November 29, 2015, http://www.ortwin-renn.de/sites/default/files/PDF/RecentPublications/PolicyBrief_IRGC_RiskGovernanceFramework_9Oct.pdf)

Many of the examples we have cited throughout this book may be termed slowly developing risks (SDRs). Whether it is a financial crisis, a pandemic or rising ocean temperatures, they generally take weeks, months, or years to evolve to the degree that they become a real problem for people, governments, risk managers, and decision makers. In order to be addressed thoughtfully, we must both learn from the lessons of history while also having a distinct long-term future orientation. An effective decision-making framework to tackle SDRs should include a focus on systemic issues, decisions that are based on facts (rather than opinions or observations), flexibility along the way (to account for new factors that may influence the process), and acting with the firm belief that recommendations that will be made are both realistic and will be acted upon.

Risk governance is all about realism, teamwork, effective communication, and the ownership of your actions and decisions. It is not supposed to be simple, fast, or straightforward. Getting it right is hard work. With the stakes so high, no organization—in either the private or the public sector—can afford to adopt a cavalier approach to the subject. The key message here is that operating in a cocoon or a silo, thinking about new challenges in old ways, and failing to have a long-term orientation will not get the job done: be collaborative, open your minds to new alternatives, involve new actors in the process, and turn the pyramid upside down.

15

Conclusion

Failing Forward[1]

Throughout this book we have explored increasingly unique attributes of risk in the twenty-first century and how decision frameworks should adapt so that organizations of all kinds may become more agile, in order to be in a position to address those risks. That does not mean, of course, that all risks *can* be effectively managed, or that there is an agile solution to every problem. Mankind has at times proven to be risk-prone and foolhardy, creating policies, procedures and thought processes that can and do in the process generate risk.

In some cases, this quality of *failing forward* has produced positive outcomes, resulting in scientific breakthroughs and the advancing of civilization. In far too many cases, however, our game of civilizational brinksmanship has continued unchecked and has claimed many a generation before us, from Columbus's 'discovery' of the New World (which claimed millions of lives) to the collapse of ancient, yet advanced societies.[2] Our proclivity as a species to make ill-fated collective choices seems to be hard wired in our DNA.

As risks have evolved from being phenomenological occurrences in the natural world, the twenty-first century is in many ways the era of man-made risk and man-stoked fires. From cyber risk—which is increasingly mutating to impact all facets of the modern economy—to terrorism and reputation risk, mounting a credible defense to these risks requires as much soft skill as it does technical risk and analytical capabilities. Moreover, twenty-first-century

---

[1] The first writer to coin the term 'failing forward' was John C. Maxwell in his book *Failing Forward* (Thomas Nelson, 2000).

[2] J. Diamond, *Collapse: How Societies Choose to Fail or Succeed* (New York: Viking Penguin, 2005).

© The Author(s) 2016
D. Wagner, D. Disparte, *Global Risk Agility and Decision Making*,
DOI 10.1057/978-1-349-94860-4_15

survival depends very much on our ability to harness risk, encourage bounded risk taking, and improving overall organizational and societal resilience. Like no time in human history, the notion of risk and resilience has taken center stage. Our individual and collective decision making are the chief actors in this drama. In our hard stance *against the gods*, it may feel like we have won, yet we find there is no one else to turn to for the answers.

The interconnectivity of the web of risks ensnaring the twenty-first century has modern man stuck as the nucleus of a delicate and, in some ways, fragile system. All the risks explored herein mutate and compound based on human action and inaction. It is therefore reasonable to assume that as individuals, organizations, and at the societal level, choices can be made that will improve our ability to thrive and, indeed, gain an upside in this era. To do so will require the evolution of corporate culture and structures, for, as we have seen in the many case studies throughout this book, many risk domains are made worse by company blind spots, managerial remoteness, or sheer ignorance regarding the presence of risk factors—none more so than the amorphous cyber and reputation risk domains.

Risk does not live in isolation, nor does it conform compliantly to the classification systems used by various risk management approaches. Rather, decision makers must view risk as a dynamic process that cannot be adequately contained with static tools. The placebo affect that can be created by believing certain risks are "covered" through traditional approaches to risk management is often more dangerous than the risk itself. The example of VW is instructive, as we have seen. Once considered a paragon of corporate governance and a leader in the automotive industry, in 2015 VW grappled with a rapidly eroding reputation and a precipitous decline in its market value due to an entirely preventable emission-rigging scandal.[3] It turns out its alleged emissions-cheat device, which was set to reduce engine output while emission tests were conducted on diesel vehicles, was also connected to the company's kill switch. No direct competitive force (perhaps other than the urge to cut corners and gain market share—all resulting from internal sources of pressure) and no discernible outside factor, caused VW executives to make these ill-fated choices. They did that entirely on their own.

In another era perhaps VW may have escaped the prying eyes of regulators and the public. Today, firms must opt in *favor* of transparency in order to remain competitive. VW's case is very much emblematic of risk in our times,

---

[3] S. Minter, "VW Scandal Lowers Germany's Brand Value by $191 Billion," *Industry Week*, October 2015, accessed November 26, 2015, http://www.industryweek.com/competitiveness/vw-scandal-lowers-germanys-brand-value-191-billion.

where climate change, reputation, and technology converged to amplify the losses to the company's shareholders, *inviting* worldwide scrutiny. Actions that were once accepted as *externality* are now treated as something punishable with monetary fines and sometimes irreversible damage to a firm's hard earned reputation.

Similarly, climate change is no longer some distant reality in a far-off and arid developing country. Catastrophic drought is a reality in California. Following Super Storm Sandy, when images of New York's Stock Exchange covered with sandbags were beamed around the world, Wall Street, Main Street and boardrooms everywhere contemplated their response to increasingly extreme natural risks. In the aftermath of the VW scandal, companies that do not leave light environmental footprints, or try to cut regulatory corners, fail to review their "green" posture at their peril. Being perceived as thoughtful about the environment is no longer simply a nice marketing motif, it should be thought of as a source of competitive advantage and survival.

Adaptability and Agility

Whether we respond to risk through fear or fiat, *risk agility* is the key attribute of the survivors. Adaptability and agility have always been the secret survival mechanism of the fittest. Many methodologies have lulled us into ignoring our instincts under the guise that we are safe and that risks are hedged. Compliance, which in some cases is merely the act of grudgingly checking boxes, is not an adequate form of risk management for the twenty-first century.

Risk agility implies a certain mastery of risk, decision making under opacity, and a level of simplicity in the face of complex, heavily interconnected systems. This is both the source of ease in implementing the approaches outlined in this book, and also one of the confounding factors. Agility, by definition, implies *nimbleness, speed*, and an *intellectual acuity* that often betray large and complex organizations.

The modern multinational is anything but agile—imagine a supertanker making a U-turn. Now imagine a supertanker guided by Francesco Schettino—the captain of the doomed *Costa Concordia*—navigating through rough seas, while facing a mutinous crew and a system failure caused by a breach. As vivid as this image is, modern corporate leaders share the waters of the market in conditions this risky, and with captains this reckless.[4] The best

---

[4] N. Squires, "*Costa Concordia* Captain Francesco Schettino to Release Book about Disaster," *The Telegraph*, June 2015. Accessed November 26, 2015, http://www.telegraph.co.uk/news/worldnews/europe/italy/11695071/Costa-Concordia-captain-Francesco-Schettino-to-release-book-about-disaster.html.

among them must make decisions that impact the lives of their crew, their missions and, in the direst of circumstances, the very survival of their ship—all with limited information, or responding to the wrong signals.

The inexorable reality is that no matter how opaque choices are, the defining attribute of a *leader* is his or her ability or obligation to choose. The defining attribute of an agile *enterprise* is the ability to align all resources to enterprise-dependent decisions. Even in the best of cases, decisions are often made applying the 80–20 rule, or some other maxim that justifies 'permitted' uncertainty in boardrooms, executive committees and government.

For many global enterprises, indecision and paralysis have crept into the decision matrix. Entire continents, such as Africa, are ring-fenced from investment under the perception that the investment climate there is too risky or illiquid. New technologies and disruptions are ceded to upstarts, and better ways of organizing for resilience and agility are left to tech companies. Often times, the best choice to make is no choice at all. For large and complex organizations, gaining momentum and responsiveness can be an exercise in futility. The perpetual motion machine of organizational behavior and the profit motive stand in the way of agility. Our natural orientation toward risk *aversion* blinds us to clear warning signs that trouble lies ahead.

The Risk-Ready Firm

In a risk-ready firm, it is everyone's business to remain in business, which requires a flat risk-response structure that leverages risk management approaches. In financial institutions, risk management is a largely backward-looking activity that is heavily quantitative and most often driven by regulatory fiat. In the non-financial world, particularly in manufacturing industry, risk management is much more qualitative, given the tangible (observational) nature of errors in a system or process. Charting a course as a hybrid between the two yields three-dimensional risk management, wherein the filter-to-noise ratio is controlled, and better, more instinctive responses can emerge.

Culturally, fear of failure and an aversion against reporting bad news must be confronted in all its guises in order for organizations to fully unlock value from risk. Withholding bad news neither makes it go away, nor does it make it better. In fact, as we have seen throughout this book, bad news tends to become amplified, and it festers with time. In the case of VW, company leaders received clear warning signs of the emissions-rigging scandal which dated back to 2007, if not earlier.[5] In the Germanwings tragedy, managers received clear signals that Andreas Lubitz, the suicidal co-pilot, was clearly a danger in the skies. Nevertheless, these risk signals were ignored, despite the relatively simple approaches to mitigate and respond to their presence.[6]

Many risk management approaches unfortunately impose far too much complexity on organizations, making them a source of risk *amplification* rather than a source of risk *abatement*. As the old adage goes, complex systems fail in complex ways. Simplicity, therefore, is a key attribute of an agile risk management framework. Some of the most enduring ways to avoid losses merely outline proscribed activities, as opposed to delineating every possible situational response. Greater simplicity is needed in risk management—and in decision-making systems—in order to keep up with the times.

The other principal weakness of traditional risk management frameworks and how they hamper agility is that they often rely on historical data to drive current understanding and future directions. The challenge of taking this approach is that it is not very effective against unprecedented, emerging, or never before seen threats. Moreover, despite all the rigor of statistical methods, for many risk domains, such as cyber and reputation risk, the data set is shallow and largely unreported. The *omerta* that follows a potentially embarrassing cyber risk or reputational exposure hampers the ability to fully understand the scope and the forces shaping these risk domains. The call to establish a central clearinghouse to capture market-wide data regarding breaches is getting louder with each high profile case.[7]

The False Positives of History

The other challenge to historical methods is the preponderance of so-called 'false positives' derived from a confirmation bias that affects individuals, teams, and organizations. The lessons of history may be consulted not for real clues about the present or future, but for a confirmation of preconceived notions. History can either be a useful guide to managing the present, or it can really

---

[5] J. Gitlin, *Report: VW Was Warned about Cheating Emissions in 2007*, Ars Technica, September 2015, accessed November 26, 2015, http://arstechnica.com/cars/2015/09/report-vw-was-warned-about-cheating-emissions-in-2007/.

[6] M. Eddy, "For Families of Germanwings Victims, Anger Simmers Through Grief," *The New York Times*, October 2015, accessed November 26, 2015, http://www.nytimes.com/2015/10/11/world/europe/for-families-of-germanwings-victims-anger-burns-through-grief.html?_r=0.

[7] K. Williams, "Manhattan DA Opens International Cyber Threat Sharing Nonprofit," *The Hill*, September 2015, accessed November 26, 2015, http://thehill.com/policy/cybersecurity/253830-manhattan-da-opens-international-cyberthreat-sharing-nonprofit.

set one off on the wrong path. The behavior of nation-states is a good example of this; either a country will behave exactly as it has and be highly predictable (such as the government of North Korea's constant antics), or it can surprise and amaze (such as Germany's willingness to throw open its doors to Syrian refugees, when many of its neighbors locked the door shut).

By the same token, preconceived notions about how the government of a country may act can defy prediction. For example, since the collapse of the Soviet Union, the world has come to expect that Vladimir Putin will seek to remind the West that Russia still matters on the global stage. Russia will therefore be expected to flex its muscles around its borders, and seek to revive an anti-U.S. alliance with China. The West could not have predicted, based on its experience since 2011, that Mr. Putin would choose to take the lead in the fight against ISIS in Syria, essentially superseding America's and the West's prior limited efforts to combat ISIS, and largely making the West irrelevant in Syria's future.

History tends to be an excellent teacher vis-à-vis great successes or great failures when it comes to decision making. It is not such a great teacher when it comes to predicting the future or taking a leap into the unknown. The behavior of complex systems such as weather, the economy, and social risk are not readily gauged by traditional approaches. Such approaches are most effective for high-frequency, reliable events. Low-frequency and rare events' statistical methods tend to suffer from "model error"—namely, the risk that the object being measured is far too complex for the methods used, or, more common still, that the person doing the measuring makes a mistake, or the model itself is too confounding. Many losses resulting from the global financial crisis where compounded by this reality. Simply put, complexity plus complexity equals complexity squared.

A Fear-Based Approach to Risk

Fear-based approaches to risk, which is how many insurance policies are sold, sometimes end up being imposed upon consumers rather than being part of deliberate, sound hedging decisions. Compulsory auto liability insurance and mandated health care policies are good examples of this part of the risk management landscape, with penalties added for good measure for non-compliance. These are, in the end, 'sold' based on fear, both of the familiar and of the unknown. Most of us know someone who has gotten into an auto accident or had a terrible illness and racked up hundreds of thousands of dollars in medical bills, yet even though we believe that it is unlikely that a similar fate will befall us, an inability to predict the future prompts us to be willing adherents to this compulsory form of risk management.

Other products of the insurance market which are not compulsory, such as political risk and trade credit insurance, promote and protect cross-border trade and investment and are clearly growth drivers. Such products go part of the way toward answering the question: What would companies do in a world without risk? A volatile country like Nigeria, which is perceived to be highly corrupt and an inherently dangerous place to do business, can become investment grade when investors or lenders utilize the right country risk management tools. Nearly all of the risks that can cause an investor to lose their capital when investing or operating in international markets can be adequately hedged this way. Moreover, these risks, and the cost of insuring them, can be readily priced into a company's economic model. It can easily be argued that not doing so is highly irresponsible—and yet, less than one percent of all trade, investment and lending utilizes political risk and trade credit insurance—not only because few decision makers pursue it, but because the insurance marketplace does not have the capacity to insure all cross-border transactions.

Instead of actively pursuing such coverage, many investors and traders either forego emerging market investments altogether, or internal investment teams contemplate assuming higher risk by making adjustments to perceived country risk in their financial discount rates. The principal challenge with this approach is that while a higher discount rate may indeed compensate for increased country risk factors, the discount rate needs to reach maturity by running the course of the investment horizon for it to be of any real value as a hedging strategy. A country-adjusted discount rate is an illiquid instrument, and many investments are lost due to unforeseen risks early on. Simply put, the higher discount or interest rate needs to accrue over time to be of any real value. Political risk insurance, on the other hand, backs into deep financial pools with highly rated private sector insurers, multilateral development banks or investment promotion agencies.

A more mundane, if controversial example of where risk serves as a catalyst for growth is the private mortgage insurance (PMI) market. This category of insurance closely parallels the concepts behind trade credit or political risk insurance, in that the insured would not have access to home ownership without the existence of the product. The typical PMI policy favors the lender and protects against borrower default. While there is some controversy surrounding this product as a "penalty on the poor," its catalytic property cannot be denied.[8]

---

[8] S. Reckard, "Lower Credit Scores Lead to Higher Mortgage Costs," *The Seattle Times*, July 2015, accessed November 26, 2015, http://www.seattletimes.com/business/real-estate/lower-credit-scores-lead-to-higher-mortgage-costs/.

Some of the best examples of catalytic risk solutions can be found in the export and trade credit domain. Export credit insurance offered through private sector providers (or one of the world's 60+ government-backed export–import banks) is a great mechanism to promote growth, particularly among small manufacturing firms.[9] The exposure to buyer defaults is a very real risk; once a ship has left port with a company's goods, it is usually on a one-way trip. Export credit programs help mitigate default risks, while at the same time offering terms of trade, such as extended repayment periods, that would otherwise not be available.

Risk Taking and Entrepreneurialism

The pursuit of risk agility implies controlled risk taking at all levels of an organization. Firms that embody risk agility are often seen as serial innovators with an enduring ability to telegraph market needs. Entrepreneurs naturally possess risk agility, for every day they tread a fine line between the success and demise of their firm or idea. Promoting entrepreneurial culture is difficult in large corporate environments, since, by default, entrepreneurs tend to favor chaos over order and traditional command and control structures. At the same time, many entrepreneurs are influencing established firms that recognize that they must innovate in order to survive.

That message has been embraced for decades by companies such as International Business Machines and GE. Increasingly, established firms across the spectrum see the value in promoting entrepreneurial culture, specifically by creating Skunk Works (made famous by Lockheed Martin) or by giving isolated teams high-priority innovative tasks. Other organizations, particularly in consumer-facing industries, are creating innovation labs tasked with imbuing Silicon Valley's culture of innovation in otherwise staid industries or business lines, or are creating corporate venturing units whose sole task is to identify, invest in and roll up innovative early stage companies. All of these approaches are great for growing market share, intellectual property and new innovations, they also work well when applying agile risk management.

For example, an organization wishing to test its resilience to a given risk or a series of unfortunate events can task a risk innovation unit with addressing the company's post mortem and outlining the most likely causes of death. However grim this task may be, it is often a useful exercise (one that is surely applied by entrepreneurs) to work backward from the end state. The chief

---

[9] D. Disparte, *Ex-Im and the Pax Americana.*

obstacle of bringing risk management to the "money-making" side of business (and the decision-making side of government) is the reality, however unfair, that risk management is often conflated with a business prevention function. Just as you cannot have a vibrant business if it does not survive, you cannot use scare tactics all the way to the boardroom.

Instead of merely paying lip service to risk management being a value-adding discipline, agile risk managers should be *embedded* in frontline teams. Risk managers have done their trade few favors by either telling people what they cannot do or by being the gloomiest people in the room. However, the real obstacles to elevating this discipline and equipping companies with able twenty-first-century risk leaders is more structural than it is about individual talent. Organizations must first acknowledge that the rules of the game have changed and that they need to be leading the pack in terms of innovation and strategy, rather than simply playing along. McDonalds figured this out too late in the game.

Whales, Rogue Traders, and Accountability

Even in cutting-edge firms such as JP Morgan, whose risk management regime has been chronicled by leading business schools, what should have been obvious structural impediments to effective risk management exacerbated otherwise preventable losses.[10] During the now-infamous "London Whale" case, JP Morgan's Chief Risk Officer reported directly to Jamie Dimon, the bank's messianic Chairman and CEO. At the same time, Ina Drew, the firm's Chief Investment Officer (CIO), commanded a $350 billion investment portfolio that was supposed to keep the liability between deposits and loans in check. The CIO's office was effectively run as a large internal hedge fund, which many observers believed pushed the limits of both structural and technical risk management, such as VaR metrics and the balance of risk-weighted assets. This created a more permissive environment, where Bruno Iksil (a.k.a. 'The Whale'), one of the bank's London-based traders, could push his trading activities to the very brink, culminating in a $6 billion dollar loss for JP Morgan in 2012.

This well-documented case provides a clear taxonomy of structural impediments to agile risk management, even in a highly regarded, well-capitalized and thoroughly regulated firm such as Chase. To begin with, Jamie Dimon

---

[10] R. Kaplan and M. Annette, JP Morgan's Loss: Bigger than "Risk Management," *Harvard Business Review*, May 2012, accessed November 26, 2015, https://hbr.org/2012/05/jp-morgans-loss-bigger-than-ri.

continues to serve as the firm's chairman and CEO. Unlike with VW, where its CEO (Martin Winterkorn) immediately resigned in disgrace following the diesel scandal, Mr Dimon remained at the helm. The same was true when Kweku Adoboli engaged in unauthorized trading that cost UBS more than $2 billion in 2012, and Sergio Ermotti remained CEO. The same also occurred again when Howard Hubler was held responsible for the largest trading loss in history ($9 billion) at Morgan Stanley in 2007, and John Mack remained CEO (Mr Hubler was granted $10 million in back pay for his trouble[11]). What kind of message does this send to investors and consumers?

Having a CEO also serve as a board of director or chairman effectively removes the separation of powers designed to keep management in check. In the case of JP Morgan, the board's risk management committee at the time of the losses also reported to Mr. Dimon, as did the CRO and CIO respectively. So, if the tone at the top is to "push risk-taking in our asset-liability management unit to the limit", few will challenge this directive. With the benefit of hindsight, Mr. Dimon described the structure of the CIO unit as "flawed, complex, poorly reviewed, poorly executed and poorly managed.[12]" This statement threw Ina Drew, the erstwhile CIO, under the bus, but Mr. Dimon failed to take a good look at himself in the mirror in the process. Sending a mixed signal to its peers and the market, JP Morgan's board awarded Mr Dimon with a 74 percent pay increase the year following the London Whale case, bringing his total compensation to $20 million in January of 2014.[13] In the preceding year, JP Morgan paid $23 billion legal bills and fines.[14]

Doing the Right Thing

Doing the right thing plays well on Main Street. It is not complicated, everyone understands what it means, and most people see inherent value in knowing the difference between right and wrong, and practicing it. That being the case, why is it, then, that so many corporate leaders are in the habit of doing exactly the opposite? Is there something intrinsic in twenty-first-century culture that also says that you should try to get away with whatever you can, or

---

[11] "How to Lose $9 Billion and Walk Away with $10 Million," *Trejdify*, February 19, 2012, accessed November 26, 2015, http://blog.trejdify.com/2012/02/how-to-lose-9-billion-and-walk-away.html.

[12] J. Schlesinger, "JPMorgan Chase: London Whale Swallows $2B," CBS News, May 10, 2012, accessed November 26, 2015, http://www.cbsnews.com/news/jpmorgan-chase-london-whale-swallows-2b/.

[13] N. Summer, "Dimon Gets a 74 Percent Raise after Billions in Fines," *Bloomberg Business*, January 24, 2014, accessed November 26, 2015, http://www.bloomberg.com/bw/articles/2014-01-24/dimon-gets-74-percent-raise-after-billions-in-fines.

[14] S. Gandel, "Dimon Gets 74% Raise After Billions in Fines," *Fortune*, January 24, 2014, accessed November 26, 2015, http://fortune.com/2013/10/11/jpmorgan-were-prepared-for-23-billion-in-legal-bills/.

you should strive to do what you want until and unless someone objects? The pressure between *maximizers* (those who want as much as they can get as fast as they can get it) and *optimizers* (those who want as much as they can have for as long as they can have it) is a central struggle in the modern economy.

Agile decision makers are champions of transparency, and doing the right thing—even if it may cost a company money. They do not just 'subscribe' to the concept, they practice it, and they imbue such values in their organizations. Not only does doing so foster trust and alignment among internal stakeholders, it can also be an advantage during times of unwanted public revelations, or bad news. Following a scandal like Toyota's sticking accelerator pedals or public debacles like BP's Gulf Oil disaster, agile leaders respond quickly and affirmatively to a crisis. They do not attempt to pass the buck of responsibility somewhere else, or to someone else.

While doing so may play well with shareholders, in the end, the long-term damage done to a firm's reputation in the process costs a lot more. What do people remember more about how BP responded to the Gulf Oil disaster—the billions of dollars it ultimately paid out (even for fraudulent claims), or the soundbite of former CEO Tony Hayward saying he "wanted his life back", after having tried to manage the crisis for a few weeks. That soundbite probably did as much damage to BP's reputation as the spill itself. To BP's credit, he was replaced shortly thereafter.

In the cases of both BP and Toyota, lasting contributions were made to the book of "what not to do following a public relations crisis". Agile firms quickly and transparently tackle bad news. The examples of Le Pain Quotidien's field mouse incident and Apple's uncharacteristically transparent supply chain report, highlighted earlier in this book, are examples of emerging best practice. Indeed, in the era of systemic cyber risk, having nothing to hide, and being perceived to have such a reputation, may very well be the best defense.

Building Trust in the Digital World

The smartest and strongest thing any leader can do to enhance a brand's reputation and, ultimately, its share value is to make the word 'trust' synonymous with the name of the brand. There is no way a company can possibly be considered to be a believer in the importance of transparency and good governance, a responsible and worthwhile steward of the environment, or an organization that believes in doing the right thing, without portraying itself as a reliable and trusted partner. There is no better way to achieve that than by exhibiting trustworthiness in everything an organization does.

Ultimately, the value of a brand can be measured by what it means in the hearts and minds of consumers, which is judged by whether a company does what it says it either has done, or is going to do. It boils down to whether a promise is kept, or not—a concept that all of our parents and grandparents instilled in us. The manner in which a company is judged to be keeping its promises is not all that different in today's digital world than it was when families gathered around the radio for their evening entertainment one hundred years ago. It is all about personal experience and word of mouth—the difference today is that the word travels at the speed of light, so the stakes are higher, because more people can instantly be reached.

> The techniques used by brands to build trust in the pre-digital age are far less salient to contemporary users, who are savvier, more skeptical, and more cynical about marketing messages. Today, consumers require more than slogans and a persuasive pitch. Delivering on the promise a brand makes has always been essential, but in today's consumer culture, fueled by social media, a discrepancy between promise and product can be instantly fatal to a brand. The emergence of digital platforms has expanded the opportunity for brands to build trust equity.[15]

Although an increasing percentage of consumers (especially younger consumers) are more inherently distrustful of big business, the very nature of the way business is done today requires that a high level of that 'trust equity' be established. Agile risk managers and decision makers understand that with consumers being more skeptical than ever before about the claims companies may make about their products, they must work harder both to establish trust, and to keep it. Among the ways they can do this are to seek and publish customer reviews in social media, and create and publish report cards about how they are doing vis-à-vis compliance with governance standards and transparency ratings. In order to be perceived as leaders in the 'trust quotient', companies should deliberately and affirmatively establish in the public domain who they are, what they stand for, and that what they actually do that is consistent with what they say they do.

A new breed of organization is emerging that is unique to the twenty-first century—the corporate activists. These are firms whose actions go well beyond value systems written on walls or kitschy corporate credos, like *do no evil*. The activist firm will often make hard choices that may deliberately erode shareholder value in the short run, but enhance overall trust equity by showing the market that they stand for something. Much like the Geneva

---

[15] J. Lee, *How Brands Built Trust in a Digital World*, hugeinc.com, August 27, 2014, accessed November 26, 2015, http://www.hugeinc.com/ideas/report/how-brands-build-trust-digitally.

Convention took on new meaning following 9/11, corporate values matter most when they may be least convenient.

Some salient examples include Apple's response to the marriage equality debate in the U.S., where it offered to relocate its staff from states that stood against Apple's views. With Tim Cook being the first openly gay CEO of a Fortune 500 company, the corporate and personal lines of the leaders' views were blurred, but much to Apple's advantage. Similarly, Starbuck's controversial *Race Together* campaign raised questions about why a barista should strike a dialogue about race relations demonstrated a move away from the traditional agnosticism that dominates corporate decisions.[16] Activist firms are optimizers of value, in that they are willing to undergo short-term losses to stand their ground on key issues. These rare glimpse of values in action have all the staying power of negative information in the market.

The Importance of Looking Beyond the Headlines

In the era of instantaneous information and the 24/7 news cycle, there are some distinct advantages to being a news junkie. Among these is that one can learn something that *could* become a sensational headline in the future, but what is required in order to discern something that is newsworthy versus something that is not is context. It is easy to focus on the latest sensational news headlines when thinking about what may negatively impact an organization's ability to operate, domestically or internationally. The problem with such an approach, however, is that there is a lot going on beyond the headlines that may have an impact on your firm directly or indirectly, in the near or distant future—you just do not know it yet.

While many risk managers and decision makers may continue to assume that the ongoing economic recovery in the U.S. (and some other countries) generally implies reduced risk in conducting business abroad, the opposite is the case in the many countries in the world which, although perhaps not in the headlines, are either experiencing heightened levels of risk or are perpetually *in* a higher state of risk. We are not simply talking about the Iraqs and Syrias of the world—but countries not otherwise normally associated with risk, such as Brazil and Thailand. A brief examination of these countries serves to remind us how different the trajectory of any given country may be, how

---

[16] D. Disparte and T. Gentry, "Corporate Activism is on the Rise," *International Policy Digest*, July 6, 2015, accessed November 26, 2015, http://www.internationalpolicydigest.org/2015/07/06/corporate-activism-is-on-the-rise/.

very different our perceptions of them may be from reality, and how little we may actually know about them. Your firm may not be doing business there, but what is happening there may be indicative of what *could* occur in your home country, or another place where you are currently operating, as a result of the trends it may represent or the nature of a government's reaction to ongoing economic, political or social challenges.

Down and Out in Brazil

As a BRIC country, people are accustomed to thinking of Brazil as an economic powerhouse which, by virtue of its geographical size, large population, and abundant natural resources, implies that its potential for growth and future prosperity seems limitless. The reality could not be more different.

In 2001, when the country was designated a BRIC, it was $200 billion in debt and spent 38 percent of its GDP servicing that debt. The following year Brazil took out a $40 billion loan from the IMF. The country's average GDP growth rate since 2000 has been under 2 percent, and Brazil has consistently underperformed its BRIC counterparts. Brazil scored poorly in the World Bank's *Doing Business* rankings (in 2015), with an overall score of 167 out of 189, and ranked in the lowest third in six of the ten categories. Its best ranking was in providing electricity (55)—and even that was under threat at the time because of persistent drought, which impacted its hydropower production capabilities, on which the country is heavily dependent. Much of the country subsequently endured several power cuts.

As is the case with many of its counterparts in Latin America, Brazil fell into the trap of export dependency on China. Approximately 80 percent of the country's basic goods exports went to China between 2000 and 2010, while less than 20 percent of its manufactured and semi-manufactured goods were purchased by China. That's fine when times are good, but as China's growth rate continued to decline, Brazil found it difficult to find buyers to take up the slack, as did many other China-dependent countries.

Another source of concern is Brazil's increasingly dilapidated infrastructure, which impacts everything from production logistics to basic transportation needs. A number of large infrastructure projects were approved during the 'boom' years under President Lula da Silva, and remain unfunded and unfinished for many years after he left office. The country's balance of payments deficit continued to rise while inflation continued to push up against the Central Bank's ceiling.

A combination of government complacency, an inadequately developed regulatory framework, and a host of infrastructure bottlenecks have prevented Brazil from achieving its full potential. Rigid labor laws, a byzantine tax system, and the government domination of long-term credit markets conspire to prevent Brazil from breaking out of its well established pattern of low economic performance.

In spite of all the hoopla over Brazil as one of the world's globalization poster boys, its worst enemy is itself. In spite of the many opportunities it has had to fix its problems and live up to its potential, it has failed to do so, time and time again. Brazil has yet to sustain mid-to-high single-digit GDP growth rates as the other BRIC countries have done, and is no better poised to do so in the second decade of the twenty-first century than it has done so far in the first. Since Brazil 'has it all' but has not managed to pull it all together, even in good times, what does that say about other countries that have far less going for them?[17]

Praying for the Tourists in Thailand

When most people think of Thailand, they think of beaches, beer, and good times. They tend not to think of coups. In the tortured modern political history of Thailand military coups have become commonplace, the result of the growing fissure between the political and economic power of the Bangkok elite and the parlous state of the rural poor. Thailand has endured 12 military coups and seven attempted coups since the absolute monarchy ended in 1932 and the army's right to intervene in political affairs has even been enshrined in law, making Thailand one of the world's most coup-prone countries.

Coups have become such a permanent component of the political landscape that they have proven to be largely irrelevant for business (that said, when the red shirts and yellow shirts battled one another in the streets in 2010, political violence ensued, and much of Bangkok's (and therefore, the country's) economy was brought to a standstill for weeks. Based on the country's economic performance over the past 40 years, coups have generally had a net positive benefit on the country's economic trajectory. Between 1995 and 2013, tourist arrivals rose from about 500,000 to 2,500,000. The country's

---

[17] Adapted from D. Wagner, "Why Brazil is Its Own Worst Enemy," *The Huffington Post*, December 23, 2014, accessed November 26, 2015, http://www.huffingtonpost.com/daniel-wagner/why-brazil-is-its-own-wor_b_6367760.html.

foreign exchange reserves rose in tandem, from just $3 billion in 1995 to $190 billion in 2011—much of it due to tourism.

Overall, the economy has performed remarkably well, given Thailand's propensity for radical political change, in spite of the government itself, which has generally done a poor job of managing all that money. Over the past 10 years, Thailand has run a budget deficit 60 percent of the time, its spending has more than doubled, and its external debt has more than tripled. Although business and consumer confidence have largely remained steady since 2010, both are beginning to weaken. And although the baht has actually strengthened fairly consistently since 2001, it is more or less at the same level it was 15 years ago.

One thing that could make a difference would be if the tourists stopped coming, prompting the exchange rate, foreign exchange reserves, and stock market to fall substantially. Perhaps, if there was enough collective pain, the red and yellow shirts would determine that it is in everyone's interest to find common ground. Given that a shared allegiance to Thailand's king has not prompted the parties to find a path forward, perhaps less money in their wallets will. It seems that ultimately the tourists are the only ones who can solve Thailand's political dilemma.[18]

The Importance of Unvarnished Opinions

What can be done from a practical perspective to address the challenges posed by operating in such countries? Naturally, it is not the risk manager's decision to head for the exit from Brazil or Thailand or any other country experiencing setbacks that have not come to the attention of decision makers, but it *is* a risk manager's responsibility to give the home office an unvarnished opinion of what conditions really are on the ground, and what may realistically be expected to occur in the future. Too often, local managers fail to do this. Many of the challenges associated with doing business abroad really are the "cost of doing business", and a firm's pricing model and risk management strategy should adequately reflect these heightened exposures. The challenge is to be able to identify the correct trigger point for when to pull the plug.

While most risks are insidious by nature, gradually eroding economic conditions, risk managers have a tendency to worry about 'spectacular' events,

---

[18] Adapted from D. Wagner, "Only the Tourists Can Save Thailand," *The Huffington Post*, July 23, 2014, accessed November 26, 2015, http://www.huffingtonpost.com/daniel-wagner/only-the-tourists-can-sav_b_5777836.html.

such as high-profile kidnappings or terrorist attacks. The reality, however, is that all processes that erode economic conditions fall under the risk manager's remit—whether they are insurable or not. The simple lesson is to consume the news and to do so with a critical eye, assuming that one day the headlines you are reading that are not on the front page may one day be there, and have a direct impact on your firm.[19]

Everyone's Opinion Is Valuable

When you have heard the prediction of an 'expert' in a given subject, have you ever thought to yourself "I could have told you that"? When 'experts' have been proven to be wrong, time after time, whether it was where a hurricane will hit land, who will win an election, or what the outcome of a civil conflict will be, have you wondered why they call themselves, or are referred to as, 'experts'? A study has been done that proves that the views of ordinary people can be more accurate than 'the experts'.

Over several years, the Good Judgement Project incorporated the views of tens of thousands of people from all walks of life who were asked to forecast global events. Many of the participants' views turned out to be more accurate than the collective judgement of widely accepted benchmarks, predictive tools and even intelligence analysts. The study proves that, sometimes, accurate forecasting does not require complicated tools, powerful computers, or expensive consultants. Rather, it requires being informed, weighing evidence from a variety of sources, having an open mind, being flexible, being willing to admit when a mistake is made, and having a willingness to change a course of direction.

What the authors of the study refer to as "Superforecasters" boils down to this:

1. Their philosophy tends to be:
   - Cautious: Recognizing that nothing is certain.
   - Humble: Because reality is infinitely complex.
   - Non-deterministic: What happens is not necessarily meant to be and does not necessarily need to happen.

---

[19] Adapted from D. Disparte and D. Wagner, "Going Beyond the Headlines with Global Risk Management," *Risk Management Magazine*, February 13, 2015. Accessed November 26, 2015, http://www.rmmagazine.com/2015/02/13/going-beyond-the-headlines-with-global-risk-management/.

2. Their abilities and thinking styles are:
   - Open-minded: Their beliefs are hypotheses that need to be tested rather than treasures to be protected.
   - Intelligent and knowledgeable: They are intellectually curious and relish mental challenges.
   - Reflective: They are naturally introspective and self-critical.
   - Numerate: They tend to be comfortable with numbers.

3. Their approach to forecasting tends to be:
   - Pragmatic: Not wedded to any idea or agenda.
   - Analytical: They consider other views.
   - Dragonfly-eyed: They value diverse views and naturally integrate them into their own.
   - Thoughtful: When facts change, they change their minds.
   - Intuitive: They tune in to their gut instincts and acknowledge cognitive and emotional biases they may have.

4. Their work ethic tends to possess:
   - A growth mindset: They believe it is possible to get better.
   - Grit: They are determined to keep at it, no matter how long it takes or how much energy they must devote to the task at hand.[20]

So, there is no real secret to what makes a good forecaster. It is a combination of a natural curiosity about the world, being informed, having an opinion, combining it with insight and instinct, and correcting course as necessary. Anyone can do that, and now there is proof that anyone who does that has a good chance of being as accurate as, or perhaps even more accurate than, all those 'experts'. Now just imagine if you were surrounded by a room full of these kinds of people, all tackling the same set of problems or challenges. That is powerful.

Globalization's Backlash

When tackling those problems together, it is important to bear in mind that globalization has not afforded the same benefits to all countries or all people, and that as the twenty-first century marches on, more of the people left

---

[20] D. Gardner and P. Tetlock, *Superforecasting: The Art and Science of Prediction* (New York: Crown Publishers, 2015), 191–192.

behind have become increasingly vocal about their concerns. This is naturally an issue for risk-agile decision makers who operate in the global arena, but it is likely to become more of a concern with time, as larger numbers of people come to believe that: (a) they have been left behind; (b) they are unlikely to become part of the one percent who have benefitted the most from globalization; and (c) in all likelihood, they have little to lose by strenuously voicing their concerns.

The global recession of 2009 certainly exacerbated the concerns of those left behind, and this occurred not just in the developing world; Anti-austerity protests have become a common feature of the landscape in Europe since then. The backlash against globalization has tended to fall into two general categories: protests against capitalism and the defense of national sovereignty. Random acts of violence have accompanied protests, but the real concern for decision makers is that millions of consumers are questioning the basic premise of globalization—that freer and all-encompassing trade and investment is a net positive for the global consumer, and that more jobs and opportunities are created with globalization than without it.

A lot of people in countries all around the world certainly do not *feel* as if globalization has had a meaningful positive impact on their lives or livelihoods, and they wonder whether, after 25 years (or so) of the modern version of globalization, they ever will. Many of them feel they are enduring an economic assault by the West, resulting in an attack on native customs and traditions, and a dependence on the rest of the world for their livelihood. They may also see 9/11 and the upheaval in the Middle East as an outgrowth of what they perceive to be an attempt by the West to basically control the world through commerce and politics.

It does not matter to these critics that the net result of globalization is that the proportion of output from the largest companies has tended to fall as challenges to their prior economic dominance have come under threat from new companies in their space, and local entities. "Incumbents could once protect themselves behind lofty barriers such as the high cost of capital, the difficulty of acquiring new technology, or the importance of close relationships with national governments. Globalization reduces the importance of all these things. Lower barriers make capital easier to raise, technology easier to buy, markets easier to reach, and ties with national governments ever less important. You no longer have to be a multinational to have the reach of one."[21] To someone on the margins, however, none of this matters.

---

[21] J. Micklethwait and A. Wooldridge, "Think Again: The Globalization Backlash," *Foreign Policy*, November 17, 2009, accessed November 26, 2015, http://foreignpolicy.com/2009/11/17/think-again-the-globalization-backlash/.

One of the things globalization has done is make it much easier for foreign leaders to declare another country the source of all its economic problems, with the U.S. naturally featuring at the top of the list. But is it Hollywood or Bollywood that is 'destroying' native culture? Is it French wine or Australian wine that puts Chilean winemakers out of business? And was it Honda or Hyundai that destroyed the local carmaker? When actually forced to look at the statistics and think about why a country has the problems it has, the conclusion often reached is that the home country has failed to get its house in order (as in the case of Brazil).

This complicates the job of the risk manager and adds an extra burden on the decision maker for those firms that operate internationally. Many companies now face a hypercompetitive environment with overly burdensome regulations (as many countries at least 'try' to appear as if they are complying with best practices according to global regulatory standards), vibrant NGOs, and supercharged consumers who cannot wait to speak their minds on social media. When faced with such a daunting climate, many will be inclined to ask themselves if it is even worth it to operate abroad.

The answer will sometimes be 'no', but as a result of globalization, those same set of challenges exist at home. In the twenty-first century the notion that one country, or for that matter one company, can be a fortress unto itself shielded from the effects of the world has been profoundly shattered. There is not a single publicly traded company that can decouple its fortunes from the performance of countries thousands of miles away, and where they have no investments or trading relationships. Similarly, the line between financial risk and sovereign risk has been blurred and made more porous with the advent of "too-big-too-fail" policies, state-sponsored enterprises, and sovereign wealth funds, whose investments often have more to do with statecraft than with investor returns.

The Rise of the Global Middle Class

The MDBs have rather different thoughts about the typical member of the growing global middle class than are found with the average business person. For the MDBs, it is the person in developing countries earning between $2 and $20 per day (comprising 70 percent of Latin Americans, 82 percent of Asians, and 85 percent of Africans[22]). For a business person, it might be the

---

[22] *Part I: Special Chapter: The Rise of Asia's Middle Class*, Key Indicators for Asia and the Pacific 2010, Asian Development Bank, 2010, accessed November 26, 2015, http://www.adb.org/publications/key-indicators-asia-and-pacific-2010.

more than 800 million Chinese who are expected to form the middle class of that country by 2030 (China's middle class numbered just 87 million in 2005 and is projected to reach 340 million in 2016).[23] According to one study, the entire global middle class may rise to 4.9 billion by 2030 (from 1.8 billion today), with 85 percent residing in Asia.[24] It is no wonder so many businesses are attempting to position themselves in Asia today.

If the statistics prove to be right, many products that Western households have been consuming for decades will be within reach of billions of new consumers, in just a few short years. What a tremendous opportunity for businesses and consumers alike. If your business operates in a mature industry or country and its intention is to continue to grow, you really will have little choice but to go global, implying that being an agile risk manager or decision maker is no longer something that *might* come in handy—it will in future be *necessary*.

However, even if these astounding projections come true, it would be foolish to imagine that your business will not experience a conflict and have a problem-free rise to glory. One of the consequences of this race for the middle-class dollar will be an inevitable rise in protectionism, economic nationalism, and a battle for a decreasing supply of resources—whether that is water, minerals, or food. Economies may not grow as rapidly as expected or predicted. Consumers may find—as they have throughout the developing world—that China's economic slowdown has trickled straight down to them. In addition, how much can someone making $5 per day possibly spend on a laptop or dishwasher, anyway? Not all of those billions of consumers will be able to afford what Westerners consider basic items, notwithstanding the acceptance in business of C.K. Prahalad's "Fortune at the Bottom of the Pyramid".

A backlash may ensue when their desire to be able to afford such items fades as they come to the realization that, in spite of globalization, the growing middle class, and rising incomes, an iPad is simply a luxury they will never be able to afford. That is what agile risk managers and decision makers should be thinking about today. Just ask all those Western companies who bulldozed their way to China in the 1990s in the firm belief that the path to riches lay there. They had no idea that what awaited them was a minimum investment of ten years of their time to become established, inevitable trouble with their joint-venture partner, and seemingly endless battles with local governments

[23] "Understanding China's Middle Class," *China Business Review*, U.S.–China Business Council, January 1, 2009, accessed November 26, 2015, http://www.chinabusinessreview.com/understanding-chinas-middle-class/.

[24] H. Kharas, *The Emerging Middle Class in Developing Countries*, OECD Development Centre, 2010, accessed November 26, 2015, http://www.oecd.org/dev/44457738.pdf.

and the courts. The land of milk and honey, it was not. Why should it be any different today, with more competitors, more regulations, and dwindling resources—all competing for a limited share of wallet?

"A significant threat to the realisation of vast growth in the global middle class is the concept of the middle-income trap, where countries that have grown to middle income levels subsequently stagnate and fail to attain advanced-country status, owing partly to growth slowdowns caused by the rising costs of wages and subsequent declines in competitiveness."[25] Brazil is a perfect example of this trap.

Another consideration, which certainly applies to the developed world as well, is rising income inequality, which is sure to become exacerbated in the developing world, with potentially serious consequences for foreign firms. While there is certainly reason to believe that many of the opportunities expected to arise as a result of the growth in the global middle class will unfold more or less as expected (with potentially stellar income and growth opportunities for global businesses), companies should also anticipate a variety of unforeseen outcomes. As income inequality rises, the level of frustration among middle- and lower-class consumers will also surely rise. The result could well be a rising propensity for social unrest and political violence—one of the unanticipated consequences of globalization.

The (Fr)Agile Balancing Act

Risk agility and the type of decision making we espouse is not about fearing the twenty-first century, although it certainly deserves deference. It is about respecting the speed with which things can fall apart as a result of unforeseen or unexpected events. As such, the basis for making bold choices should include doing something never done before, or waiting to see what happens before making a move. Agile enterprises are transparent, trustworthy, entrepreneurial and, above all, risk takers. Aversion to risk is dangerous and implies being stuck in another era and held at a standstill by the inertia of fear.

The age that we are living in—the age of globalization, instantaneous information, tremendous technological advances, seemingly constant radical political change, and unprecedented strain on critical resources—will show no remorse for the risk-averse. Those firms that embrace risk agility will be

---

[25] S. Aiyar et al. *Growth Slowdowns and the Middle Income Trap*, International Monetary Fund Working Paper WP/13/71, August 2013, accessed November 26, 2015, http://www.imf.org/external/pubs/ft/wp/2013/wp1371.pdf.

able to quickly reinevent themselves and establish frameworks and a company culture that recognizes when the enterprise is imperiled by a particular internal course of action, or by external forces. This is the very creative/destructive cycle that drives the global economy. How people, organizations, and society will thrive depends now more than ever on our individual and collective decisions.

Risk agility and decision making is mostly about common sense—remembering what your mother taught you as a child, incorporating the lessons you have learned throughout your life, and transforming it all into sensible action. It boils down to this: Be bold. Lean forward. Know more about the world. Turn the pyramid upside down. Embrace risk. Have a long-term orientation toward the future. Be thoughtful about what you are doing, and how you are doing it. Consider the viewpoints, needs and desires of others. Do the right thing. It is not just about making money or getting the job done; in the twenty-first century, it is about getting to the finishing line in one piece, and with a clear conscience.

Bibliography

10 Huge U.S. brands who profit from what Americans would call slave labor. criminal-justicedegreesguide.com. http://www.criminaljusticedegreesguide.com/features/10-huge-u-s-brands-who-profit-from-what-americans-would-call-slave-labor.html. Accessed 25 Nov 2015.

10 minutes on business continuity. PWC, 2012. http://www.pwc.com/us/en/10minutes/business-continuity-management.html. Accessed 26 Nov 2015.

15th Annual Global CEO survey. PWC, 2012. (interview of more than 1,200 CEOs from 60 countries). https://www.pwc.com/gx/en/ceo-survey/pdf/15th-global-pwc-ceo-survey.pdf. Accessed 26 Nov 2015.

2014 annual report. Volkwagen AG, 2014. http://annualreport2014.volkswagenag.com/. Accessed 24 Nov 2015.

2015 Edelman trust barometer: Global results. Edelman, 2015. http://www.edelman.com/insights/intellectual-property/2015-edelman-trust-barometer/. Accessed 25 Nov 2015.

A board perspective on enterprise risk management. 2010. McKinsey Global Institute. file:///C:/Users/daniel/Downloads/18_A_Board_perspective_on_enterprise_risk_management.pdf. Accessed 26 Nov 2015.

A G-zero world. www.foreignaffairs.com. Accessed 25 Nov 2015. https://www.foreignaffairs.com/articles/2011-01-31/g-zero-world.

A Singapore for Central America? Economist.com, July 14, 2011. http://www.economist.com/node/18959000. Accessed 24 Nov 2015.

A study of major risks: Their origins, impact and implications. Association of Insurance and Risk Managers (AIRMIC), Cass Business School, 2011, 8.

Abadie, A., and J. Gardeazabal. 2002. *The economic costs of conflict: A case study of the Basque Country.* Harvard University/NBER and the University of the Basque Country, July 2002. http://hks.harvard.edu/fs/aabadie/ecc.pdf. Accessed 25 Nov 2015.

© The Author(s) 2016
D. Wagner, D. Disparte, *Global Risk Agility and Decision Making,*
DOI 10.1057/978-1-349-94860-4

Abadie, A., and J. Gardeazabal. 2005. *Terrorism and the world economy*. Harvard University/NBER and the University of the Basque Country, October 2005. http://www.hks.harvard.edu/fs/aabadie/twep.pdf. Accessed 25 Nov 2015.

Abel, J. 2015. At least 81 percent of Major Healthcare or Health Insurance Companies had a data breach in the past two years. *Consumer Affairs*, September 4. http://www.consumeraffairs.com/news/at-least-81-of-major-healthcare-or-health-insurance-companies-had-a-data-breach-in-the-past-two-years-090415.html. Accessed on 30 Nov 2015.

Abel, D., and N. Emack-Bazelais. 2015. Boston's winter vaults to the top of snowfall records. *Boston Globe*, March 15. https://www.bostonglobe.com/metro/2015/03/15/parade-day-snow-but-snowiest-winter-record-unlikely-today/BCxfh7yPtIrxtHVzty5sPM/story.html. Accessed 25 Nov 2015.

Acharya, V. 2011. *Systemic risk and macro-prudential regulation*. NYU Stern/Center for Economic Policy Research/National Bureau of Economic Research, March 2011. http://pages.stern.nyu.edu/~sternfin/vacharya/public_html/ADB%20Systemic%20Risk%20and%20Macroprudential%20Regulation%20-%20Final%20-%20March%202011.pdf. Accessed 26 Nov 2015.

Acharya, V. et al. 2012. *Capital shortfall: A new approach to ranking and regulating systemic risks*. AEA Meetings, NYU Stern School of Business, January 7, 2012. http://pages.stern.nyu.edu/~jcarpen0/Chinaluncheon/Capital%20Shortfall.A%20New%20Approach%20to%20Ranking%20and%20Regulating%20Systemic%20Risks.pdf. Accessed 25 Nov 2015.

Adbayo, A.A. 2014. Implications of Boko Haram Terrorism on National Development in Nigeria. *Mediterranean Journal of Social Sciences* 5(16). file:///C:/Users/daniel/Downloads/3330-13111-1-PB.pdf. Accessed 25 Nov 2015.

Advancing the Science of Climate Change. 2010. *America's climate choices: Panel on advancing the science of climate change, National Research Council*. Washington, DC: The National Academies Press.

AES awarded Panama's first natural gas-fired generation plant. Press Release, Business Wire, September 11, 2015. http://www.businesswire.com/news/home/20150911005301/en/AES-Awarded-Panama%E2%80%99s-Natural-Gas-Fired-Generation-Plant. Accessed 25 Nov 2015.

Aityeh, C., and R. Blackwell. 2015. Massive Takata airbag recall: Everything you need to know, including full list of affected vehicles. *CarandDriver.com*, November 23. http://blog.caranddriver.com/massive-takata-airbag-recall-everything-you-need-to-know-including-full-list-of-affected-vehicles/#list. Accessed 24 Nov 2015.

Aiyar, S., et al. 2013. *Growth slowdowns and the middle income trap*. International Monetary Fund Working Paper WP/13/71, August 2013. http://www.imf.org/external/pubs/ft/wp/2013/wp1371.pdf. Accessed 26 Nov 2015.

Akhtar, S., et al. 2014. *Export–import Bank: Reauthorization: Frequently asked questions*. Congressional Research Service. http://fas.org/sgp/crs/misc/R43671.pdf. Accessed 26 Nov 2015.

Aleccia, J. 2009. Flu-fighting masks may help, but don't bet on it. *NBC News*, April 30. http://www.nbcnews.com/id/30464365/ns/health-cold_and_flu/t/flu-fighting-masks-may-help-dont-bet-it/#.VlYQg_mrRX8. Accessed 25 Nov 2015.

Allianz SE. 2015. *Allianz risk barometer 2015.* http://www.agcs.allianz.com/assets/PDFs/Reports/Allianz-Risk-Barometer-2015_EN.pdf. Accessed on 30 Nov 2015.

Allianz SE., and Allianz Global Corporate SE. 2015. *Allianz risk barometer: Top business risks 2015.* http://www.agcs.allianz.com/assets/PDFs/Reports/Allianz-Risk-Barometer-2015_EN.pd. Accessed 25 Nov 2015.

Altucher, J. 2011. A mouse in the salad means your business is growing. *Business Insider*, July. http://www.businessinsider.com/mouse-in-the-salad-2011-7. Accessed 24 Nov 2015.

An introduction to the IRGC risk governance framework. IRGC, October 2007. http://www.ortwin-renn.de/sites/default/files/PDF/RecentPublications/PolicyBrief_IRGC_RiskGovernanceFramework_9Oct.pdf. Accessed 29 Nov 2015.

Annual report 2014, U.S. Export–import Bank. http://www.exim.gov/sites/default/files/reports/annual/EXIM-2014-AR.pdf. Accessed 26 Nov 2015.

Anticipating criminal charges against Volkswagen. Law 360, September 24, 2015. Accessed at: http://www.law360.com/articles/706830/anticipating-criminal-charges-against-volkswagen

Argentina Foreign Direct Investment 2003–2015. Tradingeconomics.com, 2015. http://www.tradingeconomics.com/argentina/foreign-direct-investment. Accessed 25 Nov 2015.

Assange, J. 2015. Assange: What Wikileaks teaches us about how the U.S. operates. *Newsweek*, August 28. http://www.newsweek.com/emb-midnight-827-assange-what-wikileaks-teaches-us-about-how-us-operates-366364. Accessed on 30 Nov 2015.

Auerbach, D. 2014. The Sony hackers are terrorists. *Slate*, December 14. http://www.slate.com/articles/technology/bitwise/2014/12/sony_pictures_hack_why_its_perpetrators_should_be_called_cyberterrorists.html. Accessed 24 Nov 2015.

Barbati, G. 2013. World water wars: In the West Bank, water is just another conflict issue for Israelis and Palestinians. *International Business Times*, July 1. http://www.ibtimes.com/world-water-wars-west-bank-water-just-another-conflict-issue-israelis-palestinians-1340783. Accessed 25 Nov 2015.

Barney, C. 2014. *The enlightened organization.* Philadelphia: Kogan Page Limited.

Barton, C., et al. 2015. *The Caribbean has some of the world's highest energy costs—Now is the time to transform the region's energy market.* Inter-American Development Bank, November 14, 2015. http://blogs.iadb.org/caribbean-dev-trends/2013/11/14/the-caribbean-has-some-of-the-worlds-highest-energy-costs-now-is-the-time-to-transform-the-regions-energy-market/. Accessed 25 Nov 2015.

Ben-Shabat, H., et al. 2015. *2015 Global Retail e-Commerce Index.* AT Kearney. https://www.atkearney.com/consumer-products-retail/global-retail-development-index/2015. Accessed 30 Nov 2015.

Bergen, P., and S. Pandey. 2005. The madrassa myth. *New York Times*, June 14. http://www.nytimes.com/2005/06/14/opinion/the-madrassa-myth.html. Accessed 24 Nov 2015.

Berrick, C. 2008. *Transportation security administration has strengthened planning to guide investments in key aviation security programs, but more work remains.* U.S. Government Accountability Office, July 24. Accessed at: http://www.gao.gov/assets/130/120883.html

Bever, L. 2015. Former peanut plant executive faces life sentence for lethal salmonella coverup. *The Washington Post*, July 24. https://www.washingtonpost.com/news/morning-mix/wp/2015/07/24/former-peanut-plant-executive-faces-life-sentence-for-selling-salmonella-tainted-food/. Accessed 24 Nov 2015.

Bjerga, A. 2014. California draught transforms global food market. *Bloomberg Business*, August 11. http://www.bloomberg.com/news/articles/2014-08-11/california-drought-transforms-global-food-market. Accessed 25 Nov 2015.

Black, S. 2012. This Central American nation is spending roughly 50 percent of its entire economy on infrastructure. *Business Insider*, March 11. http://www.businessinsider.com/infrastructure-projects-in-panama-equivalent-to-8-trillion-in-the-us-2012-3. Accessed 25 Nov 2015.

Bliss, R.T. 2015. Shareholder value and CSR: Friends or foes? *CFO.com*, February 9. http://ww2.cfo.com/risk-management/2015/02/shareholder-value-csr-friends-foes/. Accessed 25 Nov 2015.

Bower, E., and J. Jones. 2015. American deaths in terrorism vs. gun violence in one graph. *CNN*, October 2. http://www.cnn.com/2015/10/02/us/oregon-shooting-terrorism-gun-violence/. Accessed 24 Nov 2015.

Brady, J.S. 2014. Remarks by the president in year-end press conference. *The White House, Office of the Press Secretary*, December 19. https://www.whitehouse.gov/the-press-office/2014/12/19/remarks-president-year-end-press-conference. Accessed 30 Nov 2015.

Brandimarte, W., and D. Bases. 2011. United States losses prized AAA credit rating from S&P. *Reuters*, August 7. http://www.reuters.com/article/2011/08/07/us-usa-debt-downgrade-idUSTRE7746VF20110807. Accessed 26 Nov 2015.

Breeden, A., and A. Rubin. 2015. Uber suspends UberPop in France and awaits court ruling. *New York Times*, July 3. http://www.nytimes.com/2015/07/04/technology/uber-suspends-uberpop-in-france-and-awaits-court-ruling.html. Accessed 24 Nov 2015.

Bright, J., and A. Hruby. 2015. *The next Africa: An emerging continent becomes a global powerhouse.* New York: Thomas Dunne Books.

Bright, J., and A. Hruby. 2015. *The next Africa: An emerging continent becomes a global powerhouse.* New York: St. Martin's Press.

Bringing big data to the enterprise, What is big data, IBM. https://www-01.ibm.com/software/data/bigdata/what-is-big-data.html. Accessed 27 Nov 2015.

Brown, M. 2015a. Lightning strikes kill customer data stored on Google Compute Engine (GCE) servers. *International Business Times*, August 19. http://www.

ibtimes.com/lightning-strikes-kill-customer-data-stored-google-compute-engine-gce-servers-2060314. Accessed on 30 Nov 2015.

Brown, R. 2015b. TPP? TTIP? Key trade deal terms explained. *The Brookings Institution*, May 20. http://www.brookings.edu/blogs/brookings-now/posts/2015/05/20-trade-terms-explained. Accessed 26 Nov 2015.

Brumfield, B. 2014. Moore, Oklahoma, looks back on Tornado that killed 24 one year ago. *CNN*, May 20. http://www.cnn.com/2014/05/20/us/oklahoma-moore-tornado-anniversary/. Accessed 25 Nov 2015.

Burton, K. 2014. John Paulson calls Puerto Rico Singapore of Caribbean. *Bloomberg Business*, April 24. http://www.bloomberg.com/news/articles/2014-04-24/john-paulson-says-puerto-rico-to-become-singapore-of-caribbean. Accessed 24 Nov 2015.

C&E corporate-NGO partnerships barometer 2015. candeadvisory.com. http://www.candeadvisory.com/barometer. Accessed 25 Nov 2015.

Cartography: The true size of Africa. Economist.com, November 10, 2010. http://www.economist.com/blogs/dailychart/2010/11/cartography. Accessed 24 Nov 2015.

Cat bond indices: Year in review 2012, Swiss Re, February 2013.

Certification: Professional practices. DRI International. www.drii.org. https://www.drii.org/certification/professionalprac.php. Accessed 26 Nov 2015.

CFAO group corporate profile—2015. http://www.cfaogroup.com/en/market-positioning. Accessed 24 Nov 2015.

CFAO group corporate website. http://www.cfaogroup.com. Accessed 24 Nov 2015.

Chacko, G., et al. 2004. *Bank Leu's prima cat bond fund*. Harvard Business School, December 2004. http://www.thecasesolutions.com/bank-leus-prima-cat-bond-fund-10452. Accessed 25 Nov 2015.

Chandler, D., and W.B. Werther. 2011. *Strategic corporation social responsibility*. New York: Sage Publications.

Chang, A. 2013. Why undersea internet cables are more vulnerable than you think. *Wired*, April 2. http://www.wired.com/2013/04/how-vulnerable-are-undersea-internet-cables/. Accessed 30 Nov 2015.

Chapel, B. 2015. U.S. predicted to be net energy exporter in next decade: First time since 1950s. *National Public Radio*, April 15. http://www.npr.org/sections/thetwo-way/2015/04/15/399843516/u-s-predicted-to-be-net-energy-exporter-in-next-decade-first-time-since-1950s. Accessed 25 Nov 2015.

Chatterji, A. 2013. *When corporations fail at doing good*. www.newyorker.com, August 29. http://www.newyorker.com/business/currency/when-corporations-fail-at-doing-good. Accessed 25 Nov 2015.

Chatterji, A. 2015. *Starbucks' "race together" campaign and the upside of CEO activism*. Harvard Business Review, March 24. https://hbr.org/2015/03/starbucks-race-together-campaign-and-the-upside-of-ceo-activism. Accessed 24 Nov 2015, accessed 25 Nov 2015.

Chen, A., and T. Siems. 2003. *The effects of terrorism on global capital markets*. Cox School of Business and the Federal Reserve Bank of Dallas, August 2003. http://

dev.wcfia.harvard.edu/sites/default/files/ChenSiems2004.pdf. Accessed 25 Nov 2015.

China's Alibaba breaks single's day records as sales surge. BBC News, November 11, 2015. http://www.bbc.com/news/business-34773940. Accessed 30 Nov 2015.

China's biggest brokerage citic in $166bn error. BBC News, November 25, 2015. http://www.bbc.com/news/business-34918510. Accessed 30 Nov 2015.

Christensen, C. 1997. *The innovator's dilemma: Creating and sustaining successful growth.* Boston: Harvard Business School Press.

Churchill, W. 1942a. *This is not the end (part 2),* November 1942. https://www.youtube.com/watch?v=pdRH5wzCQQw. Accessed 24 Nov 2014.

Churchill, W. 1942b. This is not the end. *YouTube,* November. https://www.youtube.com/watch?v=pdRH5wzCQQw. Accessed 24 Nov 2015.

Clear thinking needed. The Economist, November 28, 2015, 11.

Climate effects on health. Centers for Disease Control and Prevention. http://www.cdc.gov/climateandhealth/effects/. Accessed 25 Nov 2015.

CO_2 emissions. World Bank, 2015. http://data.worldbank.org/indicator/EN.ATM.CO2E.PC. Accessed 26 Nov 2015.

Cohen, J. 2013. Evidence of fabled viking navigational tool found, history, March 7. http://www.history.com/news/evidence-of-fabled-viking-navigational-tool-found. Accessed 24 Nov 2015.

Combating terrorism in the transport sector—Economic costs and benefits. Australia Department of Foreign Affairs and Trade, 2004.

Connor, T. 2015. Shoe Bomber has 'tactical regrets' over failed American Airlines Plot. *NBC News,* February 3. http://www.nbcnews.com/news/us-news/shoe-bomber-has-tactical-regrets-over-failed-american-airlines-plot-n296396. Accessed 25 Nov 2015

Cook, T. 2014. Tim cook speaks up. *Bloomberg Business,* October 30. http://www.bloomberg.com/news/articles/2014-10-30/tim-cook-speaks-up. Accessed 24 Nov 2015.

Cook, T. 2015. Tim cook: Pro-discrimination 'religious freedom' laws are dangerous. *The Washington Post,* March 29. https://www.washingtonpost.com/opinions/pro-discrimination-religious-freedom-laws-are-dangerous-to-america/2015/03/29/bdb4ce9e-d66d-11e4-ba28-f2a685dc7f89_story.html. Accessed 24 Nov 2015.

Cornwall, W. 2015. Hurricane Patricia, more Pacific storms, and 4 other signs of El Niño. *National Geographic,* October 25. http://news.nationalgeographic.com/2015/10/151023-hurricane-patricia-el-nino-extreme-weather-storms/. Accessed 24 Nov 2015.

Corporate social responsibility. Wikipedia. https://en.wikipedia.org/wiki/Corporate_social_responsibility. Accessed 25 Nov 2015.

Corrigan, T. 2015. SkyMall files for bankruptcy. *The Wall Street Journal,* January 23. http://www.wsj.com/articles/in-flight-catalog-skymall-files-for-bankruptcy-1422025308. Accessed 24 Nov 2015.

Cost of oil by country. Knoema.com, 2015. Accessed at: http://knoema.com/vyronoe/cost-of-oil-production-by-country

Country risk management. Comptroller of the Currency, Comptroller Handbook, October 2001.

Crittenden, M.R. 2013. Plan reins in biggest banks. *Wall Street Journal*, July 9. Accessed at: http://www.wsj.com/articles/SB10001424127887323336870457859 5540397183694

Cyber security: The terrorist in the data. The Economist, November 28, 2015. http://www.economist.com/news/briefing/21679266-how-balance-security-privacy-after-paris-attacks-terrorist-data. Accessed 30 Nov 2015.

Date, J., et al. 2015. *Undercover DHS tests find security failures at U.S. airports*, June 1. Accessed at: http://abcnews.go.com/ABCNews/exclusive-undercover-dhs-tests-find-widespread-security-failures/story?id=31434881

Davies J., and the Greenbiz Group. 2013. *State of the profession 2013*, January 2013. http://info.greenbiz.com/rs/greenbizgroup/images/State%20of%20the%20 Profession%202013.pdf?mkt_tok=3RkMMJWWfF9wsRonu6TAZKXonjHpfsX 74%2BkqX6axlMI%2F0ER3fOvrPUfGjI4ATMFnI%2BSLDwEYGJlv6SgFSL HEMa5qw7gMXRQ%3D. Accessed 25 Nov 2015.

Davis-Hanson, V. 2015. California is becoming a dust bowl. *Newsweek,* July 1. http://www.newsweek.com/california-becoming-dust-bowl-349255. Accessed 25 Nov 2015.

Decade of action for road safety: 2011–2020. World Health Organization, 2011. http://www.who.int/roadsafety/decade_of_action/plan/en/. Accessed 24 Nov 2015.

Definition of country risk. Wikipedia.com. http://en.wikipedia.org/wiki/Country_ risk. Accessed 25 Nov 2015.

Definition of Grey Swan. www.macmillandictionary.com, 2015. http://www.macmillandictionary.com/us/dictionary/american/grey-swan. Accessed 26 Nov 2015.

Definition of sovereign risk. Thefreedictionary.com. http://financial-dictionary.thefree-dictionary.com/sovereign+risk. Accessed 25 Nov 2015.

Devaney, T. 2014. FCC rule would lift in-flight call ban. *The Hill*, January 14. http://thehill.com/regulation/technology/195358-fcc-rule-would-lift-in-flight-call-ban. Accessed 30 Nov 2015.

DeWolf, D., and M. Mejri. 2013. Crisis communications failures: The BP case study. *International Journal of Advances in Management and Economics* 2: 54–55.

Diamond, J. 2005. *Collapse: How societies choose to fail or succeed.* New York: Viking Penguin.

Disparte, D. 2008. The weakest link: The state of humanitarian fleet management in Africa. *Monday Developments*, December. http://www.reuters.com/article/2011/08/29/us-nigeria-bombing-claim-idUSTRE77S3ZO20110829#kp6w3 qhjYCJBMEfe.97. Accessed 24 Nov 2015.

Disparte, D. 2013. *Market expansion risk and global mobility.* American Security Project, December 3. http://www.americansecurityproject.org/market-expansion-risk-and-global-mobility/. Accessed 24 Nov 2015.

Disparte, D. 2014. 3D risk management: A survivorship framework. *Risk Intelligence News Center, NYU Stern*, April 24. http://blogs.stern.nyu.edu/riskintelligence/?p=1765. Accessed 24 Nov 2015.

Disparte, D. 2015a. Ex-Im and the Pax Americana. *Huffington Post*, May 15. http://www.huffingtonpost.com/dante-disparte/exim-and-the-pax-american_b_7242152.html. Accessed 26 Nov 2015.

Disparte, D. 2015b. Half our potential: Failing forward, women and entrepreneurship. *The Huffington Post*, November 6. http://www.huffingtonpost.com/dante-disparte/half-our-potential-failin_b_8493312.html. Accessed 24 Nov 2015.

Disparte, D. 2015c. It is time for a cyber FDIC. *The Huffington Post*, April 17. http://www.huffingtonpost.com/dante-disparte/it-is-time-for-a-cyber-fd_b_7083948.html. Accessed 30 Nov 2015.

Disparte, D. 2015d. President Obama and the era of Afri-preneurship. *The Huffington Post*, August, 3. http://www.huffingtonpost.com/dante-disparte/president-obama-and-the-e_2_b_7920518.html. Accessed 24 Nov 2015.

Disparte, D. 2015e. Risk in the asset-less economy. *Huffington Post*, July 14. http://www.huffingtonpost.com/dante-disparte/risk-in-the-assetless-eco_b_7789512.html. Accessed 24 Nov 2015.

Disparte, D. 2015f. Welcome to 21st century warfare. *The Hill*, January 5. http://thehill.com/blogs/congress-blog/technology/228510-welcome-to-21st-century-warfare. Accessed 30 Nov 2015.

Disparte, D., and T. Gentry. 2015a. Corporate activism is on the rise. *International Policy Digest*, July 6. http://www.internationalpolicydigest.org/2015/07/06/corporate-activism-is-on-the-rise/. Accessed 26 Nov 2015.

Disparte, D., and T. Gentry. 2015b. The rise of corporate activism: From shareholder value to social value. *CSR Journal*, June 30. http://csrjournal.org/the-rise-of-corporate-activism-from-shareholder-value-to-social-value/. Accessed 24 Nov 2015.

Disparte, D., and T. Gentry. 2015c. The rise of corporate activism: From shareholder value to social value. *CSR Journal*, June 30. http://csrjournal.org/the-rise-of-corporate-activism-from-shareholder-value-to-social-value/. Accessed 25 Nov 2015.

Disparte, D., and D. Wagner. 2012. *Economic nationalism's impact on international business.* International Risk Management Institute, September 2012. http://www.irmi.com/articles/expert-commentary/economic-nationalisms-impact-on-international-business. Accessed 25 Nov 2015.

Disparte, D., and D. Wagner. 2015a. Entrepreneurialism and national security. *The Huffington Post*, March 25. http://www.huffingtonpost.com/daniel-wagner/entrepreneurialism-and-na_b_6935922.html. Accessed 24 Nov 2015.

Disparte, D., and D. Wagner. 2015b. Going beyond the headlines with global risk management. *Risk Management Magazine*, February 13. http://www.rmmagazine. com/2015/02/13/going-beyond-the-headlines-with-global-risk-management/. Accessed 26 Nov 2015.

Disparte, D., et al. 2014. *Moral hazard perceptions survey*. New York University, Stern School of Business.

Doing business. World Bank Group, 2015. http://www.doingbusiness.org. Accessed 26 Nov 2015.

Doing well by doing good. Nielsen, June 2014. http://www.nielsen.com/content/dam/ nielsenglobal/apac/docs/reports/2014/Nielsen-Global-Corporate-Social-Responsibility-Report-June-2014.pdf. Accessed 25 Nov 2015.

Drennan, J. 2015. Laundering the global garment industry's dirty business. *Foreign Policy*, April 24. http://foreignpolicy.com/2015/04/24/laundering-global-garment-industrys-dirty-business-rana-plaza-bangladesh-factories/. Accessed 25 Nov 2015.

Durden, T. 2015. Citi warns of "dancing", "music" and "complicated things" for the second time. *Zero Hedge*, February 21. http://www.zerohedge.com/news/2015-02-21/citi-warns-dancing-music-and-complicated-things-second-time. Accessed 24 Nov 2015.

Dutta, K. 2015. Shoe bomber Richard Reid shows no remorse after a decade in prison for failed terror atrocity. *Independent*, February 3. http://www.independent.co.uk/news/world/americas/shoe-bomber-richard-reid-shows-no-remorse-after-a-decade-in-prison-for-failed-terror-atrocity-10022074.html. Accessed 24 Nov 2015.

Ebola: Mapping the outbreak. BBC News, November 6, 2015. Accessed 25 Nov 2015.

Eddy, M. 2015. For families of Germanwings victims, Anger Simmers through grief. *New York Times*, October. http://www.nytimes.com/2015/10/11/world/europe/ for-families-of-germanwings-victims-anger-burns-through-grief.html?_r=0. Accessed 26 Nov 2015.

Edelstein, S. 2015. Hold onto your butts: These are the 10 fastest cars in the world. *Digital Trends*, May 5. http://www.digitaltrends.com/cars/fastest-cars-in-the-world-photo-gallery/. Accessed 25 Nov 2015.

Edwards, C. 2012. Lloyd's global underinsurance report. Center for Economic and Business Research/Lloyd's. https://www.lloyds.com/~/media/files/news%20 and%20insight/360%20risk%20insight/global_underinsurance_report_311012. pdf. Accessed 25 Nov 2015.

Edwards, C., et al. 2012. *Lloyd's global underinsurance report*. Lloyd's of London, August. https://www.lloyds.com/~/media/files/news%20and%20insight/360%20 risk%20insight/global_underinsurance_report_311012.pdf. Accessed 25 Nov 2015.

Egan, M. 2015. Saudi Arabia to run out of cash in less than 5 years. *CNN Money*, October 25. http://money.cnn.com/2015/10/25/investing/oil-prices-saudi-arabia-cash-opec-middle-east/. Accessed 24 Nov 2015.

Elliott, L., and E. Pilkington. 2015. New Oxfam report says half of global wealth held by the 1%. *The Guardian*, January 19. http://www.theguardian.com/business/2015/jan/19/global-wealth-oxfam-inequality-davos-economic-summit-switzerland. Accessed 24 Nov 2015.

Enders, W., and T. Sandler. 1996. Terrorism and foreign direct investment in Spain and Greece. *Kyklos* 49(3).

Enders, W., et al. 2006. The impact of transnational terrorism on U.S. FDI. Create Homeland and Security Center, Paper 55. http://research.create.usc.edu/cgi/viewcontent.cgi?article=1035&context=published_papers. Accessed 25 Nov 2015.

Energy access expansion in Haiti. World Bank Energy Summary, April 4, 2014. http://www.worldbank.org/en/results/2014/04/04/energy-access-expansion-in-haiti. Accessed 24 Nov 2015.

Energy security in the Caribbean. The American Security Project, March 4, 2015. http://www.americansecurityproject.org/energy-security-in-the-caribbean/. Accessed 25 Nov 2015.

Enron annual report 1999. Enron, 1999. http://picker.uchicago.edu/Enron/EnronAnnualReport1999.pdf. Accessed 26 Nov 2015.

Enron annual report 2000. Enron, 1999. http://picker.uchicago.edu/Enron/EnronAnnualReport2000.pdf. Accessed 26 Nov 2015.

Evans, M., et al. 2010. Volcanic ash cloud: British Airways Fly in the face of ban. *Telegraph*, April 18. http://www.telegraph.co.uk/travel/travelnews/7605305/Volcanic-ash-cloud-British-Airways-fly-in-the-face-of-ban.html. Accessed 24 Nov 2015.

Facebook online newsroom. Facebook corporate website, www.facebook.com, November 23, 2015. http://newsroom.fb.com/company-info/. Accessed 30 Nov 2015.

Federal Aviation Administration. *Advisory for Egypt Sinai Peninsula*. https://www.faa.gov/air_traffic/publications/us_restrictions/media/FDC_5-9155_Egypt-Sinai_Advisory_NOTAM.pdf. Accessed 24 Nov 2015.

Ferris, E., and M. Solís. 2013. *Earthquake, tsunami, meltdown—The triple disaster's impact on Japan, impact on the world*. Brookings Institution, March 11. http://www.brookings.edu/blogs/up-front/posts/2013/03/11-japan-earthquake-ferris-solis. Accessed 24 Nov 2015.

Ferro, S. 2015. A bunch of people wanted to punish the government for bailing out AIG. *Business Insider*, June 15. http://www.businessinsider.com/hank-greenberg-aig-lawsuit-win-2015-6. Accessed 26 Nov 2015.

Finkle, J., and B. Woodall. 2015. Researcher says can hack GM's OnStar app, open vehicle, start engine. *Reuters*, July 30. http://www.reuters.com/article/2015/07/30/gm-hacking-idUSL1N10A3XK20150730#BAV0OC0vC9YCwDty.97. Accessed 30 Nov 2015.

Forbes. *Forbes JC Penney profile*. Accessed on the Forbes website 24 Nov 2015.

Ford, M. 2015. Japan Curtails its pacifist stance. *The Atlantic*, September 19. http://www.theatlantic.com/international/archive/2015/09/japan-pacifism-article-nine/406318/. Accessed 24 Nov 2015.

Ford earnings: New products drive record third quarter profit. Fool.com, November 1, 2015. Accessed at: http://www.fool.com/investing/general/2015/11/01/ford-earnings-new-products-drive-record-3rd-quarte.aspx

Foreign direct investment—Net (BoP—U.S. dollar) in Pakistan. Tradingeconomics. com, 2015. http://www.tradingeconomics.com/pakistan/foreign-direct-investment-net-bop-us-dollar-wb-data.html. Accessed 25 Nov 2015.

Foreign exchange rate: U.S. dollar and Pakistani rupee. 10-year Currency Converter, Bank of Canada, 2015. http://www.bankofcanada.ca/rates/exchange/10-year-converter/. Accessed 25 Nov 2015.

Fortune and Reuters Editors. 2015. Petrobras takes $17 billion hit on scandal, promises 'normality'. *Fortune*, April 23. http://fortune.com/2015/04/23/petrobras-takes-17-billion-hit-on-scandal-promises-normality. Accessed 24 Nov 2015.

Fowler, D. 2013. The Toyota way. *Simplicable*, August 14. http://management.simplicable.com/management/new/toyota-way. Accessed 24 Nov 2015.

Fox news poll: 2016 matchups. *Fox News*, November 20, 2015. http://www.foxnews.com/politics/interactive/2015/11/20/fox-news-poll-2016-matchups-syrian-refugees/. Accessed 26 Nov 2015.

Gage, D. 2012. The venture capital secret: 3 out of 4 start-ups fail. *The Wall Street Journal*, September 20. http://www.wsj.com/articles/SB10000872396390443720204578004980476429190. Accessed 24 Nov 2015.

Gandel, S. 2014. Dimon gets 74% raise after billions in fines. *Fortune*, January 24. http://fortune.com/2013/10/11/jpmorgan-were-prepared-for-23-billion-in-legal-bills/. Accessed 26 Nov 2015.

Gandel, S. 2015. Lloyd's CEO: Cyber attacks cost companies $400 billion every year. *Fortune*, January 23. http://fortune.com/2015/01/23/cyber-attack-insurance-lloyds/. Accessed 30 Nov 2015.

Gardner, D., and P. Tetlock. 2015. *Superforecasting: The art and science of prediction.* New York: Crown Publishers.

Gass, N. 2015. Obama's damage control team goes corporate. *Politico*, June 9. http://www.politico.com/story/2015/06/obama-white-house-staff-corporate-jobs-118786. Accessed 24 Nov 2015.

GDP data. Google, 2015. http://www.google.ca/publicdata/explore?ds=d5bncppjof8f9_&met_y=ny_gdp_mktp_cd&idim=country:ARG&dl=en&hl=en&q=argentina+gdp. Accessed 25 Nov 2015.

GDP forecasts 2015. 2015. International Monetary Fund. http://www.imf.org/external/pubs/ft/weo/2015/update/02/. Accessed 24 Nov 2015.

Ghose, T. 2015. Why monster storm 'Juno' will be so snowy. *LiveScience*, January 26. http://www.livescience.com/49571-northeastern-snowstorm-juno-causes.html. Accessed 25 Nov 2015.

Gitlin, J. 2015. Report: VW was warned about cheating emissions in 2007. *Ars Technica*, September 2015. http://arstechnica.com/cars/2015/09/report-vw-was-warned-about-cheating-emissions-in-2007/. Accessed 26 Nov 2015.

Global net FDI flows: 2008–2009. United Nations Conference on Trade and Development, World Investment Report 2010, 2010. http://unctad.org/en/Docs/wir2010ch1_en.pdf. Accessed 25 Nov 2015.

Global risks 2015. World Economic Forum, 2015. http://www.weforum.org/reports/global-risks-report-2015. Accessed 24 Nov 2015.

Global terrorism database. National Consortium for the Study of Terrorism and Reponses to Terrorism (START), 2015. http://www.start.umd.edu/gtd/. Accessed 24 Nov 2015.

Global Terrorism Index 2014. Institute for Economics and Peace, 2014. http://economicsandpeace.org/wp-content/uploads/2015/06/Global-Terrorism-Index-Report-2014.pdf. Accessed 25 Nov 2015.

Global Terrorism Index 2015. Institute for economics and peace, 2015. http://economicsandpeace.org/wp-content/uploads/2015/11/Global-Terrorism-Index-2015.pdf. Accessed 25 Nov 2015.

Goldman, D. 2015. Cheap oil puts the house of Saud at risk. *Asia Times*, October 22. Accessed at: http://atimes.com/2015/10/cheap-oil-puts-the-house-of-saud-at-risk/.

Goldsmith, S., and D. Wagner. 2010. A new era for PNG. *Project Finance International*, May 19.

Goodpaster, K.E., and J.B. Matthews. 1982. Can a corporation have a conscience? *Harvard Business Review*, January. https://hbr.org/1982/01/can-a-corporation-have-a-conscience. Accessed 25 Nov 2015.

Grauer, Y. 2015. Security news this week: Turns out baby monitors are wildly easy to hack. *Wired*, September 5. http://www.wired.com/2015/09/security-news-week-turns-baby-monitors-wildly-easy-hack/. Accessed 30 Nov 2015.

Green insurance. Insurance Information Institute, November, 2015. http://www.iii.org/article/green-insurance. Accessed 25 Nov 2015.

Greider, W. 2010. The AIG scandal. *The Nation*, August 6. http://www.thenation.com/article/aig-bailout-scandal/. Accessed 26 Nov 2015.

Gresser, E. 2014. 60 export credit agencies operate worldwide. *Progressive Economy*, September 17. http://www.progressive-economy.org/trade_facts/60-export-credit-agencies-operate-worldwide/. Accessed 26 Nov 2015.

Grey Swans: Transformation of risk in an interconnected world. PWC, 2013.

Guidance on supervisory interaction with financial institutions on risk culture. Financial Stability Board, April 7, 2014. http://www.financialstabilityboard.org/2014/04/140407/. Accessed 24 Nov 2015.

Gunther, M. 2015. Under pressure: Campaigns that persuaded companies to change the world. *The Guardian*, February 9. http://www.theguardian.com/sustainable-business/2015/feb/09/corporate-ngo-campaign-environment-climate-change. Accessed 25 Nov 2015.

Hackett, C. 2015a. *5 facts about the Muslim population in Europe*. Pew Research Center, November 17. http://www.pewresearch.org/fact-tank/2015/11/17/5-facts-about-the-muslim-population-in-europe/. Accessed 24 Nov 2015.

Hackett, R. 2015b. How much do data breaches cost big companies? Shockingly little. *Fortune*, March 27. http://fortune.com/2015/03/27/how-much-do-data-breaches-actually-cost-big-companies-shockingly-little/. Accessed 30 Nov 2015.

Hajzler, C. Expropriation of foreign direct investments: Sectoral patterns from 1993 to 2006.

Haldane, A. 2012. *The dog and the frisbee*. Bank of England, August 31, 2012 (adapted from a speech given by Mr. Haldane. http://www.bankofengland.co.uk/publications/Pages/speeches/2012/596.aspx. Accessed 24 Nov 2015.

Hallman, B. 2012. Four years since Lehman brothers, 'Too big to fail' banks, now even bigger, fight reform. *Huffington Post*, September 15. http://www.huffingtonpost.com/2012/09/15/lehman-brothers-collapse_n_1885489.html. Accessed 24 Nov 2015.

Harish, S.P., and S.P. Santosh. 2013. Affects of globalization and limitations of CSR. *IOSR Journal of Humanities and Social Science* 14(4): 46.

Hartwig, R. 2015. *Terrorism risk and insurance program: Renewed and restructured*. Insurance Information Institute, April 2015. http://www.iii.org/sites/default/files/docs/pdf/paper_triastructure_2015_final.pdf. Accessed 24 Nov 2015.

Hassan, S.S. 2015. Analysis: Reko Diq's billion dollar mystery. *Dawn*, January 23. http://www.dawn.com/news/1158808. Accessed 25 Nov 2015.

Hass-Edersheim, E. 2010. The BP culture's role in the Gulf oil crisis. *Harvard Business Review*, June 8. https://hbr.org/2010/06/the-bp-cultures-role-in-the-gu. Accessed 24 Nov 2015.

Henning, E., and H. Varnholt. 2015. Volkswagen assesses emissions scandal's impact on its finances. *The Wall Street Journal*, October 4. http://www.wsj.com/articles/volkswagen-evaluating-emissions-scandals-impact-on-companys-finances-1443980626. Accessed 25 Nov 2015.

Higham, S. 2015. Longtime USAID contractor embroiled in scandal fires top managers, others. *Washington Post*, February 20. https://www.washingtonpost.com/news/federal-eye/wp/2015/02/20/longtime-usaid-contractor-embroiled-in-scandal-fires-top-managers-others/. Accessed 26 Nov 2015.

Horn, H. 2015. The staggering scale of Germany's refugee project. *The Atlantic*, September 12. Accessed at: http://www.theatlantic.com/international/archive/2015/09/germany-merkel-refugee-asylum/405058/

Hornyak, T. 2015. Hack to cost Sony $35 million in IT repairs. *CSO*, February 4. http://www.csoonline.com/article/2879444/data-breach/hack-to-cost-sony-35-million-in-it-repairs.html. Accessed 30 Nov 2015.

How to lose $9 billion and walk away with $10 million. *Trejdify*, February 19, 2012. http://blog.trejdify.com/2012/02/how-to-lose-9-billion-and-walk-away.html. Accessed 26 Nov 2015. http://www.iata.org/pressroom/pr/pages/2012-12-06-01.

aspx. Accessed 24 Nov 2015, http://www.wsj.com/articles/SB100014240529702 03833004577249434081658686

Hufbauer, G.C., et al. 2007. *Economic sanctions reconsidered.* 3rd ed. Peterson Institute for International Relations, November 2007, 23.

Huntington, S. 1996. *The clash of the civilizations and the remaking of the world order,* 121. New York: Simon & Schuster.

Hussey, A. 2013. Algiers: A city where France is the promised land—And still the enemy. *The Guardian,* January 26. http://www.theguardian.com/world/2013/jan/27/algeria-france-colonial-past-islam. Accessed 24 Nov 2015.

Independent expert committee makes forward-looking recommendations on foot & mouth disease. Press Release, The Royal Society of Edinburgh, 2002. https://www.royalsoced.org.uk/209_Independentexpertcommitteemakesforwardlooking recommendationsonFootMouthDisease.html. Accessed 26 Nov 2015.

Insurance-linked securities: Catastrophe bonds, sidecars and life insurance securitization. National Association of Insurance Commissioners, September 3, 2015. http://www.naic.org/cipr_topics/topic_insurance_linked_securities.htm. Accessed 25 Nov 2015.

Insurance-linked securities: Market update volume XVI. Swiss Re, July 2011.

Insurance-linked securities: Market update, volume XIX. Swiss Re, July, 2013.

Irvine, C., and T. Parfitt. 2013. Kremlin returns to typewrriters to avoid computers leaks. *The Telegraph,* July 11. http://www.telegraph.co.uk/news/worldnews/europe/russia/10173645/Kremlin-returns-to-typewriters-to-avoid-computer-leaks.html. Accessed 30 Nov 2015.

Izadi, E. 2015. Paris tries to fight smog by banning half its cars from the roads. *The Washington Post,* March 23. https://www.washingtonpost.com/news/worldviews/wp/2015/03/23/paris-tries-to-fight-smog-by-banning-half-its-cars-from-the-roads/. Accessed 25 Nov 2015.

Jackson, B.A. 2007. *Economically targeted terrorism.* Rand Center for Terrorism Risk Management Policy. http://www.rand.org/content/dam/rand/pubs/technical_reports/2007/RAND_TR476.pdf. Accessed 25 Nov 2015.

Javers, E. 2011. Citigroup tops list of banks who received federal aid. *CNBC,* March 16. http://www.cnbc.com/id/42099554. Accessed 24 Nov 2015.

JC Penney Company, Inc. corporate website. JC Penney Company, Inc. www.jcpenny.com. Accessed 24 Nov 2015.

Jie, Y., and L. Wei. 2015. Apple seeks to launch Apple Pay in China by February. *The Wall Street Journal,* November 24. http://www.wsj.com/articles/apple-pay-may-launch-in-china-by-february-sources-say-1448340287. Accessed 30 Nov 2015.

John Hancock Life Insurance. Johnhancockinsurance.com. https://www.johnhancock-insurance.com/life/John-Hancock-Vitality-Program.aspx. Accessed 25 Nov 2015.

Johnston, C. 2015. Berlin anti-TTIP trade deal protest attracts hundreds of thousands. *The Guardian,* October 10. http://www.theguardian.com/world/2015/oct/10/berlin-anti-ttip-trade-deal-rally-hundreds-thousands-protesters. Accessed 26 Nov 2015.

Judge gives preliminary approval of $8 million settlement over Sony Hack. *Time*, November 26, 2015. http://time.com/4127711/sony-hack-security-legal-settlement/. Accessed 30 Nov 2015.

Just good business. economist.com, January 17, 2008. http://www.economist.com/node/10491077. Accessed 25 Nov 2015.

Kanter, J., et al. 2010. Toyota has a pattern of slow responses on safety issues. *New York Times*, February 6. http://www.nytimes.com/2010/02/07/business/global/07toyota.html?pagewanted=all&_r=0. Accessed 24 Nov 2015.

Kaplan, R., and M. Annette. 2012. JP Morgan's loss: Bigger than "risk management". *Harvard Business Review*, May. https://hbr.org/2012/05/jp-morgans-loss-bigger-than-ri. Accessed 26 Nov 2015.

Kaplan, R., et al. 2012. JP Morgan's loss: Bigger than "risk management". *HBR Blog Network*, May 23. https://hbr.org/2012/05/jp-morgans-loss-bigger-than-ri. Accessed 24 Nov 2015.

Katz, I., et al. 2014. MetLife to file first lawsuit over systemic-risk label. *Bloomberg Business*, January 13. http://www.bloomberg.com/news/articles/2015-01-13/metlife-sues-over-too-big-to-fail-designation-by-u-s-regulators. Accessed 24 Nov 2015.

Katzenstein, P.J., et al. 2006. *Anti-Americanism in world politics*. Cornell: Cornell University Press.

Kawasaki, J., and S. Herath. 2010. *Thailand's rice farmers adapt to climate change*. Our World, United Nations University, November 19. https://collections.unu.edu/eserv/UNU:1581/journal-issaas-v17n2-02-kawasaki_herath.pdf. Accessed 25 Nov 2015.

Keegan, P. 2015. Here's what really happened at that company that set a $70,000 minimum wage, Inc. *Slate*, October 23. http://www.slate.com/blogs/moneybox/2015/10/23/remember_dan_price_of_gravity_payments_who_gave_his_employees_a_70_000_minimum.html. Accessed 24 Nov 2015.

Kelly, E. 2015. Obama signs funding bill averting government shutdown—For now. *USA Today*, September 30. http://www.usatoday.com/story/news/2015/09/30/senate-approves-funding-bill-avert-government-shutdown/73032366/. Accessed 26 Nov 2015.

Khan, I. 2010. Flood brings chaos back to Pakistan's Swat Valley. *New York Times*, August 19. http://www.nytimes.com/2010/08/20/world/asia/20swat.html?_r=0. Accessed 25 Nov 2015.

Kharas, H. 2010. *The emerging middle class in developing countries*. OECD Development Centre. http://www.oecd.org/dev/44457738.pdf. Accessed 26 Nov 2015.

Kimmelman, M. 2009. Footprints of Pieds-Noirs reach deep into France. *New York Times*, March 5. http://www.nytimes.com/2009/03/05/world/europe/05iht-kimmel.4.20622745.html?_r=0. Accessed 24 Nov 2015.

Kleinman, A. 2014. NSA: Tech companies knew about PRISM the whole time. *The Huffington Post*, March 20. http://www.huffingtonpost.com/2014/03/20/nsa-prism-tech-companies_n_4999378.html. Accessed 30 Nov 2015.

Kluver, R. 2014. Social media wouldn't have changed Tiananmen Square. *U.S. News and World Report*, June 4. http://www.usnews.com/opinion/articles/2014/06/04/revisiting-tiananmen-square-after-25-years-in-the-age-of-social-media. Accessed 30 Nov 2015.

Kopan, T. 2014. Cybercrime costs $575 billion yearly. *Politico*, June 9. http://www.politico.com/story/2014/06/cybercrime-yearly-costs-107601. Accessed 30 Nov 2015.

Kopecki, D., and L. Woellert. 2008. *Moody's, S&P employees doubted ratings, e-mails say (update2), Bloomberg*. October 22. Accessed at: http://www.bloomberg.com/apps/news?pid=newsarchive&sid=a2EMlP5s7iM0

Kopytoff, V. 2015. Apple: The first $700 billion company. *Fortune*, February 10. http://fortune.com/2015/02/10/apple-the-first-700-billion-company/. Accessed 24 Nov 2015.

Korn, B., and B.Y.M. Tham. 2013. Why we could easily have another flash crash. *Forbes*, August 9. http://www.forbes.com/sites/deborahljacobs/2013/08/09/why-we-could-easily-have-another-flash-crash/. Accessed 30 Nov 2015.

Kotter, J.P. 2007. Leading change: Why transformation efforts fail. *Harvard Business Review*, January. https://hbr.org/2007/01/leading-change-why-transformation-efforts-fail. Accessed 26 Nov 2015.

Kovensky, J. 2014. Chief happiness officer is the latest, creepiest job in corporate America. *New Republic*, July 23. Accessed at: http://www.newrepublic.com/article/118804/happiness-officers-are-spreading-across-america-why-its-bad

Kramer, A. 2009. Ikea tries to build public case against Russian corruption. *New York Times*, September 11. http://www.nytimes.com/2009/09/12/business/global/12ikea.html. Accessed 24 Nov 2015.

Krayenbuehl, T. 1985. *Country risk: Assessment and monitoring*. Cambridge: Woodhead-Faulkner.

Kunkle, F. 2015. FAA, airlines still working to resume normal air traffic after major glitch. *The Washington Post*, August 16. https://www.washingtonpost.com/local/trafficandcommuting/faa-airlines-still-working-to-resume-normal-air-traffic-after-major-glitch/2015/08/16/2f973a48-442c-11e5-846d-02792f854297_story.html. Accessed 24 Nov 2015.

Kytle, B., and J.G. Ruggie. 2005. *CSR as risk management*. Harvard University/JFK School of Government, Working paper 10, March 2005, 9.

La Monica, P. 2015. Can JC Penney and Sears avoid RadioShack's fate? *CNN Money*, February 10. http://money.cnn.com/2015/02/10/investing/radioshack-jcpenney-sears/. Accessed 24 Nov 2015.

Larson, C. *Rates of lung cancer rising steeply in Smoggy Beijing*. Bloomberg.com. http://www.bloomberg.com/bw/articles/2014-02-28/rates-of-lung-cancer-rising-steeply-in-smoggy-beijing. Accessed 25 Nov 2015.

Leahy, J. 2015. Petrobras scandal lays bare Brazil's political fragilities. *Financial Times*, March 11. http://www.ft.com/cms/s/0/0fdb4796-c6f8-11e4-9e34-00144feab7de. html#axzz3sQhRNpeW. Accessed 24 Nov 2015.

Leber, R. 2014. This is what our hellish world will look like after we hit the global warming tipping point. *New Republic*, December 21. https://newrepublic.com/article/120578/global-warming-threshold-what-2-degrees-celsius-36-f-looks. Accessed 25 Nov 2015.

Lee, J. 2014. How brands built trust in a digital world. *hugeinc.com*, August 27. http://www.hugeinc.com/ideas/report/how-brands-build-trust-digitally. Accessed 26 Nov 2015.

Lemos, R. 2015. Sony Pegs initial cyber-attack losses at $35 million, February 2. http://www.eweek.com/security/sony-pegs-initial-cyber-attack-losses-at-35-million.html. Accessed 30 Nov 2015.

Leopold, G., and K. Wafo. 1998. *Political risk and foreign direct investment*. Faculty of Economics and Statistics, University of Konstanz. http://kops.uni-konstanz.de/bitstream/handle/123456789/12070/161_1.pdf?sequence=1. Accessed 25 Nov 2015.

Levin, C. Senator, Chairman, et al. 2013. *JP Morgan Chase Whale Trade: A case history of derivatives risks and abuses*. Permanent Subcommittee on Investigations, U.S. Senate Hearing, March 15, 2013.

Li, Q. 2005. Does democracy promote or reduce transnational terrorist incidents? *The Journal of Conflict Resolution* 49(2).

Li, Q. 2006. Political violence and foreign direct investment. *Research in Global Strategic Management* 12.

Li, Q., and D. Schaub. 2004. Economic globalization and transnational terrorism. *The Journal of Conflict Resolution* 48(2).

List of countries and dependencies by population. Wikipedia. https://en.wikipedia.org/wiki/List_of_countries_and_dependencies_by_population. Accessed 24 Nov 2015.

List of countries by GDP (nominal). Wikipedia. https://en.wikipedia.org/wiki/List_of_countries_by_GDP_(nominal). Accessed 24 Nov 2015.

Living conditions in Europe. Eurostat, ec.europa.eu, 2014. http://ec.europa.eu/eurostat/documents/3217494/6303711/KS-DZ-14-001-EN-N.pdf/d867b24b-da98-427d-bca2-d8bc212ff7a8. Accessed 25 Nov 2015.

Lloyd's City Risk Index: 2015–2025—Executive summary. *Lloyd's of London*, 2015. http://www.lloyds.com/cityriskindex/files/8771-city-risk-executive-summary-aw.pdf. Accessed 25 Nov 2015.

Long, C. 2012. Puerto Rico is America's Greece. *Reuters*, March 8. http://blogs.reuters.com/muniland/2012/03/08/puerto-rico-is-americas-greece/. Accessed 25 Nov 2015.

Longworth, A., et al. 2012. The sustainability executive: Profile and progress. *PWC*, October. http://www.pwc.com/us/en/corporate-sustainability-climate-change/

publications/sustainability-executive-profile-and-progress.html. Accessed 25 Nov 2015.

Lopez, L. 2014. Puerto Rico keeps the lights on, but debt crisis far from over. *Reuters*, August 15. http://www.reuters.com/article/2014/08/15/us-usa-puertorico-utility-insight-idUSKBN0GF0C320140815. Accessed 24 Nov 2015.

Lowder, B. 2015. Obama on marriage equality: America should be very proud. *Slate*, June 26. http://www.slate.com/blogs/outward/2015/06/26/obama_on_supreme_court_gay_marriage_decision_america_should_be_very_proud.html. Accessed 24 Nov 2015.

Macaskill, J. 2012. Rivals hover over JP Morgan as farce threatens to turn into tragedy. *Euromoney* June. http://www.euromoney.com/Article/3039413/Rivals-hover-over-JPMorgan-as-farce-threatens-to-turn-into-tragedy.html. Accessed 24 Nov 2015.

MacBride, E. 2015. The story behind Abraaj Group's stunning rise in global private equity. *Forbes*, November 4. http://www.forbes.com/sites/elizabethmacbride/2015/11/04/the-story-behind-abraajs-stunning-rise/. Accessed 24 Nov 2015.

Maddock, M. 2012. If you have to fail—And you do—Fail forward. *Forbes,* October 10. http://www.forbes.com/sites/mikemaddock/2012/10/10/if-you-have-to-fail-and-you-do-fail-forward/. Accessed 24 Nov 2015.

Markel, M. 2014. How the Tylenol murders of 1982 changed the way we consume medication. *PBS Newshour*, September 29. http://www.pbs.org/newshour/updates/tylenol-murders-1982/. Accessed 24 Nov 2015.

Marshall, C. 2014. Massive seawall may be needed to keep New York City dry. *Scientific American,* May 5. http://www.scientificamerican.com/article/massive-seawall-may-be-needed-to-keep-new-york-city-dry/. Accessed 25 Nov 2015.

Martin, R. 2013. Rethinking the decision factory. *Harvard Business Review*, October. https://hbr.org/2013/10/rethinking-the-decision-factory. Accessed 24 Nov 2015.

Maxwell, J. 2000. *Failing forward.* New York: Nelson Business.

McCay, T. 2015. The TSA has been letting a lot of people with links to terrorism work at airports. *news.mic*, June 9. Accessed at: http://mic.com/articles/120393/the-tsa-has-been-letting-a-lot-of-people-with-links-to-terrorism-work-at-airports

McGrath, C. 2011. *Mubarak regime shuts down Internet in futile attempt to stop protests. The Electronic Intifada*, January 28. https://electronicintifada.net/content/mubarak-regime-shuts-down-internet-futile-attempt-stop-protests/9794. Accessed 24 Nov 2015.

McNichol, T. 2011. Be a jerk: The worst business lesson from the Steve Jobs biography. *The Atlantic*, November 28. http://www.theatlantic.com/business/archive/2011/11/be-a-jerk-the-worst-business-lesson-from-the-steve-jobs-biography/249136/. Accessed 24 Nov 2015.

Merchant, N. 2014. How to invent the future. *Harvard Business Review*, October 17. https://hbr.org/2014/10/how-to-invent-the-future. Accessed 24 Nov 2015.

Merelli, A. The 30-year-old agreement that symbolizes European Unification is fraying at the edges. *Quartz*, September 13. http://qz.com/501064/the-30-year-old-agreement-that-symbolizes-european-unification-is-fraying-at-the-edges/. Accessed 24 Nov 2015.

Micklethwait, J., and A. Wooldridge. 2009. Think again: The globalization backlash. *Foreign Policy*, November 17. http://foreignpolicy.com/2009/11/17/think-again-the-globalization-backlash/. Accessed 26 Nov 2015.

Miller, Z. 2011. Starbucks CEO takes 'no campaign donations' pledge to the public with full page NYT Ad. *Business Insider*, September 4. http://www.businessinsider.com/starbucks-ceo-takes-no-campaign-donations-pledge-to-the-public-2011-9. Accessed 24 Nov 2015.

Miller, G. 2015. California's Uber ruling could erase billions. *Wall Street Daily*, June 26. http://www.wallstreetdaily.com/2015/06/26/uber-california-ruling/. Accessed 24 Nov 2015.

Milliken, K. 2015. The Yanomami are great observers of nature. *Survival International*. http://www.survivalinternational.org/articles/3162-yanomami-botanical-knowledge. Accessed 24 Nov 2015.

Minkel, J.R. 2008. The 2003 Northeast blackout—Five years later. *Scientific American*, August 13. http://www.scientificamerican.com/article/2003-blackout-five-years-later/. Accessed 30 Nov 2015.

Minter, S. 2015. VW scandal lowers Germany's brand value by $191 billion. *Industry Week*, October. http://www.industryweek.com/competitiveness/vw-scandal-lowers-germanys-brand-value-191-billion. Accessed 26 Nov 2015.

Moore, S. 2012. Ship accidents sever data cables off East Africa. *The Wall Street Journal*, February 28. http://www.wsj.com/articles/SB10001424052970203833004577249434081658686. Accessed 30 Nov 2015.

Moore schools destroyed in tornado poorly built, civil engineer says. Associated Press, Tulsa World, February 22, 2014. http://www.tulsaworld.com/news/local/moore-schools-destroyed-in-tornado-poorly-built-civil-engineer-says/article_a00b291d-3b1c-5fe5-a4dd-9d575c5a1323.html. Accessed 25 Nov 2015.

Morales, E. 2015. Puerto Rico's soaring cost of living, from giant electric bills to $5 cornflakes. *The Guardian*, July 12. http://www.theguardian.com/world/2015/jul/12/puerto-rico-cost-of-living. Accessed 25 Nov 2015.

Most Muslims 'desire democracy'. BBC News, February 27, 2008. http://news.bbc.co.uk/2/hi/americas/7267100.stm. Accessed 24 Nov 2015.

Mshelizza, I. 2011. Islamist Sect Boko Haram claims Nigerian U.N. bombing. *Reuters*, August 29. http://www.reuters.com/article/2011/08/29/us-nigeria-bombing-claim-idUSTRE77S3ZO20110829#kp6w3qhjYCJBMEfe.97. Accessed 24 Nov 2015.

Muir, D., and A. Castellano. 2013. Oklahoma tornado: 2 devastated elementary schools had no safe rooms. *ABC News*, May 22. http://abcnews.go.com/US/oklahoma-tornado-devastated-elementary-schools-safe-rooms/story?id=19230427. Accessed 25 Nov 2015.

Muslim-Western tensions persist. Pew Research Center, July 21, 2011. http://www.pewglobal.org/2011/07/21/muslim-western-tensions-persist/. Accessed 24 Nov 2015.

Mustoe, H. 2015. VW and the never-ending cycle of corporate scandals. *BBC News*, October 26. http://www.bbc.com/news/business-34572562. Accessed 24 Nov 2015.

Mydans, S. 2011. Flood defenses are overrun in Bangkok. *New York Times*, October 25. http://www.nytimes.com/2011/10/26/world/asia/flood-waters-in-bangkok-shut-domestic-airport.html. Accessed 25 Nov 2015.

Nakashima, E., and J. Warrick. 2012. Stuxnet was work of U.S. and Israeli experts, officials say. *Washington Post*, June 2. https://www.washingtonpost.com/world/national-security/stuxnet-was-work-of-us-and-israeli-experts-officials-say/2012/06/01/gJQAlnEy6U_story.html. Accessed 30 Nov 2015.

National flood insurance program. National Association of Insurance Commissioners, 2015. http://www.naic.org/cipr_topics/topic_nfip.htm. Accessed 25 Nov 2015.

Natural disasters: Counting the costs. *Economist.com*, March 21, 2011. http://www.economist.com/blogs/dailychart/2011/03/natural_disasters. Accessed 24 Nov 2015.

New York University, *Volatility lab systemic rankings*, 2015. http://vlab.stern.nyu.edu/welcome/risk/. Accessed 24 Nov 2015.

Newport, F. 2002. Gallup poll of the Islamic world. *Gallup*, February 26. http://www.gallup.com/poll/5380/gallup-poll-islamic-world.aspx. Accessed 24 Nov 2015.

Niziol, T. 2015. The science behind naming winter storms at the weather channel. *The Weather Channel*, October 13. http://www.weather.com/news/news/science-behind-naming-winter-storms-weather-channel-20140121. Accessed 25 Nov 2015.

Nohria, N. 2015. You're not as virtuous as you think. *Washington Post*, October. https://www.washingtonpost.com/opinions/youre-not-as-virtuous-as-you-think/2015/10/15/fec227c4-66b4-11e5-9ef3-fde182507eac_story.html. Accessed 26 Nov 2015.

Nuccitelli, D. 2013. Is climate change humanity's greatest-ever risk management failure? *The Guardian*, August 22. http://www.theguardian.com/environment/climate-consensus-97-per-cent/2013/aug/23/climate-change-greatest-risk-management-failure. Accessed 25 Nov 2015.

O'Brien, M. 2012. Meet the most indebted man in the world. *The Atlantic,* November 2. http://www.theatlantic.com/business/archive/2012/11/meet-the-most-indebted-man-in-the-world/264413/. Accessed 24 Nov 2015.

O'Reilly, A. 2014. Plagued by violence, bad economy, Puerto Rico rings in 2014 with bang, 13 murders in 5 days. *Fox News Latino*, January 8. http://latino.foxnews.com/latino/news/2014/01/08/plagued-by-violence-bad-economy-puerto-rico-rings-in-2014-with-bang-13-murders/. Accessed 25 Nov 2015.

OECD and the G20: Monitoring investment and trade measures. www.oecd.org. http://www.oecd.org/g20/topics/trade-and-investment/g20.htm. Accessed 25 Nov 2015.

Oppel, R. 2004. Word for word/energy hogs; Enron traders on Grandma Millie and making out like bandits. *New York Times,* June 13. http://www.nytimes.com/2004/06/13/weekinreview/word-for-word-energy-hogs-enron-traders-grandma-millie-making-like-bandits.html. Accessed 24 Nov 2015.

Pakhomov, N., and D. Wagner. 2015. How Western energy sanctions against Russia have backfired. *Russia Direct,* June 23. http://www.russia-direct.org/opinion/how-western-energy-sanctions-russia-have-backfired. Accessed 25 Nov 2015.

Pakin, N. 2013. The case against Dodd-Frank act's living wills: Contingency planning following the financial crisis. *Berkeley Business Law Journal.* http://scholarship.law.berkeley.edu/cgi/viewcontent.cgi?article=1093&context=bblj. Accessed 24 Nov 2015.

Pakistan floods, Thomson Reuters Foundation information website on Pakistan flooding, April 8, 2013. http://www.trust.org/spotlight/Pakistan-floods-2010. Accessed 25 Nov 2015.

Panama orders power rationing as drought continues. BBC News, May 8, 2013. http://www.bbc.com/news/world-latin-america-22449328. Accessed 24 Nov 2015.

Part I: Special chapter: The rise of Asia's middle class. Key indicators for Asia and the Pacific 2010, Asian Development Bank, 2010. http://www.adb.org/publications/key-indicators-asia-and-pacific-2010. Accessed 26 Nov 2015.

Perez, I. 2013. Climate change and rising food prices heightened Arab Spring. *Scientific American,* March 4. http://www.scientificamerican.com/article/climate-change-and-rising-food-prices-heightened-arab-spring/. Accessed 25 Nov 2015.

PetroCaribe and the Caribbean: Single point of failure. The Economist, October 4. http://www.economist.com/news/americas/21621845-venezuelas-financing-programme-leaves-many-caribbean-countries-vulnerable-single-point. Accessed 24 Nov 2015.

Philips, M. 2012. Knight capital shows how to lose $440 million in 30 minutes. *Bloomberg Business,* August, 2. http://www.bloomberg.com/bw/articles/2012-08-02/knight-shows-how-to-lose-440-million-in-30-minutes. Accessed 24 Nov 2015.

Plummer, B. 2013. Climate change will open up surprising new Artic shipping routes. *The Washington Post,* March 5. https://www.washingtonpost.com/news/wonk/wp/2013/03/05/climate-change-will-open-up-surprising-new-arctic-shipping-routes/. Accessed 25 Nov 2015.

Prospects weekly: Protectionism muted, FDI plummets in 2009, global oil demand now rising. blogs.worldbank.org. http://blogs.worldbank.org/prospects/prospects-weekly-protectionism-muted-fdi-plummets-in-2009-global-oil-demand-now-rising. Accessed 25 Nov 2015.

Ranking Caribbean countries by population density. Caribbean Journal, October 22, 2013. http://caribjournal.com/2013/10/22/ranking-caribbean-countries-by-population-density/#. Accessed 25 Nov 2015.

Reckard, S. 2015. Lower credit scores lead to higher mortgage costs. *The Seattle Times*, July. http://www.seattletimes.com/business/real-estate/lower-credit-scores-lead-to-higher-mortgage-costs/. Accessed 26 Nov 2015.

Resolution plans, Board of Governors of the Federal Reserve System, U.S. Federal Reserve, November 26, 2015, http://www.federalreserve.gov/bankinforeg/resolution-plans.htm. (Living wills or resolutions plans for 293 financial institutions are available at the Federal Reserve's website).

Rice, D. 2015. Patricia tops list of the world's strongest storms. *USA Today*, October 24. http://www.usatoday.com/story/weather/2015/10/23/hurricane-patricia-strongest-hurricane/74461754/. Accessed 25 Nov 2015.

Right wing extremism in Europe. Friedrich-Ebert-Stiftung Forum, 2013. http://library.fes.de/pdf-files/dialog/10031.pdf. Accessed 25 Nov 2015.

Risk resilience: Reckoning with a new era of threats. PWC, 2013.

Romero, S. 2012. Brazil bars oil workers from leaving after spill. *New York Times*, March 18. http://www.nytimes.com/2012/03/19/business/energy-environment/brazil-bars-17-at-chevron-and-transocean-from-leaving-after-spill.html. Accessed 24 Nov 2015.

Rosen, C. 2015. Amy Pascal leaving Sony Pictures role to launch own production company at studio. *The Huffington Post*, February 2. http://www.huffingtonpost.com/2015/02/05/amy-pascal-sony_n_6622920.html. Accessed 24 Nov 2015.

Roy, A. 2011. How Mitt Romney's health-care experts helped design Obamacare. *Forbes*, October 11. http://www.forbes.com/sites/aroy/2011/10/11/how-mitt-romneys-health-care-experts-helped-design-obamacare/. Accessed 24 Nov 2015.

Rubin, R. 2015. U.S. companies are stashing $2.1 trillion overseas to avoid taxes. *Bloomberg*, March 4. http://www.bloomberg.com/news/articles/2015-03-04/u-s-companies-are-stashing-2-1-trillion-overseas-to-avoid-taxes. Accessed 26 Nov 2015.

Ruzhinskaya, T.I. 2011. Russian oil and gas industry's investment potential and problems. *IPSIonline.com*, December. http://www.ispionline.it/it/documents/Analysis_90_2011.pdf. Accessed 25 Nov 2015.

Sabi, A. 2015. Saudi Arabia's deficit problem. *Global Risk Insights*, October 29. http://globalriskinsights.com/2015/10/saudi-arabias-deficit-problem/. Accessed 25 Nov 2015.

Safety report 2015 edition. International Civil Aviation Organization, 2015. http://www.icao.int/safety/Documents/ICAO_Safety_Report_2015_Web.pdf. Accessed 24 Nov 2015.

Satler, A. 1932. The future of economic nationalism. *Foreign Affairs,* October. https://www.foreignaffairs.com/articles/1932-10-01/future-economic-nationalism. Accessed 25 Nov 2015.

Savage, C. 2012. Election to decide future interrogation methods in terrorism cases. *New York Times*, September 27. http://www.nytimes.com/2012/09/28/us/politics/election-will-decide-future-of-interrogation-methods-for-terrorism-suspects.html?_r=0. Accessed 24 Nov 2015.

Schlesinger, J. 2012. JPMorgan Chase: London Whale swallows $2B. *CBS News*, May 10. http://www.cbsnews.com/news/jpmorgan-chase-london-whale-swallows-2b/. 26 Nov 2015.

Schmid, A., et al. 1988. *Political terrorism: A new guide to actors, authors, concepts, data bases, theories, and literature*. Amsterdam: North Holland, Transaction Books.

Schneider, K. 2015. Panama's hydropower development defined by fierce resistance and tough choices. *Circle of Blue*, February 13. http://www.circleofblue.org/waternews/2015/world/panamas-hydropower-development-defined-fierce-resistance-tough-choices/. Accessed 24 Nov 2015.

Schollhammer, H. 1974. *Locational strategies of multinational firms*. Los Angeles: Pepperdine University.

Schwartz, P.N. 2014. Crimea's strategic value to Russia. *Center for Strategic and International Studies*, March 18. http://csis.org/blog/crimeas-strategic-value-russia. Accessed 25 Nov 2015.

Sciutto, J. 2015. OPM Government data breach impact 21.5 million. *CNN Politics*, July 10. http://www.cnn.com/2015/07/09/politics/office-of-personnel-management-data-breach-20-million/. Accessed 30 Nov 2015.

Security expert said he hacked plan controls midflight. Chicago Tribune, May 18, 2015. http://www.chicagotribune.com/news/nationworld/ct-flight-hacking-investigation-20150518-story.html. Accessed 30 Nov 2015.

SFA Statement on the UNCLOS Arbitral Proceedings against China. Philippines Department of Foreign Affairs, January 22, 2015. http://www.dfa.gov.ph/newsroom/unclos. Accessed 25 Nov 2015.

Shaikh, S. 2011. Mohamed Bouazizi: A fruit seller's legacy to the Arab people. *CNN*, December 17. http://www.cnn.com/2011/12/16/world/meast/bouazizi-arab-spring-tunisia/. Accessed 26 Nov 2015.

Sheftalovich, Z. 2015. UberPop suspended in Paris. *Politico*, July 3. http://www.politico.eu/article/uberpop-suspended-in-france/. Accessed 24 Nov 2015.

Sidel, R., et al. 2008. WaMu is seized, sold off to J.P. Morgan, in largest failure in U.S. Banking history. *Wall Street Journal*, September 26. http://www.wsj.com/articles/SB122238415586576687. Accessed 26 Nov 2015.

Simmons, C. 2015. Firms' discovery fee request is trimmed in 9/11 litigation. *New York Law Journal*, November 2. http://www.newyorklawjournal.com/id=1202741208985/Firms-Discovery-Fee-Request-Is-Trimmed-in-911-Litigation?slreturn=20151024090609. Accessed 24 Nov 2015.

Simon, J.D. 1992. Political risk analysis for international banks and multinational enterprises. In *Country risk analysis: A handbook*, ed. R.L. Solberg. London: Routledge.

Slaterry, B., et al. 2014. Sink the Jones Act: Restoring America's competitive advantage in maritime-related industries. *The Heritage Foundation*, May 22. http://www.heritage.org/research/reports/2014/05/sink-the-jones-act-restoring-americas-competitive-advantage-in-maritime-related-industries. Accessed 25 Nov 2015.

Smith, D. 2014. MasterCard takes a full page ad in The New York Times to show off Apple Pay and The iPhone 6. *Business Insider*, September 12. http://www.businessinsider.com/mastercard-new-york-times-full-page-ad-shows-apple-pay-and-the-iphone-6-2014-9. Accessed 30 Nov 2015.

Smith-Park, L., and N. Elbagir. 2013. Somalia famine killed 260,000 people, report says. *CNN*, May 2. http://www.cnn.com/2013/05/02/world/africa/somalia-famine/. Accessed 25 Nov 2015.

Squires, N. 2015. Costa Concordia Captain Francesco Schettino to release book about disaster. *The Telegraph*, June. http://www.telegraph.co.uk/news/worldnews/europe/italy/11695071/Costa-Concordia-captain-Francesco-Schettino-to-release-book-about-disaster.html. Accessed 26 Nov 2015.

Stampler, L. 2015. Salesforce CEO gave an employee $50,000 to help leave Indiana. *Time*, April 2. http://time.com/3768955/salesforce-boycott-indiana-religious-freedom/. Accessed 24 Nov 2015.

Staub, R. 2007. The heart of leadership: 12 practices of courageous leader. North Carolina: *Staub Leadership Consultants,* January 2007.

Stiglitz, J., and L.J. Bilmes. 2008. *The three trillion dollar war: The true cost of the Iraq conflict.* New York: W.W. Norton & Company. http://threetrilliondollarwar.org. Accessed 25 Nov 2015.

Stokes, B., et al. 2015. Global concern about climate change, broad support for limiting emissions. *Pew Research Center*, November 5. http://www.pewglobal.org/2015/11/05/global-concern-about-climate-change-broad-support-for-limiting-emissions/. Accessed 25 Nov 2015.

Stout, A., and D. Wagner. 2015. Why there is no stopping the funding of the IS. *The Huffington Post*, June 5. http://www.huffingtonpost.com/daniel-wagner/why-there-is-no-stopping-_b_7518012.html. Accessed 25 Nov 2015.

Summer, N. 2014. Dimon gets a 74 percent raise after billions in fines. *Bloomberg Business*, January 24. http://www.bloomberg.com/bw/articles/2014-01-24/dimon-gets-74-percent-raise-after-billions-in-fines. Accessed 26 Nov 2015.

Supplier responsibility. Apple.com, 2015. http://www.apple.com/supplier-responsibility/. Accessed 25 Nov 2015.

Sytas, A., and A. Croft. 2014. Ukraine crisis will be game changer for NATO. *Reuters*, May 18. http://www.reuters.com/article/2014/05/18/us-ukraine-crisis-nato-insight-idUSBREA4H01V20140518#RfbjufyjAAHIZeri.97. Accessed 24 Nov 2015.

Tabaka, M. 2014. *Why Richard Branson thinks unlimited vacation time is awesome— And you should too,* Inc., October 6. http://www.inc.com/marla-tabaka/richard-branson-s-unlimited-vacation-policy-will-it-work-for-your-business.html. Accessed 24 Nov 2015.

Tabuchi, H. 2014. Takata saw and hid risk in airbags in 2004 former workers say. *New York Times,* November 6. http://www.nytimes.com/2014/11/07/business/airbag-maker-takata-is-said-to-have-conducted-secret-tests.html. Accessed 24 Nov 2015.

Taleb, N. 2012. *Antifragile: Things that gain from disorder.* New York: Random House.

TalkTalk hack to cost up to £35 million. BBC News, November 11, 2015. http://www.thefrontierpost.com/article/351796/talktalk-hack-to-cost-up-to-35-million/. Accessed 30 Nov 2015.

Taylor, D. 2007. *Brand vision: How to energize your team to drive business growth,* p. 222. Hoboken: Wiley.

The 10 poorest states in America. CNBC, 2013. http://www.cnbc.com/2013/09/27/the-10-poorest-states-in-america.html. Accessed 25 Nov 2015.

The charter of the Export–import Bank of the United States. U.S. Export–import Bank. http://www.exim.gov/sites/default/files//newsreleases/Updated_2012_EXIM_Charter_August_2012_Final.pdf. Accessed 26 Nov 2015.

The facts about EXIM Bank. U.S. Export–import Bank. http://www.exim.gov/about/facts-about-ex-im-bank. Accessed 26 Nov 2015.

The history of the automobile. About.com. http://inventors.about.com/library/weekly/aacarsgasa.htm. Accessed 25 Nov 2015.

The LIBOR scandal: The rotten heart of finance. Economist.com, July 7, 2013. http://www.economist.com/node/21558281. Accessed 25 Nov 2015.

The most intense hurricanes in the United States 1851–2004. National Hurricane Center, 2015. http://www.nhc.noaa.gov/pastint.shtml. Accessed 25 Nov 2015.

The world goes to town. The Economist, May 3, 2007. http://www.economist.com/node/9070726. Accessed 24 Nov 2015.

The world population and the top ten countries with the highest population. www.internetworldstats.com, 2015. http://www.internetworldstats.com/stats8.htm. Accessed 26 Nov 2015.

The world's most powerful people. Forbes.com. http://www.forbes.com/powerful-people/list/. Accessed 24 Nov 2015.

Thomas, D. 2015. Teenager arrested in TalkTalk hack. *Financial Times,* November 25. http://www.ft.com/cms/s/0/fdc801ae-936e-11e5-bd82-c1fb87bef7af.html#axzz3t0wJ8n13. . Accessed 30 Nov 2015.

Thompson, J. 2013. Texas Governor Rick Perry spends thousands on ads to Poach Missouri Businesses. *Channel 41 KSHB Kansas City,* August 26. http://www.kshb.com/news/local-news/texas-governor-rick-perry-spends-thousands-on-ads-to-poach-missouri-businesses. Accessed 26 Nov 2015.

Thompson, C. 2014. Does 'don't be evil' still apply to Google? *CNBC,* August 19. http://www.cnbc.com/2014/08/19/does-dont-be-evil-still-apply-to-google.html. Accessed 24 Nov 2015.

Throwing money at the street. The Economist, March 10, 2011. http://www.economist.com/node/18332638. Accessed 24 Nov 2015.

Tichy, N. 2014. JC Penney and the terrible cost of hiring an outsider CEO. *Fortune*, November 13. http://fortune.com/2014/11/13/jc-penney-ron-johnson-ceo-succession/. Accessed 24 Nov 2015.

Timeline of airline bombing attacks. Wikipedia. https://en.wikipedia.org/wiki/Timeline_of_airliner_bombing_attacks. Accessed 23 Nov 2015.

Top 8 reasons why transformations fail. *Morgan Franklin Consulting*, Core Confidence, Volume 1, (2014). http://www.morganfranklin.com/core-confidence/article/top-8-reasons-why-transformations-fail. Accessed 26 Nov 2015.

Touryalai, H. 2012. Knight capital: The ideal way to screw up on Wall Street. *Forbes*, August 6. http://www.forbes.com/sites/halahtouryalai/2012/08/06/knight-capital-the-ideal-way-to-screw-up-on-wall-street/. Accessed 24 Nov 2015.

Tran, M. 2008. Colombia apologizes over use of Red Cross symbol in Betancourt rescue. *The Guardian*, July 16. http://www.theguardian.com/world/2008/jul/16/colombia. Accessed 25 Nov 2015.

Trendsetter barometer. PWC, 2012. http://www.pwc.com/us/en/press-releases/2012/private-company-optimism-and-revenue.html. Accessed 26 Nov 2015.

Tshabalala, S. 2015. Nigeria is fining MTN $1,000 per illegal SIM card even though customers generate just $5 a month. *Quartz Africa*, October 26. http://qz.com/533041/africas-largest-mobile-network-is-being-fined-5-2-billion-for-flouting-nigerias-sim-card-rules/. Accessed 30 Nov 2015.

Tupaz, E., and D. Wagner. 2013. China, the Philippines, and the Rule of Law. *The Huffington Post*, January 23. http://www.huffingtonpost.com/daniel-wagner/china-philippines-rule-law_b_2533736.html. Accessed 25 Nov 2015.

Turner, G. 2015. HSBC threatens to leave London, again. *Wall Street Journal*, April 24. http://blogs.wsj.com/moneybeat/2015/04/24/hsbc-threatens-to-leave-london-again/. Accessed 24 Nov 2015.

Two degree Celsius climate change target 'utterly inadequate', expert argues. Science News, March 27, 2015. http://www.sciencedaily.com/releases/2015/03/150327091016.htm. Accessed 25 Nov 2015.

U.S. plans to accept 10,000 Syrian refugees next year. BBC News, September 10, 2015. Accessed at: http://www.bbc.com/news/world-us-canada-34215920.

Understanding China's middle class. *China Business Review*, U.S.-China Business Council, January 1, 2009. http://www.chinabusinessreview.com/understanding-chinas-middle-class/. Accessed 26 Nov 2015.

Unemployment statistics. Eurostat, ec.europa.eu, September 2015. http://ec.europa.eu/eurostat/statistics-explained/index.php/Unemployment_statistics. Accessed 25 Nov 2015.

Volcker Rule, December 10, 2013, Federal Deposit Insurance Corporation. https://www.fdic.gov/regulations/reform/volcker/rule.html. Accessed 24 Nov 2015.

Volkswagen AG. CNBC website. http://data.cnbc.com/quotes/VLKAF/tab/2. Accessed 24 Nov 2015.

Volkswagen U.S. corporate website. Volkswagen AG, October 2015. www.vw.com.

Volkswagen was warned years ago about illegal emissions tricks: Report. Chicago Tribune, September 27, 2015. http://www.chicagotribune.com/business/ct-volkswagen-scandal-20150927-story.html. Accessed 25 Nov 2015.

VW AG website on corporate values, Volkswagen AG. http://www.volkswagenag.com/content/vwcorp/content/en/human_resources.html. Accessed 24 Nov 2015.

Wagner, D. 2000. *Defining political risk.* International Risk Management Institute, October. https://www.irmi.com/articles/expert-commentary/defining-political-risk. Accessed 25 Nov 2015.

Wagner, D. 2004a. *Project financiers' and insurers' roles in promoting social responsibility in the developing world.* International Risk Management Institute, January. https://www.irmi.com/articles/expert-commentary/promoting-social-responsibility-in-the-developing-world. Accessed 25 Nov 2015.

Wagner, D. 2004b. *The implications of recurring terrorism for business.* International Risk Management Institute, May. http://www.irmi.com/articles/expert-commentary/the-implications-of-recurring-terrorism-for-business. Accessed 25 Nov 2015.

Wagner, D. 2006. *Bolivia's larger message.* International Risk Management Institute, June. https://edit.irmi.com/articles/expert-commentary/bolivia's-larger-message. Accessed 25 Nov 2015.

Wagner, D. 2009. The boardroom vacuum. *Risk Management Magazine*, December. http://cf.rims.org/Magazine/PrintTemplate.cfm?AID=4020. Accessed 25 Nov 2015.

Wagner, D. 2010a. *Is country risk really rising?* International Risk Management Institute, July. http://www.irmi.com/articles/expert-commentary/is-country-risk-really-rising. Accessed 25 Nov 2015.

Wagner, D. 2010b. *Pakistan's message to foreign investors.* International Risk Management Institute, February 12. https://www.irmi.com/articles/expert-commentary/expropriation-pakistan's-message-to-foreign-investors. Accessed 25 Nov 2015.

Wagner, D. 2011a. *How political change in MENA is affecting country risk analysis.* International Risk Management Institute, April 21. https://www.irmi.com/articles/expert-commentary/how-political-change-in-the-middle-east-and-north-africa-is-affecting-country-risk-analysis. Accessed 25 Nov 2015.

Wagner, D. 2011b. *Managing political risk in the new normal.* International Risk Management Institute, January 21. https://www.irmi.com/articles/expert-commentary/managing-political-risk-in-the-new-normal. Accessed 25 Nov 2015.

Wagner, D. 2012a. Managing country risk. New York: CRC Press.

Wagner, D. 2012b. Argentina's expropriation and the lessons of history. *The Huffington Post*, April 17. http://www.huffingtonpost.com/daniel-wagner/argentinas-expropriation-_b_1431288.html. Accessed 25 Nov 2015.

Wagner, D. 2012c. *Common sense political forecasting*. International Risk Management Institute, February 24. https://www.irmi.com/articles/expert-commentary/common-sense-political-forecasting. Accessed 25 Nov 2015.

Wagner, D. 2012d. *Managing the media in times of crisis*, www.smashwords.com. https://www.smashwords.com/books/view/252158. Accessed 26 Nov 2015.

Wagner, D. 2012e. *What the Euro crisis implies about managing country risk*. International Risk Management Institute, August 17. https://www.irmi.com/articles/expert-commentary/euro-crisis-and-country-risk. Accessed 25 Nov 2015.

Wagner, D. 2013. Europe's rising social and political risks. *The Huffington Post*, March 11. http://www.huffingtonpost.com/daniel-wagner/europes-rising-social-and_b_2853191.html. Accessed 25 Nov 2015.

Wagner, D. 2014a. *Anticipating country risk*. International Risk Management Institute, June 2014. https://www.irmi.com/articles/expert-commentary/anticipating-country-risk. Accessed 26 Nov 2015.

Wagner, D. 2014b. Only the tourists can save Thailand. *The Huffington Post*, July 23. http://www.huffingtonpost.com/daniel-wagner/only-the-tourists-can-sav_b_5377836.html. Accessed 26 Nov 2015.

Wagner, D. 2014c. Why Brazil is its own worst enemy. *The Huffington Post*, December 23. http://www.huffingtonpost.com/daniel-wagner/why-brazil-is-its-own-wor_b_6367760.html. Accessed 26 Nov 2015.

Wagner, D. 2015a. Do sanctions work? *The Huffington Post*, May 4. http://www.huffingtonpost.com/daniel-wagner/do-sanctions-work_b_7191464.html. Accessed 25 Nov 2015.

Wagner, D. 2015b. The impact of energy on China/Myanmar relations. *The Huffington Post*, April 14. http://www.huffingtonpost.com/daniel-wagner/the-impact-of-energy-of-c_b_7061318.html. Accessed 25 Nov 2015.

Wagner, D. 2015c. The need for country risk management. *Risk Management Magazine*, June 2. http://www.rmmagazine.com/2015/06/02/the-need-for-country-risk-management/. Accessed 25 Nov 2015.

Wagner, D. 2015d. What the IS teaching the West about social media. *The Huffington Post*, March 23. http://www.huffingtonpost.com/daniel-wagner/what-the-islamic-state-is-teaching-the-west-about-social-media_b_6918384.html. Accessed 25 Nov 2015.

Wallace, G. 2014. Biggest auto recalls ever. *CNN Money*, May 27. Accessed at: http://money.cnn.com/2014/05/27/autos/biggest-auto-recalls/.

Washington Post, Stuxnet.

Watts, J. 2015. Brazil in crisis mode as ruling party sees public trust rapidly dissolving. *The Guardian*, March 17. http://www.theguardian.com/world/2015/mar/17/brazil-crisis-petrobas-scandal-dilma-rousseff-protests. Accessed 24 Nov 2015.

Wealth without workers, workers without wealth. Economist.com, October 4. http://www.economist.com/news/leaders/21621800-digital-revolution-bringing-sweeping-change-labour-markets-both-rich-and-poor. Accessed 24 Nov 2014.

Weather Time Series, 2015, National Oceanic and Atmospheric Administration, National Centers for Environmental Information. http://www.ncdc.noaa.gov/cag/time-series/global/globe/land_ocean/ytd/7/1880-2015. Accessed 25 Nov 2015.

What caused the flash crash? One big, bad trade. The Economist, October 1, 2010. http://www.economist.com/blogs/newsbook/2010/10/what_caused_flash_crash. Accessed 30 Nov 2015.

What is CSR? United Nations Industrial Development Organization, 2015. http://www.unido.org/en/what-we-do/trade/csr/what-is-csr.html. Accessed 25 Nov 2015.

Wheatley, M. 2015. *Hidden costs of Sony's data breach will add up for years. experts say,* February 20. http://siliconangle.com/blog/2015/02/20/hidden-costs-of-sonys-data-breach-will-add-up-for-years-experts-say/. Accessed 30 Nov 2015.

Williams, M.T. 2010. *Uncontrolled risk*. New York: McGraw-Hill.

Williams, K. 2015a. Manhattan DA opens international cyber threat sharing non-profit. *The Hill*, September. http://thehill.com/policy/cybersecurity/253830-manhattan-da-opens-international-cyberthreat-sharing-nonprofit. Accessed 26 Nov 2015.

Williams, P. 2015b. Massive NSA phone data collection to cease. *NBC News*, November 28. http://www.nbcnews.com/news/us-news/massive-nsa-phone-data-collection-cease-n470521. Accessed 30 Nov 2015.

Wilson, A. 2006. Ford overhauls way forward plan: Cuts include 14,000 white collar jobs. *Autoweek*, September 14. http://autoweek.com/article/car-news/ford-overhauls-way-forward-plan-cuts-include-14000-white-collar-jobs. Accessed 24 Nov 2015.

Winston, A. 2014a. GE is avoiding hard choices about ecomagination. *Harvard Business Review*, August 1. https://hbr.org/2014/08/ges-failure-of-ecomagination. Accessed 25 Nov 2015.

Winston, A. 2014b. GE is avoiding hard choices about ecomagination. *hbr.org*, August 1. https://hbr.org/2014/08/ges-failure-of-ecomagination. Accessed 25 Nov 2015.

Wooldridge, M. 2014. World still learning from Ethiopian famine. *BBC News*, November 29. http://www.bbc.com/news/world-africa-30211448. Accessed 25 Nov 2015.

World development indicators (GDP), Google. http://www.google.ca/publicdata. Accessed 24 Nov 2015.

World investment report 2014, United Nations Conference on Trade and Development, 2014.

World population prospects, United Nations, 2015. http://esa.un.org/unpd/wpp/publications/files/key_findings_wpp_2015.pdf. Accessed 25 Nov 2015.

World urbanization prospects, United Nations, UN Department of Economic and Social Affairs, 2014. http://esa.un.org/unpd/wup/highlights/wup2014-highlights. pdf. Accessed 25 Nov 2015.

Yarrow, J. 2015a. Silicon Valley is 'incredibly white and male'and there's a 'sort of pride' about that fact, says Silicon Valley culture reporter. *Business Insider*, April 4. http://www.businessinsider.com/silicon-valley-is-incredibly-white-and-male-2015-4. Accessed 24 Nov 2015.

Yarrow, J. 2015b. Silicon Valley is 'incredibly white and male' and there's a 'sort of pride' about that fact, says Silicon Valley culture reporter. *Business Insider*, April 4. http://www.businessinsider.com/silicon-valley-is-incredibly-white-and-male-2015-4. Accessed 24 Nov 2015.

Zetter, K. 2015. Feds say that banned researcher commandeered a plane. *Wired*, May 15. http://www.wired.com/2015/05/feds-say-banned-researcher-commandeered-plane/. Accessed 24 Nov 2015.

Zimmerman, K.A. 2015. Hurricane Katrina: Facts, damage and aftermath. *Live Science*, August 27. http://www.livescience.com/22522-hurricane-katrina-facts. html. Accessed 25 Nov 2015.

Zuckerman, E. 2011. The first twitter revolution? *Foreign Policy*, January 15. http://foreignpolicy.com/2011/01/15/the-first-twitter-revolution-2/. Accessed 30 Nov 2015.

Zuil, L. 2009. AIG's title as world's largest insurer gone forever. *Insurance Journal*, April 29. http://www.insurancejournal.com/news/national/2009/04/29/100066. htm. Accessed 26 Nov 2015.

Zurcher, A. 2014. Rush Limbaugh and his 'black Bond' outrage. *BBC News*, December 29. http://www.bbc.com/news/blogs-echochambers-30594460. Accessed 24 Nov 2015.

Index

© The Author(s) 2016
D. Wagner, D. Disparte, *Global Risk Agility and Decision Making*,
DOI 10.1057/978-1-349-94860-4